AF389226

High Resolution Active Optical Remote Sensing Observations of Aerosols, Clouds and Aerosol-Cloud Interactions and Their Implication to Climate

High Resolution Active Optical Remote Sensing Observations of Aerosols, Clouds and Aerosol-Cloud Interactions and Their Implication to Climate

Editors

Simone Lolli

Kai Qin

James Campbell

Sheng-Hsiang (Carlo) Wang

MDPI • Basel • Beijing • Wuhan • Barcelona • Belgrade • Manchester • Tokyo • Cluj • Tianjin

Editors
Simone Lolli
Italian National Research
Council (CNR)
Italy
SSAI-NASA
USA

Kai Qin
China University of Mining
and Technology
China

James Campbell
Naval Research Laboratory
USA

Sheng-Hsiang (Carlo) Wang
National Central University
Taiwan

Editorial Office
MDPI
St. Alban-Anlage 66
4052 Basel, Switzerland

This is a reprint of articles from the Special Issue published online in the open access journal *Remote Sensing* (ISSN 2072-4292) (available at: https://www.mdpi.com/journal/remotesensing/special_issues/HRAO_rs).

For citation purposes, cite each article independently as indicated on the article page online and as indicated below:

LastName, A.A.; LastName, B.B.; LastName, C.C. Article Title. *Journal Name* **Year**, *Article Number*, Page Range.

ISBN 978-3-03943-601-9 (Hbk)
ISBN 978-3-03943-602-6 (PDF)

Cover image courtesy of Kai Qin.

Contents

About the Editors

Simone Lolli is a permanent senior scientist for the Italian National Research Council (CNR) and scientific Co-PI of the NASA MPLNET Lidar Network. He is also an adjunct professor at Kent State University (KSU, Florence Campus) . In the last ten years, Simone has developed significant experience in applied atmospheric remote sensing techniques, both from space and from the ground, to study aerosol, cloud and precipitation's optical, geometrical and microphysical properties, together with their interaction with climate, meteorology, air quality and global earth–atmosphere energy budget. He also participated in several field campaigns in Southeast Asia in the frame of 7-SEAS NASA mission. Simone holds a Ph.D. degree in Physics from the Ecole Polytechnique, Palaiseau, France, and an M.Sc. degree in Physics from the University of Florence, Italy. He has lived in four countries and speaks Italian, English, French, German, Italian, Portuguese and Spanish. His colleagues describe him as detail focused, organized, and goal oriented.

Kai Qin received his PhD from, and is a professor at as well as the deputy director of the Department of Remote Sensing & GIS, China University of Mining and Technology. He earned the Ph.D. degree in 2013 in Remote Sensing from the China University of Mining and Technology (Beijing). He is currently a professor of CUMT. His research interest focuses on atmospheric environment studies using ground and satellite-based observations. He has been responsible for an AERONET site and a MAX-DOAS site in Xuzhou, eastern China, since 2013. He serves as the secretary of ISPRS Working Group on Remote Sensing of Atmospheric Environment Since 2016. He is an official member of Sentinel-5 Precursor validation team. He has published more than 50 peer-reviewed papers. He is the Topic Editor and Guest Editor of Remote Sensing. He was a visiting scholar at Hong Kong Polytechnic University in May 2015 and a visiting scholar at the Remote Sensing Technology Institute, German Aerospace Center, from March 2016 to February 2017. He has been involved in several research projects, e.g., Natural Science Foundation of China (NSFC).

James Campbell is the Head of the Atmospheric Properties and Effects Section (Code 7544) for the Marine Meteorology Division at the Naval Research Laboratory (NRL) in Monterey, CA. His research experience is based on atmospheric remote sensing and thermodynamics. The primary focus of his work is the characterization of cloud and aerosol physical properties using lidar (light detection and ranging) instruments. He is the Principal Investigator on a current NRL project evaluating the influence that visibly thin cirrus clouds exert on climate radiative balance. He played a primary support role in the development of the federated worldwide Micro Pulse Lidar Network (MPLNET) for the NASA Earth Observing System, serving now as an MPLNET science team member. He further contributes to the NASA Cloud Aerosol Lidar with the Orthogonal Polarization satellite lidar science working group. Dr. Campbell has conducted field research on five continents, from Antarctica to the Maritime Continent of Southeast Asia.

Sheng-Hsiang (Carlo) Wang received his BS and MS degree from the National Central University (NCU), Taiwan, in 1999 and 2001. During 2008 and 2011, Dr. Sheng-Hsiang Wang was an exchange scholar (post doc) of National Central University and the University of Maryland Collage Park, and worked on-site in NASA/GSFC. His research work was strongly related to many groups in GSFC, i.e., SMARTLabs, AERONET, and MPLNET. In 2012, Dr. Wang returned to NCU for a faculty position to continue his research related to aerosol and radiation. His major research topics include: (1) aerosol and radiation interactions; (2) aerosol optics measurements and remote sensing; (3) satellite remote sensing for air pollution; and (4) numerical model development and application. He has authored or co-authored 60 published journal papers.

Editorial

Editorial for Special Issue "High Resolution Active Optical Remote Sensing Observations of Aerosols, Clouds and Aerosol–Cloud Interactions and Their Implication to Climate"

Simone Lolli [1,2,*], Kai Qin [3], James R. Campbell [4] and Sheng-Hsiang Wang [5]

[1]   CNR-IMAA, Contrada S. Loja, 85050 Tito Scalo (PZ), Italy
[2]   Department of Physics, Kent State University (Florence Campus), Kent, OH 44240, USA
[3]   School of Environment Science and Spatial Informatics, CUMT, Xuzhou 221116, China; qinkai@cumt.edu.cn
[4]   Naval Research Laboratory, Monterey, CA 93943, USA; james.campbell@nrlmry.navy.mil
[5]   Department of Atmospheric Sciences, National Central University, Taoyuan City 32001, Taiwan; carlo@g.ncu.edu.tw
*   Correspondence: simone.lolli@cnr.it

Received: 2 July 2020; Accepted: 4 July 2020; Published: 7 July 2020

Abstract: This Special Issue contains twelve publications that, through different remote sensing techniques, investigate how the atmospheric aerosol layers and their radiative effects influence cloud formation, precipitation and air-quality. The investigations are carried out analyzing observations obtained from high-resolution optical devices deployed on different platforms as satellite and ground-based observational sites. In this editorial, the published contributions are taken in review to highlight their innovative contribution and research main findings.

Keywords: lidar; aerosols; remote sensing; precipitation; wind lidar; air-pollution; radiative effects

---

New observations of atmospheric aerosols and clouds and, eventually, their mute interaction on sub-km, sub-diurnal scales, enabled by active optical remote sensing methodologies, are fundamental to assessing their role on climate and on Earth–Atmosphere radiative budget. The received submissions reflect the state-of-the-art of the active optical remote sensing instruments for determining the vertical and horizontal distribution of clouds and aerosols throughout the atmospheric column.

In [1], a new technology is developed to retrieve the vertically-resolved atmospheric precipitation intensity through a synergy between measurements from the National Aeronautics and Space Administration (NASA) micro-pulse Lidar network (MPLNET), an analytical model solution and ground-based disdrometer measurements. In [2], the aerosol optical depth (AOD) from Terra-Moderate Resolution Imaging Spectroradiometer (MODIS) High-quality flag Collections 6 and 6.1 (C6 and C6.1) retrievals were retrieved from dark-target (DT), deep-blue (DB) and merged DT and DB (DTB) level–2 AOD products for verification against Aerosol Robotic Network (AERONET) Version 3 Level 2.0 AOD data obtained from 2004–2014 for three sites located in the Beijing–Tianjin–Hebei (BTH) region. In [3], the authors, based on lidar and aircraft soundings, investigated the features of the convective boundary layer height and determined the thresholds of the environmental relative humidity (RH) corresponding to the observed convective boundary layer heights over Southeast China from October 2017 to September 2018. In [4], the spatial–temporal distribution of dust aerosols over East Asia was investigated using Cloud–Aerosol Lidar and Infrared Pathfinder Satellite Observations (CALIPSO) retrievals (01/2007–12/2011) from the perspective of the frequency of dust occurrence (FDO), dust top layer height (TH) and profiles of aerosol subtypes. The results put in evidence that a typical dust belt was generated from the dust source regions (the Taklimakan and Gobi Deserts), in the latitude range of 25–45° N and reaching eastern China, Japan and Korea and, eventually, the Pacific Ocean. In [5], an improved above low-level cloud–aerosol (ACA) identification and retrieval methodology was developed to provide a new global view of the ACA distribution by combining three-channel Cloud–Aerosol Lidar with Orthogonal Polarization (CALIOP) observations. The new method can reliably identify and retrieve both thin and dense ACA layers, providing consistent results between the day- and night-time retrieval of ACAs. Then, new four-year (2007 to 2010) global ACA datasets were built, and new seasonal mean views of global ACA occurrence, optical depth, and geometrical thickness were presented and analyzed. In [6], the authors carried out the retrieval of the aerosol properties through sunphotometer observation data from March 2012 to February 2014 in Kunming, China, speculating possible causes about seasonal variations. In [7], the authors presented a proof-of-concept algorithm to automatically detect precipitation from lidar measurements obtained from the National Aeronautics and Space Administration micro-pulse Lidar network (MPLNET). In [8], the authors, utilizing the satellite observations and reanalysis data, investigated the effects of Black Carbon on the climate over the Tibetan Plateau, finding that the emissions intensify the East Asian Summer monsoon. In [9], the authors employed the wind profiling observations from the fine-time-resolution radar wind profiler (RWP), together with hourly ground-level $PM_{2.5}$ measurements, to explore the wind features in the planetary boundary layer (PBL) and their association with aerosols in Beijing for the period from December 1, 2018, to February 28, 2019. In [10], the authors developed a prototype of a homemade portable no-blind zone lidar system designed to map the three-dimensional distribution of aerosols based on a dual-field-of-view receiver system. This innovative lidar prototype has a spatial resolution of 7.5 m and a time resolution of 30 s. In [11], the author developed a Three-Dimensional Real-time Atmospheric Monitoring System to measure and analyze the vertical profiles of horizontal wind speed and direction, vertical wind velocity as well as aerosol backscatter obtained from lidar measurements. The system was applied to Hong Kong, a highly dense city with complex topography, during each season and including hot-and-polluted episodes (HPEs) in 2019. In [12], the authors conducted Doppler lidar measurements in 2019 to reveal the characteristics of typical daytime turbulent mixing processes in the convective boundary layer over Hong Kong. The authors assessed the contribution of cloud-radiative cooling on turbulent mixing and determined the altitudinal dependence of the contribution of surface heating and vertical wind shear to turbulent mixing.

References

1. Lolli, S.; D'Adderio, L.; Campbell, J.; Sicard, M.; Welton, E.; Binci, A.; Rea, A.; Tokay, A.; Comerón, A.; Barragan, R.; et al. Vertically Resolved Precipitation Intensity Retrieved through a Synergy between the Ground-Based NASA MPLNET Lidar Network Measurements, Surface Disdrometer Datasets and an Analytical Model Solution. *Remote Sens.* **2018**, *10*, 1102. [CrossRef]
2. Bilal, M.; Nazeer, M.; Nichol, J.; Qiu, Z.; Wang, L.; Bleiweiss, M.; Shen, X.; Campbell, J.; Lolli, S. Evaluation of Terra-MODIS C6 and C6.1 Aerosol Products against Beijing, XiangHe, and Xinglong AERONET Sites in China during 2004–2014. *Remote Sens.* **2019**, *11*, 486. [CrossRef]
3. Liu, D.; Zhao, T.; Boiyo, R.; Chen, S.; Lu, Z.; Wu, Y.; Zhao, Y. Vertical Structures of Dust Aerosols over East Asia Based on CALIPSO Retrievals. *Remote Sens.* **2019**, *11*, 701. [CrossRef]
4. Zhang, W.; Deng, S.; Luo, T.; Wu, Y.; Liu, N.; Li, X.; Huang, Y.; Zhu, W. New Global View of Above-Cloud Absorbing Aerosol Distribution Based on CALIPSO Measurements. *Remote Sens.* **2019**, *11*, 2396. [CrossRef]
5. Liu, Y.; Tang, Y.; Hua, S.; Luo, R.; Zhu, Q. Features of the Cloud Base Height and Determining the Threshold of Relative Humidity over Southeast China. *Remote Sens.* **2019**, *11*, 2900. [CrossRef]
6. Wang, H.; Zhang, C.; Yu, K.; Tang, X.; Che, H.; Bian, J.; Wang, S.; Zhou, B.; Liu, R.; Deng, X.; et al. Aerosol Optical Radiation Properties in Kunming (the Low–Latitude Plateau of China) and Their Relationship to the Monsoon Circulation Index. *Remote Sens.* **2019**, *11*, 2911. [CrossRef]
7. Lolli, S.; Vivone, G.; Lewis, J.; Sicard, M.; Welton, E.; Campbell, J.; Comerón, A.; D'Adderio, L.; Tokay, A.; Giunta, A.; et al. Overview of the New Version 3 NASA Micro-Pulse Lidar Network (MPLNET) Automatic Precipitation Detection Algorithm. *Remote Sens.* **2020**, *12*, 71. [CrossRef]

8. Luo, M.; Liu, Y.; Zhu, Q.; Tang, Y.; Alam, K. Role and Mechanisms of Black Carbon Affecting Water Vapor Transport to Tibet. *Remote Sens.* **2020**, *12*, 231. [CrossRef]
9. Zhang, Y.; Guo, J.; Yang, Y.; Wang, Y.; Yim, S. Vertical Wind Shear Modulates Particulate Matter Pollutions: A Perspective from Radar Wind Profiler Observations in Beijing, China. *Remote Sens.* **2020**, *12*, 546. [CrossRef]
10. Wang, J.; Liu, W.; Liu, C.; Zhang, T.; Liu, J.; Chen, Z.; Xiang, Y.; Meng, X. The Determination of Aerosol Distribution by a No-Blind-Zone Scanning Lidar. *Remote Sens.* **2020**, *12*, 626. [CrossRef]
11. Yim, S. Development of a 3D Real-Time Atmospheric Monitoring System (3DREAMS) Using Doppler LiDARs and Applications for Long-Term Analysis and Hot-and-Polluted Episodes. *Remote Sens.* **2020**, *12*, 1036. [CrossRef]
12. Huang, T.; Yim, S.; Yang, Y.; Lee, O.; Lam, D.; Cheng, J.; Guo, J. Observation of Turbulent Mixing Characteristics in the Typical Daytime Cloud-Topped Boundary Layer over Hong Kong in 2019. *Remote Sens.* **2020**, *12*, 1533. [CrossRef]

Article

Determination of Lidar Ratio for Major Aerosol Types over Western North Pacific Based on Long-Term MPLNET Data

Sheng-Hsiang Wang [1,2,*], Heng-Wai Lei [1], Shantanu Kumar Pani [1], Hsiang-Yu Huang [1], Neng-Huei Lin [1,2], Ellsworth J. Welton [3], Shuenn-Chin Chang [4,5] and Yueh-Chen Wang [1]

[1]   Department of Atmospheric Sciences, National Central University, Taoyuan 32001, Taiwan; chris@atm.ncu.edu.tw (H.-W.L.); skpani@cc.ncu.edu.tw (S.K.P.); hsiangyu.huang@g.ncu.edu.tw (H.-Y.H.); nhlin@cc.ncu.edu.tw (N.-H.L.); stonyyc@g.ncu.edu.tw (Y.-C.W.)
[2]   Center for Environmental Monitoring and Technology, National Central University, Taoyuan 32001, Taiwan
[3]   Goddard Space Flight Center, NASA, Greenbelt, MD 20771, USA; Ellsworth.J.Welton@nasa.gov
[4]   Environmental Protection Administration, Taipei 10042, Taiwan; scchang@epa.gov.tw
[5]   School of Public Health, National Defense Medical Center, Taipei 11490, Taiwan
*   Correspondence: carlo@g.ncu.edu.tw

Received: 10 July 2020; Accepted: 19 August 2020; Published: 26 August 2020

Abstract: East Asia is the most complex region in the world for aerosol studies, as it encounters a lot of varieties of aerosols, and aerosol classification can be a challenge in this region. In the present study, we focused on the relationship between aerosol types and aerosol optical properties. We analyzed the long-term (2005–2012) data of vertical profiles of aerosol extinction coefficients, lidar ratio (S_p), and other aerosol optical properties obtained from a NASA Micro-Pulse Lidar Network and Aerosol Robotic Network site in northern Taiwan, which frequently receives Asian continental outflows. Based on aerosol extinction vertical profiles, the profiles were classified into two types: type 1 (single-layer structure) and type 2 (two-layer structure). Fall season (October–November) was the prevailing season for the Type 1, whereas type 2 mainly happened in spring (March–April). In type 1, air masses normally originated from three regional sectors, i.e., Asia continental (AC), Pacific Ocean (PO), and Southeast Asia (SA). The mean S_p values were 39 ± 17 sr, 30 ± 12 sr, and 38 ± 18 sr for the AC, PO, and SA sectors, respectively. The S_p results suggested that aerosols from the AC sector contained dust and anthropogenic particles, and aerosols from the PO sector were most likely sea salts. We further combined the EPA dust event database and backward trajectory analysis for type 2. Results showed that S_p was 41 ± 14 sr and 53 ± 21 sr for dust storm and biomass-burning events, respectively. The S_p for biomass-burning events in type 2 showed two peaks patterns. The first peak occurred within range of 30–50 sr corresponding to urban pollutant, and the second peak occurred within range of 60–80 sr in relation to biomass burning. Finally, our study summarized the S_p values for four major aerosol types over northern Taiwan, viz., urban (42 ± 18 sr), dust (34 ± 6 sr), biomass-burning (69 ± 12 sr), and oceanic (30 ± 12 sr). Our findings provide useful references for aerosol classification and air pollution identification over the western North Pacific.

Keywords: ground based remote sensing; aerosols optical properties; lidar ratio; aerosol type

---

1. Introduction

Atmospheric aerosols play a crucial role in governing the regional-to-global climate change. They influence the Earth-atmosphere energy budget directly by absorbing and scattering incoming (shortwave) and outgoing terrestrial (longwave) radiation ([1]) and indirectly by modifying cloud microphysical properties ([2]). However, uncertainties in their compositions, characteristics, size distributions, concentrations, and vertical distributions throughout the atmospheric column make the exact quantification of their overall impact challenging. Several types of aerosols originate from different sources, and their physical, chemical, and optical characteristics also vary significantly ([3]). Such variations lead to large uncertainties in aerosol radiative forcing estimations. Information about vertical distributions of aerosols and their optical properties is of prime importance and crucial for quantifying the accurate radiative forcing ([4–6]). However, vertical aerosol distributions are unevenly distributed over the globe ([7–11]). Due to wide spatial and temporal variability, the understanding of the vertical structure of aerosols is still very limited [8].

The knowledge on vertical distributions—mostly gained worldwide—was from in-situ probing using rocket and balloon-borne instrumentations (e.g., [12,13]), ground-based lidar (light detection and ranging) measurements (e.g., [14–23]), and satellite remote-sensing ([24,25]). In-situ measurements on airplanes are sparse and passive remote-sensing only provides a coarse estimation of vertical aerosol distribution. However, lidar (an active remote-sensing instrument) is the most prominent tool for aerosol profiling in terms of the optical properties of aerosols ([26]). Lidar was used to characterize the vertical profiles during several field experiments, such as FIRE (First ISCCP Regional Experiment; [27]), INDOEX (Indian Ocean Experiment; [28,29]), SAFARI-2000 (Southern African Regional Science Initiative; [30]), ACE-2 (Aerosol Characterization Experiment; [31]); BASE-ASIA (Biomass-Burning Aerosols in South-East Asia: Smoke Impact Assessment; cf. http://smartlabs.gsfc.nasa.gov/; [32]), and 7-SEAS (Seven South East Asian Studies; [33]). Moreover, the variations of vertical profiles on the regional scale, such as in major cities, e.g., Los Angeles, Paris, Tokyo, and Hong Kong, have been extensively investigated by using lidar datasets ([22] and references therein).

In addition to sunphotometer (e.g., [34,35]) and passive satellite products (e.g., [36]), lidar measurements can also be used to classify different aerosol types (e.g., [37]). Measurements from depolarization lidars (ground and space-borne) can be used to separate between different aerosol types (e.g., [38,39]), and nowadays even for the calculation of aerosol number concentrations in the atmosphere ([26,40]). Among lidar measurements, lidar ratio (or extinction-to-backscatter ratio; S_p) particularly contains the information about aerosol types and provides insight into the size and absorption of aerosol particles. S_p generally depends on various factors, such as size, shape, and refractive index of aerosols (e.g., [19]) and relative humidity (RH). This can be computed as a function of wavelength of the laser light if the composition, shape, and size of aerosols are known ([40]). Spectral analyses of S_p combined with information from AERONET (Aerosol Robotic Network) inversion products serve to estimate aerosol type (e.g., [41]). Measurements of S_p combined with aerosol depolarization also serve for aerosol typing (e.g., [42–45]). Such information is also used in CALIPSO (Cloud-Aerosol Lidar and Infrared Pathfinder Satellite Observations) satellite data to classify the aerosol type through the use of inversion and processing S_p. This aspect has been greatly useful for observation and modelling for the global distribution of aerosols.

In recent years, Asian pollution (e.g., haze) has attracted worldwide attention. Aerosols mainly from anthropogenic activities in East Asia are recognized as an important source of regional and global pollution ([29,46]). Moreover, depending on meteorological conditions, this pollution also affects the air quality over downwind areas through long-range transport. Some studies reported that the aerosol types in East Asia (i.e., Taiwan) were large and highly variable in seasonal trends, influenced by different aerosol delivery mechanisms, and also impacted the radiation and air quality ([47]). Long-range transport of anthropogenic emissions from mainland China, dust events from the desert regions of northern China and Mongolia, and biomass-burning from peninsular Southeast Asia significantly impact the air quality over northern Taiwan ([23,47,48]). Over the past several years, most of the air quality studies over northern Taiwan relied on sampling, ground telemetry equipment to monitor observations, and model simulations (e.g., [49–53]). However, those studies are not sufficient to enable one to understand the impacts of air pollutants with accuracy due to the lack of understanding of vertical pollutant distributions. In addition to monitoring the transmitted light reaching the height of the spatial structure of atmospheric pollutants, the novelty of surface telemetry tool technology is to overcome the lack of information on the vertical profile.

Therefore, this study quantitatively analyzed vertical aerosol distributions in the lower troposphere by examining the aerosol extinction profiles derived from MPL (micro-pulse lidar) measurements over a rural location in northern Taiwan. Long-term (2005–2012) data of vertical aerosol profiles and column-integrated aerosol optical properties from MPL and AERONET (Aerosol Robotic Network) observations, respectively, were derived and used in this study. The classifications of aerosol types were made on the basis of S_p value, vertical aerosol profiles, and columnar optical properties of aerosols. The S_p values for several major air pollutants that affect northern Taiwan were determined and further compared with previous studies, to identify the pollutant phase between S_p corresponding numerical relationships. The major issues studied in depth in the present study were (a) seasonal variation of optical properties and vertical distribution of aerosols; (b) single and two-layered structures of aerosols; (c) the typical S_p values associated with different aerosol types, particularly over northern Taiwan.

2. Methodology

2.1. Site Description

Aerosol observation data used in this current study were acquired at the National Central University (NCU), Taoyuan city (24.97°N, 121.18°E; 133 m above sea level, a.s.l.) in northern Taiwan. This site is located at the western edge of Taoyuan city, situated 50 km south of Taipei, the national capital of Taiwan, and serves as a rural site with no significant near-source emissions. The northern part of Taiwan belongs to subtropical climate zone. The site stays normally under the influence of south-westerly Asian monsoon in summer (June–August) and north-easterly monsoon in late autumn to winter (October–February). The weather over northern Taiwan is normally cloudy and humid during the summer, but relatively cloud-free and dry during the winter ([23]). Several studies have reported that northern Taiwan is positioned on the pathway of continental Asian outflow to the west Pacific during the pollution outbreaks ([54–58]).

2.2. Ground-Based Remote Sensing Observations

MPL is an effective instrument to provide both high vertical and temporal-resolution aerosol distribution, mainly based on the principle of elastic back-scattering of the emitted low-power high repeatability, eye-safe laser [59]. It consists of a solid crystal (ND: YLF, neodymium-doped yttrium lithium fluoride crystals) laser which emits radiation at 1064 nm wavelength and its second harmonic generation is at 527 nm wavelength (green light) with a pulse repetition rate of 2500 Hz. The receiver section consists of a telescope (Cassegrain-type) with a coaxial optical lens diameter of 20 cm, and the back scattered light detected by the telescope is made to fall on an avalanche photodiode (a semiconductor photomultiplier tube) to count the back scattered photons, and finally records the number of photons per second into the computer. Data are recorded in the format of a time resolution of 1 min, spatial resolution of 75 m. The NCU MPL system is a member of the NASA Micro-Pulse Lidar Network (MPLNET; http://mplnet.gsfc.nasa.gov; [60]) project. The instrument maintenance, calibration and data processing were following well to the MPLNET portal.

Direct-sun measurements of aerosol optical depth (AOD; τ) at different spectral wavelengths (440, 500, 675, 870, and 1020 nm), recorded by a Cimel sun–sky radiometer at EPA-NCU site, were obtained from AERONET (http://aeronet.gsfc.nasa.gov). A more in-depth description of AERONET data can be found elsewhere ([61]). In the present study, we used AERONET Level 2 and MPLNET beta level 2a data ([23]) for analysis and discussion. Measurements from co-located AERONET and MPLNET can provide accurate aerosol products such as the aerosol backscattering coefficient profile and lidar ratio (S_p) at 527 nm (hereafter all S_p calculated in this study refer to this wavelength). The S_p was obtained through the best agreement of AERONET Level 2 AOD and MPL vertically integrated AOD retrieved by Fernald's method ([62]). This parameter contains the information of aerosol type in optical properties sense, which varied with the laser's wavelength, aerosol particle size distribution, and composition (i.e., physicochemical properties of refractive index) significantly ([63,64]). It is noted that the data used in this study only represent clear-sky conditions.

2.3. Source Identification Through Back Trajectory Analysis

In order to identify the source origin of aerosol transport, 5-day back trajectory calculations were carried out using the Hybrid Single-Particle Lagrangian Integrated Trajectory Version 4 (HYSPLIT; Draxler and Rolph, 2003) model. HYSPLIT is an air parcel trajectory and dispersion model maintained by the National Oceanic and Atmospheric Administration (NOAA) Air Resources Laboratory and this model uses wind fields to trace air parcel transport through the atmosphere ([65,66]). In its backward mode, HYSPLIT integrates back in time the path of travel of an air parcel arriving at a receptor location defined by horizontal and vertical coordinates at a given time. Meteorological input data for the back-trajectory calculation were from the NCEP Global Data Assimilation System (GDAS) at $1° \times 1°$ resolution.

3. Results

3.1. Long-Term Data Analysis of Aerosol Optical Properties

The long-term optical properties of aerosols at NCU station (EPA-NCU as the official name shown in AERONET and MPLNET) are shown in Figure 1. Figure 1a shows the monthly variations in AOD at 500 nm (τ_{500}). The observed annual average τ_{500} was 0.41 ± 0.25 (range of 0.2–1). High values (greater than annual average) of τ_{500} were observed from February to May, and low values (less than or equal to annual average) during the remaining months. The standard deviation of τ_{500} was about 0.2 to 0.3, indicating a clue on the high variability of columnar aerosol loading. High τ_{500} values were found to be similar to the nearby AERONET station (i.e., Taipei_CWB; 25°N, 121°E; 26 m a.s.l.; ≈25 km northeast of EPA-NCU site) in Taipei city, and low values of τ_{500} were similar to a high-altitude background AERONET site at mountain Lulin (23.51°N, 120.92°E; 2862 m a.s.l.; ≈100 km southeast of EPA-NCU) in central Taiwan. This indicates relatively higher aerosol loading over the study region and made it a complex aerosol environment. The monthly mean τ_{500} was the highest in March (0.72 ± 0.28), followed by April (0.58 ± 0.22). This was mainly due to the long-range transport of biomass-burning from Indochina region and dust aerosols from arid region in China [23]. It was further confirmed that the EPA-NCU stations have frequent long-range transport events, and to strengthen the transport mechanism, the vertical profiles of aerosol extinction coefficient and trajectories for origin of aerosols were analyzed.

The long-term analysis of the angstrom exponent between 440 nm and 870 nm ($\alpha_{440/870}$), which stands for aerosol particle sizes (higher value for smaller particle and vice versa), has been plotted in Figure 1b. Annual average of $\alpha_{440/870}$ was 1.26 ± 0.11 (ranged from 1.34 to 1.05). The maximum value ≈1.4 was found in February followed by the low in April (1.12 ± 0.32), and the minimum was found in September (1.05 ± 0.42). The smallest size of particle observed in February and March is due to the long-rang transported biomass-burning aerosols in the free atmosphere ([67]). Higher values of τ_{500} in April with lower values of $\alpha_{440/870}$ indicated the presence of dust aerosols due to Asian dust transport from the Gobi Desert.

Figure 1c shows the monthly S_p averaged from MPLNET data set during 2005–2012. The annual average S_p was 47 ± 21 sr. S_p was the highest in March (54 ± 23 sr), followed by 51 ± 23 sr in April, and the lowest value was observed during May (37 ± 17 sr). As aforementioned, the S_p is mainly dependent on aerosol physicochemical properties and can be further interpreted to aerosol types.

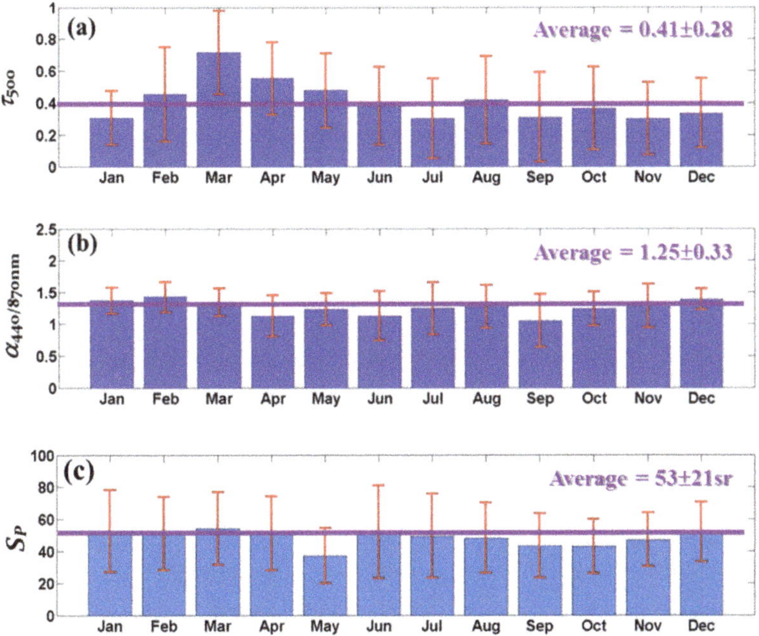

Figure 1. Monthly averaged (**a**) τ_{500}, (**b**) $\alpha_{440/870}$, and (**c**) S_p at EPA-NCU site during 2005–2012. Vertical red bars indicate ±1 standard deviation from the mean.

3.2. Monthly Aerosol Extinction Profiles

Figure 2 shows the vertical profile of aerosol extinction coefficient for each month averaged for the period of 2005–2012. As is clearly seen in Figure 3, the extinction coefficient extended from the surface to ≈6 km during March, April, and May, whereas the extinction coefficient was almost zero at ≈4 km during other months. A small aerosol layer at 2–4 km was observed over the region during March. This type of aerosol layer is mainly attributed to the dry convicting lift of air pollutants from far-off regions and subsequent horizontal transport of aerosols ([23,68]). On average, the vertical aerosol distribution follows the scale height, higher aerosol extinction near the surface and decreases as height increases.

3.3. Aerosol Vertical Distribution, Source Region, and Optical Properties

According to the discussion in Section 3.2, the air masses in the PBL and free atmosphere could be from different source origins and imply different aerosol types. In order to clearly define the impacts of vertical distribution on the optical properties (i.e., τ_{500}, $\alpha_{440/870}$, and S_p), we divided the classification of vertical distribution into two categories based on [23], i.e., single-layer (type 1) and two-layer (type 2) aerosol structures of the aerosol extinction coefficient.

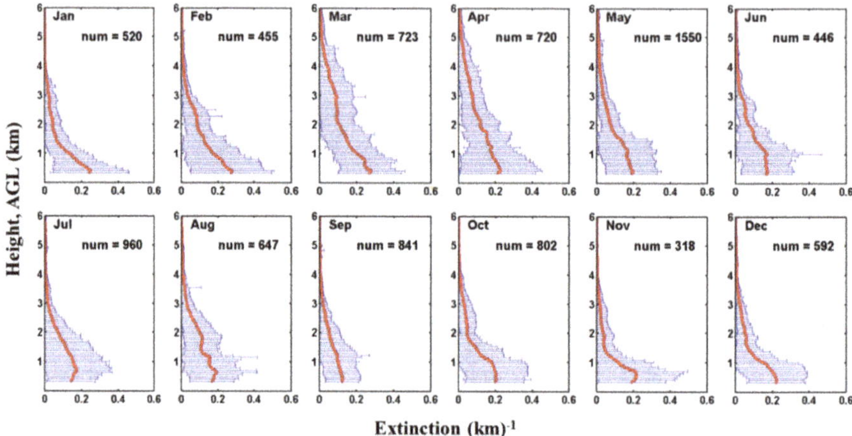

Figure 2. Vertical profiles of monthly averaged aerosol extinction along with ±1 standard deviation during 2005–2012. The number of profiles for monthly average is denoted in the figure.

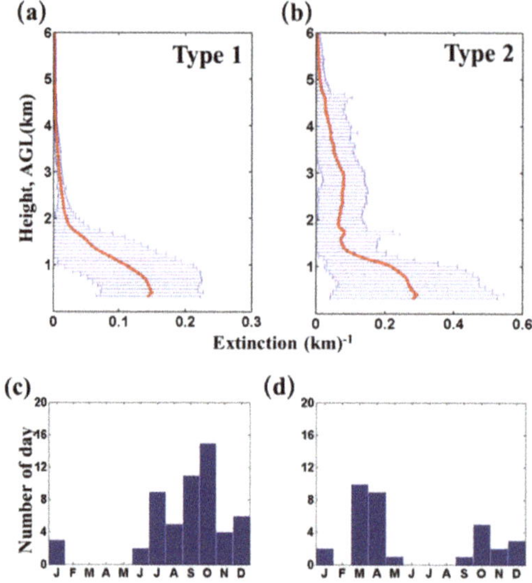

Figure 3. Averaged aerosol extinction profiles for (**a**) single-layer structure (type 1) and (**b**) two-layer structure (type 2). (**c**,**d**) indicate the number of days in the month corresponding to type 1 and type 2, respectively.

3.3.1. Characteristics of Single-Layer Aerosol Structure

The averaged aerosol extinction along with one standard deviation for single aerosol layer structure (type 1) is shown in Figure 3a. The aerosols are concentrated in the lower troposphere within 2 km altitude and the averaged extinction coefficient value is less than 0.2 km^{-1} at the near surface. This confinement of aerosols was attributed to boundary layer dynamics and inversion layer below free atmosphere. The total number of days corresponding to type 1 was 55 days (total number of data points = 649) and there were a maximum of 15 days in October, followed by a total of 11 days in September (Figure 4c). The single-layered structure of aerosols mainly occurred in the autumn and winter, especially in September and October, as shown in Figure 3c.

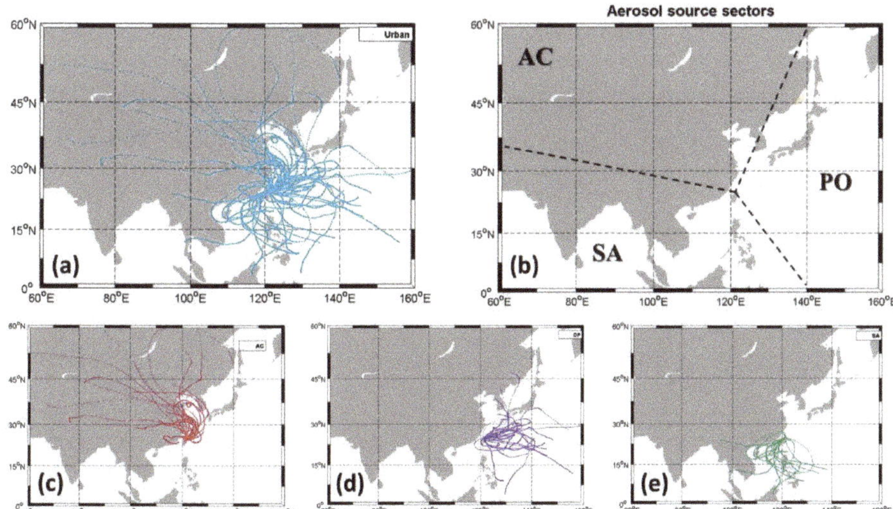

Figure 4. (**a**) Five-day back trajectories for all cases of single-layer aerosol structure (type 1). (**b**) Aerosol source origins at EPA-NCU from (**c**) Asia continental (AC), (**d**) Pacific Ocean (PO), and (**e**) Southeast Asia (SA).

In order to explore the possible sources and types of aerosols, the backward trajectory simulations were employed. Five-day back trajectories (Figure 4a) made at a height of 500 m represent the occurrence height of maximum aerosol extinction. The back trajectories for type 1 showed disorder and originated from all the directions. We further divided the total area into three regions (Figure 4b), viz., the Asia continental region (AC; Mongolia, Tibet, Xinjiang, North China and Central China region, Japan, and South Korea), the Pacific Ocean region (PO; Northwestern Pacific), and the Southeast Asia region (SA; Southeast Asia and southern China). Back trajectory analysis showed that a total of 28 day (number of data points = 458), 12 day (number of data points = 59), and 21 day (number of data points = 154) air masses were coming from AC, PO, and SA regions to EPA-NCU site, respectively (Figure 4c,e). It is important to note that few trajectories passed through more than one region, which resulted in double counting of data numbers in Table 1.

Table 1. Aerosol optical properties for single-layer aerosol structure (type 1) with respect to different source regions. Data were averaged for 2005–2012 collected at the MPLNET EPA-NCU site. AERONET derived ω_{440}, g_{440}, n_r, n_i stand for single-scattering albedo at 440 nm wavelength, asymmetry factor at 440 nm wavelength, and real part and imaginary part of refractive index, respectively.

Aerosol Optical Property	Type 1	Source Region of Type 1		
		PO	AC	SA
Data Number	649	59	458	154
τ_{500}	0.21 ± 0.1	0.19 ± 0.1	0.24 ± 0.11	0.19 ± 0.08
$\alpha_{440/870}$	1.29 ± 0.30	1.24 ± 0.31	1.34 ± 0.19	1.32 ± 0.32
S_p [sr]	39 ± 17	30 ± 12	39 ± 16	38 ± 17
ω_{440}	0.93 ± 0.02	0.96 ± 0.03	0.91 ± 0.06	0.91 ± 0.07
g_{440}	0.77 ± 0.03	0.77 ± 0.02	0.72 ± 0.04	0.72 ± 0.03
n_r	1.46 ± 0.06	1.46 ± 0.05	1.45 ± 0.06	1.46 ± 0.06
n_i	0.0055 ± 0.002	0.0033 ± 0.0037	0.0031 ± 0.0037	0.0054 ± 0.0036

± denotes the one standard deviation.

Combining information of $\alpha_{440/870}$ and S_p can help us to distinguish the aerosol types. Figure 5a shows the scatter plot of $\alpha_{440/870}$ and S_p with code denoting τ_{500} together with the histograms representing spread of samples and color magnitude of for type 1. The τ values were in the range of 0.1 to 0.6, with an average value of 0.21; $\alpha_{440/870}$ was between 0.6 and 1.8, with an average value of 1.26 ± 0.31. The S_p is concentrated between 30 and 60 sr with an average of 39 sr. Previous studies reported that the S_p values of (37 ± 9) sr and (33 ± 6) sr were for dust and marine aerosols, respectively [69].

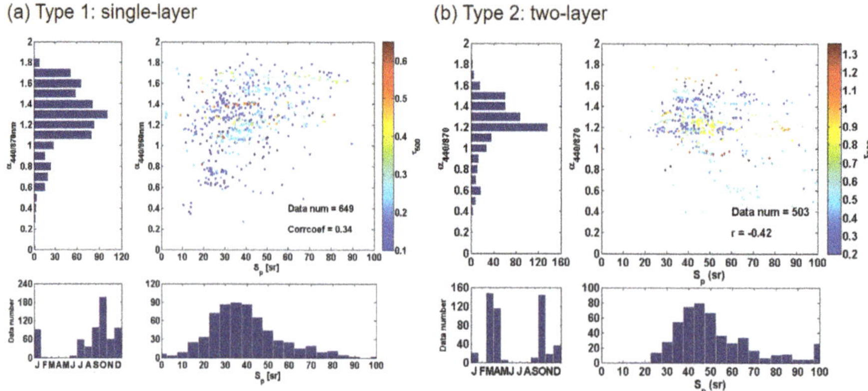

Figure 5. Scatter plot and data spread distribution histograms of aerosol optical properties for (**a**) single-layer (type 1) and (**b**) two-layer (type 2) versions.

Table 1 presents the aerosol optical properties of type 1, and for the values in different source sectors. The τ_{500}, $\alpha_{440/870}$, and S_p for AC region were 0.24 ± 0.11, 1.34 ± 0.19, and (39 ± 16) sr, respectively. The real part (n_r = 1.45) of the refractive index was small and relatively close to the urban aerosol refractive index (n_r = 1.46). The τ_{500}, $\alpha_{440/870}$, and S_p for PO source region were 0.19 ± 0.1, 1.24 ± 0.31, and (30 ± 12) sr, respectively. The moderate high value of $\alpha_{440/870}$ suggests that the mixture of coarse mode sea salt aerosol and fine mode urban aerosol. The similar value of S_p about 33 ± 6 sr was also reported by [28] in the Indian Ocean. For the SA region, the average τ_{500} and $\alpha_{440/870}$ were 0.19 ± 0.08 and 1.32 ± 0.32, respectively, indicating the abundant fine mode particle aerosol compared to other source regions. The average of S_p was 38 ± 17 sr in the SA region; however, the data histogram (not shown) exhibited two distinguished peaks: the first peak ranged between 20 and 40 sr, whereas the second peak ranged between 45 and 60 sr. The median value of the first peak was about 30 sr corresponding to marine type aerosols, while the second peak was similar to the results from local pollutants. The complex of emission sources (i.e., marine, ship, local Taiwan island) in the SA region might need further study in the future when more data becomes available. Regarding to the AERONET derived single-scattering albedo at 440 nm wavelength (ω_{440}) and asymmetry factor at 440 nm wavelength (g_{440}), aerosols from AC and SA regions show similar values and suggest a moderate absorption compared to that from PO regions. The result implies the segment of source sector with backward trajectory may not be sufficient to aerosol classification for majority of data.

3.3.2. Characteristics of Two-Layer Aerosol Structure

Two-layer structured vertical profiles have been classified as type 2 ([23]). In this type, we found another peak with an aerosol extinction coefficient at 2–4 km in addition to the extinction coefficient peak that appears at the near surface. Lower than 2 km, the aerosol vertical profile has a similar impact as type 1, but at the 2–4 km range, the average extinction coefficient is 0.25 km^{-1} in type 2, whereas it is only less than 0.1 km^{-1} in type 1. The two-layered structure mainly occurs in spring and winter seasons, and a few days in other months, as shown in Figure 3d.

The τ_{500}, $\alpha_{440/870}$, and S_p for all observations (total number of data points = 503) under the spread and histograms for two-layer aerosol structure (type 2) are shown in Figure 5b. From the Figure 5b it can be observed that τ_{500} is between 0.1 and 1.5, with an average of 0.51 ± 0.26; $\alpha_{440/870}$ is between 0.34 and 1.85, with an average of 1.23 ± 0.26; and S_p ranges between 30 and 80 sr with an average of 52 ± 23 sr.

These two-layered aerosol structures high-up and low-down are caused by the different delivery mechanisms; therefore, they can be tracked by the use of air mass back trajectories analysis in the HYSPLIT model (e.g., [23]). In addition, according to a database (https://airtw.epa.gov.tw/CHT/Forecast/Sand.aspx) provided by the Taiwan Environmental Protection Agency (EPA) which determines the date for a sandstorm event, we can classify dust transport as taking place down low. The dust and biomass-burning cases for two-layer aerosol structure are defined as follows:

1. Two-layer aerosol structure (dust case): Sandstorm pollution recorded by the EPA on the same day and the main source of air mass from the AC region (mainly from the elevated regions over northern China and Mongolia and covered the longest distance along China's coast). Those mineral dust particles usually transport near the surface behind of a frontal system, and occasionally intrude to higher levels via frontal dynamics.

2. Two-layer aerosol structure (biomass-burning case): Transport of springtime biomass-burning emissions from the source regions over SA (comprising Cambodia, Laos, Myanmar, and Thailand). Those aerosols mainly transport in free atmosphere and arrive in Taiwan at higher elevation.

3.4. Dust Case

From the vertical profile of the aerosol extinction coefficient, it can be seen that dust mainly transports in the height region of 500–2000 m (Figure 6a). We carefully checked the airflow source with backward trajectories; it was found that low in the atmosphere (500 m), the aerosol sources were from the northern desert regions of China. In addition, we also found that three dust events (on 18 March, 28 March, and 25 November in 2005) had influences on biomass-burning transport from Indochina through the backward trajectory analysis at 2500 m. A recent publication by [70] also demonstrated a long-range transport event with concurrent dust and biomass-burning aerosol layers based on EPA-NCU lidar observation. More study related to the coexisting of dust and biomass-burning over the west Pacific can be explored in the future.

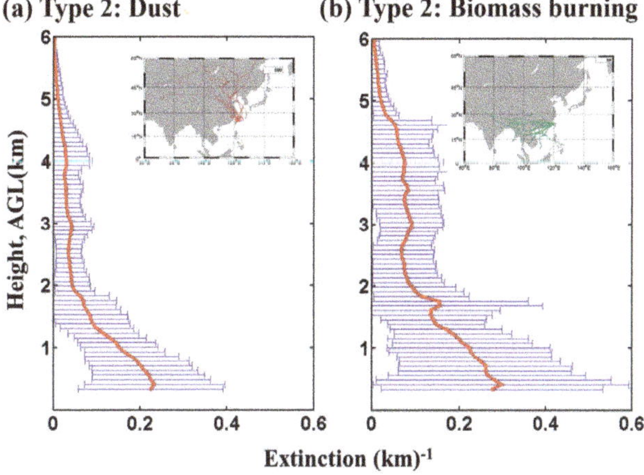

Figure 6. Averaged aerosol extinction profiles for (**a**) type 2 dust source and (**b**) type 2 biomass-burning source. In the subplots, the respective air mass backward trajectories for dust (ended at the height of 500 m) and for biomass-burning (ended at the height of 3 km) cases are illustrated.

The τ_{500} for the dust events over the years was at about 0.45 ± 0.16 with the averaged $\alpha_{440/870}$ of about 1.10 ± 0.24. Coarse particle size was higher when compared to other regions, and S_p was found in the range of 20 to 60 sr with an average of 40 ± 16 sr (Figure 7a), and followed the normal Gaussian distribution with the median value of 40 sr. The estimation S_p of dust aerosols in African region was 37 ± 9 sr, and it also found that it changed in the course of dust transportation when mixed with urban pollutants ([71]).

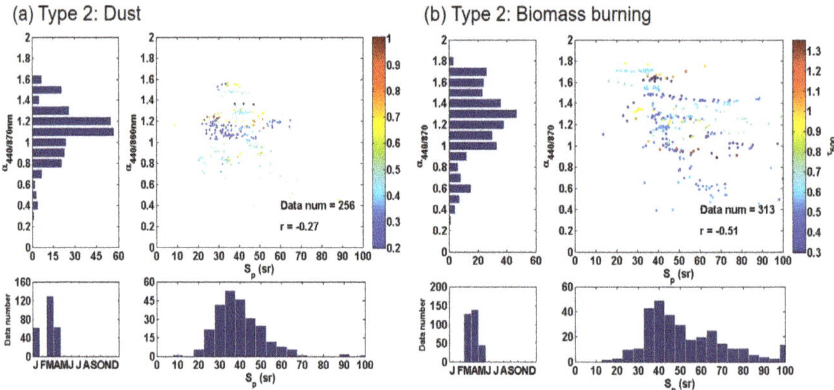

Figure 7. Scatter plot and data spread distribution histograms of aerosol optical properties for (**a**) type 2 dust cases and (**b**) type 2 biomass-burning cases.

3.5. Biomass-Burning Case

The vertical distribution of aerosol extinction coefficient shows a significant secondary peak 0.1 km^{-1} in the height range of 3–4 km approximately (Figure 6b). Based on the back trajectory at the height of 3000 m above sea level, we found that the air mass sources were from the SA region (taken over 22 days; total number of data points = 313). The analyzed averaged optical characteristics for biomass-burning were τ_{500} of 0.58 ± 0.20, $\alpha_{440/870}$ of 1.22 ± 0.33, and S_p of 53 ± 21 sr (Table 2). The distribution of S_p shows double peaks, the first peak in the range of 20 to 50 sr and the second peak in the range of 60 to 70 sr (Figure 7b). Compared to the mean S_p value of 47 ± 21 sr in northern Taiwan, it was found that the first peak could be linked to mixture of anthropogenic and biomass-burning aerosols. When the air by the burning of biomass and other outside influences affects the area, it gives rise in S_p to a more apparent second peak (60 to 70 sr) due to high absorption of light by the high altitude aerosols. The similar value of S_p as 63 ± 10 sr for African savannah biomass-burning was also reported by [17].

Table 2. Same as Table 1 but for aerosol optical properties of a two-layered aerosol structure (type 2) with respect to different emission sources. Due to the layer being decoupled, the dust and biomass-burning layer can co-occur, which causes the double count issue of data number.

Aerosol Optical Property	Type 2	Emission Source of Type 2	
		Dust	Biomass Burning
Data Number	503	256	313
τ_{500}	0.51 ± 0.22	0.45 ± 0.16	0.58 ± 0.20
$\alpha_{440/870}$	1.23 ± 0.26	1.10 ± 0.24	1.22 ± 0.32
S_p [sr]	52 ± 23	40 ± 16	53 ± 21
ω_0	0.93 ± 0.02	0.96 ± 0.03	0.93 ± 0.02
g	0.77 ± 0.03	0.77 ± 0.02	0.76 ± 0.02
n_r	1.46 ± 0.06	1.46 ± 0.05	1.48 ± 0.03
n_i	0.0055 ± 0.0020	0.0033 ± 0.0037	0.0054 ± 0.0036

4. Discussion

In this section, we further discuss the S_p of urban, dust, and biomass-burning aerosols representing the subtropical region of East Asia based on aforementioned analysis with additional estimations. In the previous section, based on statistics of long-term (2005–2012) EPA-NCU MPL dataset, we found that the average value of S_p is 47 ± 21 sr. The results report a mixture of dust and biomass-burning aerosols. In order to quantify the S_p value of urban aerosols, we selected cases with two constraints: (1) type 1 vertical profile and only from AC source region, (2) no dust events. We estimated the mean S_p of urban aerosols to be 42 ± 18 sr, which is in agreement with the value of 45 ± 10 suggested by [19] (see Table 3). For oceanic aerosols, we suggested that the S_p value can be estimated from PO source region in type 1, which is 30 ± 12 sr. This value is very similar to the values (32–33) provided by [72]. It should be noted that a higher value observed in our study may have been due to the mixing of local urban aerosols during the transport to northern Taiwan.

Table 3. Comparison of S_p values (unit in sr) in the literature.

Aerosol type	This Study	Literature	References
Urban	42 ± 18	45 ± 10	[19]
Oceanic	30 ± 12	33 ± 6, 32 ± 6	[72]
Dust	34 ± 6	37 ± 9, 35 ± 5	[71,73]
Biomass-burning	69 ± 12	63 ± 10	[17]

Regarding dust aerosol, we selected cases that met the criteria of type 1 classification and EPA dust event dates. Only one dust event was recorded on 29 January 2007, and comprises 37 profiles. As shown in Figure 8a, the dust plume was transported near the surface and constrained within 1.4 km height. The maximum aerosol extinction coefficient was 0.2 ± 0.04 km^{-1} observed at 1 km. Figure 8b shows the scattering plot of S_p and $\alpha_{440/870}$ for all type 1 data points, in which red dots highlight the dust signature. The $\alpha_{440/870}$ was about 1.05, suggesting a slightly coarser particle for transported dust when compared to the mean $\alpha_{440/870}$ value of 1.29 for type 1 at EPA-NCU station. The number distribution of S_p for dust data is shown in Figure 8c. The S_p ranged between 25 and 50 sr with an average of 34 ± 6 sr. The value shows good agreement with previous studies ([71,73]) that measured dust particles (Table 3). It is worth mentioning that a large portion of dust events may have a mixture of anthropogenic aerosols during the transport. Those mixed events can have a higher lidar ratio of up to 40 sr, as described in dust cases of type 2 profile.

The transport of biomass-burning aerosols to Taiwan is often characterized by upper-level transport, which makes it difficult to separate the biomass-burning S_p value from the total columnar mean S_p value (as obtained from the EPA-NCU MPLNET). The histogram of S_p number distribution for biomass-burning cases in type 2 shows a bimodal distribution (Figure 7b), where the first peak ranged from 30 to 50 sr and the second peak ranged from 60 to 80 sr. In type 2, the aerosol type contributing to the surface layer can be urban, oceanic, or dust, whereas biomass-burning aerosols are attributed to the upper level. Large portions of aerosols are suspended within the surface layer. As listed in Table 3, the mean S_p values for urban, oceanic, and dust are between 30–42 sr, and correspond to the first peak of Figure 7b. That suggests the domination of surface layer transport. In other words, the second peak of Figure 7b may represent the domination of upper-level transport, which is mainly attributed to biomass-burning aerosols. The mode S_p of the second peak was 69 ± 12 sr, representing biomass-burning aerosols originating from Indochina. Our result shows a little higher value compared to the literature value of 63 ± 10 sr for African biomass-burning ([17]).

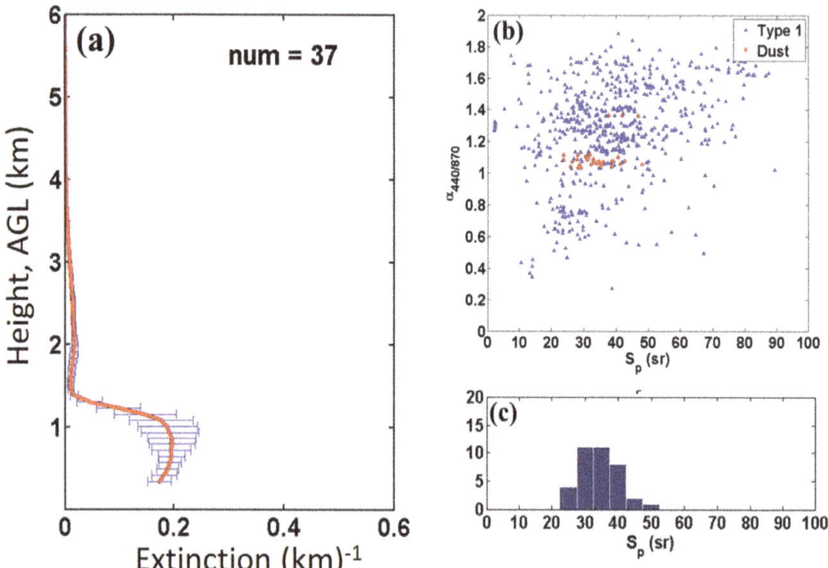

Figure 8. Single-layer dust transfer cases: (**a**) aerosol extinction profiles, (**b**) scatter plots of S_p and $\alpha_{440/870}$ for type 1 (blue triangles) and dust (red diamonds) events, and (**c**) histogram of S_p number distribution.

5. Conclusions

Lidar ratio (S_p) is a distinctive aerosol optical property and can be used for aerosol type classification, which is important for pollution source identification and radiative forcing assessment. This study analysed the long-term (2005–2012) observations of MPLNET and AERONET from a rural site (EPA-NCU) in northern Taiwan located in the western North Pacific. Those observations, consisting of aerosol optical characteristics, extinction coefficient profiles, and lidar ratios, in conjunction with air mass trajectories, were further used to investigate the relationship between aerosol types and aerosol optical properties over the region. The important findings are as follows:

- The long-term average values of τ_{500}, $\alpha_{440/870}$, and S_p were found to be 0.41 ± 0.28, 1.25 ± 0.33, and (47 ± 21) sr, respectively.

- The highest τ_{500} (0.72 ± 0.28) and S_p (54 ± 23 sr) values in the month of March were primarily attributed to the long-range transport of biomass-burning aerosols from Indochina.

- Two types of aerosol structure were classified based on the vertical cross-sections of aerosols. Type 1 aerosol structure (near-surface aerosol transport mainly prevails during October–November) showed low values τ_{500} and S_p and mainly originated from the AC region (τ_{500}: 0.24 ± 0.11; S_p: 39 ± 8 sr), PO region (0.19 ± 0.10; 30 ± 12 sr), and SA region (0.19 ± 0.11; 38 ± 17 sr).

- Type 2 aerosol transport (mainly during March–April at an altitude of 3 to 6 km) was associated with high τ_{500} (0.51 ± 0.22), $\alpha_{440/870}$ (1.23 ± 0.26), and S_p (52 ± 23 sr).

- For type 2 dust case, the estimated τ_{500}, $\alpha_{440/870}$, and S_p were found to be 0.45 ± 0.16, 1.10 ± 0.24, and (40 ± 16) sr, respectively. For type 2 biomass-burning case, the estimated τ_{500}, $\alpha_{440/870}$, and S_p were found to be 0.58 ± 0.20, 1.22 ± 0.33, and (53 ± 21) sr, respectively.

- S_p values, for four major aerosol types over northern Taiwan, were estimated to be 42 ± 18 sr (urban), 34 ± 6 sr (dust), 69 ± 12 sr (biomass-burning), and 30 ± 12 sr (oceanic).

The findings of this study offer a reference for future attempts to monitor the sources of aerosol over Taiwan and western North Pacific. Results of S_p value provide useful references for aerosol classification and air pollution identification in this region. This kind of study is also important for resolving aerosol-perturbed atmospheric circulation (e.g., [74–76]) and temporal changes in regional air quality (e.g., [77,78]). Nevertheless, due to the principle of Mie elastic lidar, the S_p value only represents a columnar integrated number by using a retrieval method. The newly available lidar ratio datasets with data fusion technology, such as AERONET version 3 product (e.g., [34,79]), MPLNET version 3 depolarization ratio product, Raman lidar ([72]), and high spectral resolution lidar (e.g., [42–45]) do also give an aspect of aerosol classification. More observation and data analyses are needed to further explore the relationship between aerosol types and aerosol optical properties over the region.

Author Contributions: Conceptualization, S.-H.W.; methodology, H.-W.L., H.-Y.H., and Y.-C.W.; writing—original draft preparation, S.-H.W.; writing—review and editing, S.K.P.; project administration, N.-H.L. and E.J.W.; funding acquisition, S.-C.C. All authors have read and agreed to the published version of the manuscript.

Funding: This work is funded by the Taiwan Environmental Protection Administration under grant EPA-107-FA11-03-A071 and by the Ministry of Science and Technology under grant 108-2111-M-008-025.

Acknowledgments: The NASA MPLNET and AERONET are funded by the NASA Earth Observing System and Radiation Sciences Program. The authors gratefully acknowledge the NOAA Air Resources Laboratory for the provision of the HYSPLIT transport and dispersion model used in this publication. We are also grateful to the National Center for High-performance Computing for computer time and facilities.

Conflicts of Interest: The authors declare no conflict of interest.

References

1. Charlson, R.J.; Schwartz, S.E.; Hales, J.M.; Cess, R.D.; Coakley, J.D.; Hansen, J.E.; Hofmann, D.J. Climate forcing by anthropogenic aerosols. *Science* **1992**, *255*, 423–430. [CrossRef] [PubMed]

2. Penner, J.E.; Charlson, R.J.; Hales, J.M.; Laulainen, N.S.; Leifer, R.; Novakov, T.; Ogren, J.; Radke, L.F.; Schwartz, S.E.; Travis, L. Quantifying and minimizing uncertainty of climate forcing by anthropogenic aerosols. *Bull. Am. Meteorol. Soc.* **1994**, *75*, 375–400. [CrossRef]

3. Intergovernmental Panel on Climate Change. *Climate Change 2007, Working Group I Report—The Physical Science Basis*; Solomon, S., Qin, D., Manning, M., Marquis, M., Averyt, K., Tignor, M.M.B., LeRoy Miller, H., Jr., Chen, Z., Eds.; Cambridge University Press: New York, NY, USA, 2007.

4. Kaufman, Y.J.; Tanre, D.; Gordon, H.R.; Nakajima, T.; Lenoble, J.; Frouin, R.; Grassl, H.; Hermann, B.M.; King, M.D.; Teillet, P.M. Passive Remote Sensing of Tropospheric Aerosol and Atmospheric Correction for the Aerosol Effect. *J. Geophys. Res.* **1997**, *102*, 16815–16830. [CrossRef]

5. Yu, H.B.; Chin, M.; Winker, D.M.; Omar, A.H.; Liu, C.K.; Diehl, T. Global View of Aerosol Vertical Distributions from CALIPSO Lidar Measurements and GOCART Simulations: Regional and Seasonal Variations. *J. Geophys. Res.* **2010**, *115*, D00H30. [CrossRef]

6. Haywood, J.M.; Ramaswamy, V. Global sensitivity studies of the direct radiative forcing due to anthropogenic sulfate and blackcarbon aerosols. *J. Geophys. Res.* **1998**, *103*, 6043–6058. [CrossRef]

7. Gadhavi, H.; Jayaraman, A. Airborne lidar study of the vertical distribution of aerosols over Hyderabad, an urban site in central India, and its implication for radiative forcing calculations. *Ann. Geophys.* **2006**, *24*, 2461–2470. [CrossRef]

8. Chew, B.N.; Campbell, J.; Hyer, E.J.; Salinas, S.V.; Reid, J.S.; Welton, E.J.; Holben, B.N.; Liew, S.C. Relationship between aerosol optical depth and particulate matter over Singapore: Effects of aerosol vertical distributions. *Aerosol Air Qual. Res.* **2016**, *16*, 2818–2830. [CrossRef]

9. Wang, S.-H.; Welton, E.J.; Holben, B.N.; Tsay, S.-C.; Lin, N.-H.; Giles, D.; Stewart, S.A.; Janjai, S.; Nguyen, X.A.; Hsiao, T.-C. Vertical distribution and columnar optical properties of springtime biomass-burning aerosols over Northern Indochina during 2014 7-SEAS campaign. *Aerosol Air Qual. Res.* **2015**, *15*, 2037–2050. [CrossRef]

10. Hee, W.S.; San Lim, H.; Jafri, M.Z.M.; Lolli, S.; Ying, K.W. Vertical profiling of aerosol types observed across monsoon seasons with a Raman lidar in Penang Island, Malaysia. *Aerosol Air Qual. Res.* **2016**, *16*, 2843–2854. [CrossRef]

11. Muñoz Magnino, R.; Alcafuz, R.I. Variability of urban aerosols over Santiago, Chile: Comparison of surface PM10 concentrations and remote sensing with ceilometer and lidar. *Aerosol Air Qual. Res.* **2012**, *12*, 8–19. [CrossRef]

12. Jayaraman, A.; Subbaraya, B.H. In situ measurements of aerosol extinction profiles and their spectral dependencies at tropospheric levels. *Tellus* **1993**, *45B*, 473–478. [CrossRef]

13. Ramachandran, S.; Jayaraman, A. Balloon-borne study of the upper tropospheric and stratospheric aerosols over a tropical station in India. *Tellus* **2003**, *55B*, 820–836. [CrossRef]

14. He, Q.S.; Li, C.C.; Mao, J.T.; Lau, A.K.H. A study on aerosol extinction-to-backscatter ratio with combination of micro-pulse lidar and MODIS over Hong Kong. *Atmos. Chem. Phys.* **2006**, *6*, 3243–3256. [CrossRef]

15. Devara, P.C.S.; Raj, P.E.; Pandithurai, G. Aerosol-profile measurements in the lower troposphere with four-wavelength bistatic argon-ion lidar. *Appl. Opt.* **1995**, *34*, 4416–4425. [CrossRef]

16. Parameswaran, K.; Rajan, R.; Vijayakumar, G.; Rajeev, K.; Moorthy, K.K.; Nair, P.R.; Satheesh, S.K. Seasonal and long term variations of aerosol content in the atmospheric mixing region at a tropical station on the Arabian sea-coast. *J. Atmos. Sol. Terr. Phys.* **1998**, *60*, 17–25. [CrossRef]

17. Welton, E.J.; Voss, K.J.; Gordon, H.R.; Maring, H.; Smirnov, A.; Holben, B.N.; Schmid, B.; Livingston, J.M.; Russell, P.B.; Durkee, P.A.; et al. Ground-based lidar measurements of aerosols during ACE-2: Lidar description, results, and comparisons with other ground-based and airborne measurements. *Tellus* **2000**, *52B*, 636–651. [CrossRef]

18. Chen, W.B.; Kuze, H.; Uchiyama, A.; Suzuki, Y.; Takeuchi, N. One-year observation of urban mixed layer characteristics at Tsukuba, Japan using a micro pulse lidar. *Atmos. Environ.* **2001**, *35*, 4273–4280. [CrossRef]

19. Campbell, J.R.; Welton, E.J.; Spinhirne, J.D.; Ji, Q.; Tsay, S.; Piketh, S.J.; Barenbrug, M.; Holben, B.N. Micropulse lidar observations of tropospheric aerosols over northeastern South Africa during the ARREX and SAFARI 2000 dry season experiments. *J. Geophys. Res.* **2003**, *108*, 8497. [CrossRef]

20. Schmid, B.; Ferrare, R.; Flynn, C.; Elleman, R.; Covert, D.; Strawa, A.; Welton, E.; Turner, D.; Jonsson, H.; Redemann, J. How well do state-of-the-art techniques measuring the vertical profile of tropospheric aerosol extinction compare? *J. Geophys. Res.* **2006**, *111*, D05S07. [CrossRef]

21. Hayasaka, T.; Satake, S.; Shimizu, A.; Sugimoto, N.; Matsui, I.; Aoki, K.; Muraji, Y. Vertical distribution and optical properties of aerosols observed over Japan during the Atmospheric Brown Clouds–East Asia Regional Experiment 2005. *J. Geophys. Res.* **2007**, *112*, D22S35. [CrossRef]

22. He, Q.; Li, C.; Mao, J.; Lau, A.K.-H.; Chu, D.A. Analysis of aerosol vertical distribution and variability in Hong Kong. *J. Geophys. Res.* **2008**, *113*, D14211. [CrossRef]

23. Wang, S.-H.; Lin, N.-H.; Chou, M.-D.; Tsay, S.-C.; Welton, E.J.; Hsu, N.C.; Giles, D.M.; Liu, G.-R.; Holben, B.N. Profiling transboundary aerosols over Taiwan and assessing their radiative effects. *J. Geophys. Res.* **2010**, *115*, D00K731. [CrossRef]

24. Kent, G.S.; Trpte, C.R.; Lucker, P.L. Long-term Stratospheric Aerosol and Gas Experiment I and II measurements of upper tropospheric aerosol extinction. *J. Geophys. Res.* **1998**, *103*, 28863–28874. [CrossRef]

25. Spinhirne, J.D.; Palm, S.P.; Hart, W.D.; Hlavka, D.L.; Welton, E.J. Cloud and aerosol measurements from GLAS: Overview and initial results. *Geophys. Res. Lett.* **2005**, *32*, L22S03. [CrossRef]

26. Mamouri, R.-E.; Ansmann, A. Potential of polarization lidar to provide profiles of CCN- and INP-relevant aerosol parameters. *Atmos. Chem. Phys.* **2016**, *16*, 5905–5931.

27. Sassen, K.; Cho, B.S. Subvisual-Thin Cirrus LIDAR Dataset for Satellite Verification and Climatological Research. *J. Appl. Meteorol.* **1992**, *31*, 1275–1285. [CrossRef]

28. Welton, E.J.; Voss, K.J.; Flatau, P.J.; Markowicz, K.; Campbell, J.R.; Spinhirne, J.D.; Gordon, H.R.; Johnson, J.E. Measurements of aerosol vertical profiles and optical properties during INDOEX 1999 using micropulse lidars. *J. Geophys. Res.* **2002**, *107*, 8019. [CrossRef]

29. Pelon, J.; Flamant, C.; Chazette, P.; Leon, J.-F.; Tanre, D.; Sicard, M.; Satheesh, S.K. Characterization of aerosol spatial distribution and optical properties over the Indian Ocean from airborne LIDAR and radiometry during INDOEX'99. *J. Geophys. Res.* **2002**, *107*, 8029. [CrossRef]

30. McGill, M.J.; Hlavka, D.L.; Hart, W.D.; Welton, E.J.; Campbell, J.R. Airborne lidar measurements of aerosol optical properties during SAFARI-2000. *J. Geophys. Res.* **2003**, *108*, 8493. [CrossRef]

31. Flamant, C.; Pelon, J.; Chazette, P.; Trouillet, V.; Quinn, P.K.; Frouin, R.; Bruneau, D.; Leon, J.-F.; Bates, T.; Johnson, J.; et al. Airborne lidar measurements of aerosol spatial distribution and optical properties over the Atlantic Ocean during an European pollution outbreak of ACE-2. *Tellus* **2000**, *52B*, 662–677. [CrossRef]

32. Tsay, S.C.; Hsu, N.C.; Lau, W.K.M.; Li, C.; Gabriel, P.M.; Ji, Q.; Holben, B.N.; Judd Welton, E.; Nguyen, A.X.; Janjai, S.; et al. From BASE-ASIA toward 7-SEAS: A satellite-surface perspective of boreal spring biomass-burning aerosols and clouds in Southeast Asia. *Atmos. Environ.* **2013**, *78*, 20–34. [CrossRef]

33. Wang, S.-H.; Tsay, S.-C.; Lin, N.-H.; Chang, S.-C.; Li, C.; Welton, E.J.; Holben, B.N.; Hsu, N.C.; Lau, K.-M.; Lolli, S.; et al. Origin, transport, and vertical distribution of atmospheric pollutants over the northern South China Sea during 7-SEAS/Dongsha experiment. *Atmos. Environ.* **2013**, *78*, 124–133. [CrossRef]

34. Shin, S.-K.; Tesche, M.; Noh, Y.; Müller, D. Aerosol-type classification based on AERONET version 3 products. *Atmos. Meas. Tech.* **2019**, *12*, 3789–3803. [CrossRef]

35. Siomos, N.; Fountoulakis, I.; Natsis, A.; Drosoglou, T.; Bais, A. Automated Aerosol Classification from Spectral UV Measurements Using Machine Learning Clustering. *Remote Sens.* **2020**, *12*, 965. [CrossRef]

36. Chen, Q.-X.; Huang, C.-L.; Yuan, Y.; Mao, Q.-J.; Tan, H.-P. Spatiotemporal Distribution of Major Aerosol Types over China Based on MODIS Products between 2008 and 2017. *Atmosphere* **2020**, *11*, 703. [CrossRef]

37. Nicolae, D.; Vasilescu, J.; Talianu, C.; Binietoglou, I.; Nicolae, V.; Andrei, S.; Antonescu, B. A neural network aerosol-typing algorithm based on lidar data. *Atmos. Chem. Phys.* **2018**, *18*, 14511–14537. [CrossRef]

38. Ansmann, A.; Seifert, P.; Tesche, M.; Wandinger, U. Profiling of fine and coarse particle mass: Case studies of Saharan dust and Eyjafjallajökull/Grimsvötn volcanic plumes. *Atmos. Chem. Phys.* **2012**, *12*, 9399–9415. [CrossRef]

39. Kim, M.-H.; Omar, A.H.; Tackett, J.L.; Vaughan, M.A.; Winker, D.M.; Trepte, C.R.; Hu, Y.; Liu, Z.; Poole, L.R.; Pitts, M.C.; et al. The CALIPSO version 4 automated aerosol classification and lidar ratio selection algorithm. *Atmos. Meas. Tech.* **2018**, *11*, 6107–6135. [CrossRef]

40. Georgoulias, A.K.; Marinou, E.; Tsekeri, A.; Proestakis, E.; Akritidis, D.; Alexandri, G.; Zanis, P.; Balis, D.; Marenco, F.; Tesche, M.; et al. A first case study of CCN concentrations from spaceborne lidar observations. *Remote Sens.* **2020**, *12*, 1557. [CrossRef]

41. Ackermann, J. The extinction-to-backscattering ratio oftropospheric aerosol: A numerical study. *J. Atmos. Ocean. Technol.* **1998**, *15*, 1043–1050. [CrossRef]

42. Perez-Ramirez, D.; Whiteman, D.N.; Veselovskii, I.; Colarco, P.; Korenski, M.; da Silva, A. Retrievals of aerosol single scattering albedo by multiwavelength lidar measurements: Evaluations with NASA Langley HSRL-2 during discover-AQ field campaigns. *Remote Sens. Environ.* **2019**, *222*, 144–164. [CrossRef]

43. Burton, S.P.; Ferrare, R.A.; Hostetler, C.A.; Hair, J.W.; Rogeres, R.R.; Obland, M.D.; Obland, M.D.; Butler, C.F.; Cook, A.L.; Harper, D.B.; et al. Aerosol classification using airborne High Spectral Resolution Lidar measurements—Methodology and examples. *Atmos. Meas. Tech.* **2012**, *5*, 73–98. [CrossRef]

44. Burton, S.P.; Ferrare, R.A.; Vaughan, M.A.; Omar, A.H.; Rogers, R.R.; Hostetler, C.; Hair, J.W. Aerosol classification from airborne HSRL and comparisons with the CALIPSO vertical feature mask. *Atmos. Meas. Tech.* **2013**, *6*, 1397–1412. [CrossRef]

45. Burton, S.P.; Vaughan, M.A.; Ferrare, R.A.; Hostetler, C.A. Separating mixtures of aerosol types in airborne High Spectral Resolution Lidar data. *Atmos. Meas. Tech.* **2014**, *7*, 419–436. [CrossRef]

46. Streets, D.G.; Bond, T.C.; Carmichael, G.R.; Fernandes, S.D.; Fu, Q.; He, D.; Klimont, Z.; Nelson, S.M.; Tsai, N.Y.; Wang, M.Q.; et al. An inventory of gaseous and 10 primary aerosol emissions in Asia in the year 2000. *J. Geophys. Res.* **2003**, *108*, 8809. [CrossRef]

47. Wenig, M.; Spichtinger, N.; Stohl, A.; Held, G.; Beirle, S.; Wagner, T.; Jahne, B.; Platt, U. Intercontinental transport of nitrogen oxide pollution plumes. *Atmos. Chem. Phys.* **2003**, *3*, 387–393. [CrossRef]

48. Chiang, C.-W.; Das, S.K.; Nee, J.-B. An iterative calculation to derive extinction-to-backscatter ratio based on lidar measurements. *J. Quant. Spectrosc. Radiat. Transf.* **2008**, *109*, 1187–1193. [CrossRef]

49. Lin, N.-H.; Tsay, S.C.; Maring, H.B.; Yen, M.C.; Sheu, G.R.; Wang, S.H.; Chi, K.H.; Chuang, M.T.; Ou-Yang, C.F.; Fu, J.S.; et al. An overview of regional experiments on biomass burning aerosols and related pollutants in Southeast Asia: From BASE-ASIA and the Dongsha Experiment to 7-SEAS. *Atmos. Environ.* **2013**, *78*, 1–19. [CrossRef]

50. Chen, W.-N.; Tsao, C.C.; Nee, J.B. Tropospheric and Lower Stratospheric Temperature Measurement by Rayleigh Lidar. *J. Atmos. Sol. -Terr. Phys.* **2004**, *66*, 39–49. [CrossRef]

51. Chen, W.-N.; Tsai, F.-J.; Chou, C.C.-K.; Chang, S.-Y.; Chen, I.-W.; Chen, J.-P. Cases studies of Asian Dust in the Free Atmosphere by Raman Depolarization Lidar at Taipei, Taiwan. *Atmos. Environ.* **2007**, *41*, 7698–7714. [CrossRef]

52. Chen, W.-N.; Tsai, F.-J.; Chou, C.C.-K.; Chang, S.-Y.; Chen, Y.-W.; Chen, J.-P. Cases Study of Relationship Between Water-soluble Ca^{2+} and Lidar Depolarization Ratio for Spring Aerosol in the Boundary Layer. *Atmos. Environ.* **2007**, *41*, 1440–1455.

53. Chen, W.-N.; Chen, Y.-W.; Chou, C.C.-K.; Chang, S.-Y.; Lin, P.-H.; Chen, J.-P. Columnar optical properties of tropospheric aerosol by combined lidar and sunphotometer measurements at Taipei, Taiwan. *Atmos. Environ.* **2009**, *43*, 2700–2716. [CrossRef]

54. Tsai, J.-H.; Hsu, Y.-C.; Yang, J.-Y. The relationship between volatile organic profiles and emission sources in ozone episode region, a case study in southern Taiwan. *Sci. Total Environ.* **2004**, *328*, 131–142. [CrossRef] [PubMed]

55. Omar, A.H.; Winker, D.M.; Vaughan, M.A.; Hu, Y.; Trepte, C.R.; Ferrare, R.A.; Lee, K.-P.; Hostetler, C.A.; Kittaka, C.; Rogers, R.R.; et al. The CALIPSO Automated Aerosol Classification and Lidar Ratio Selection Algorithm. *J. Atmos. Ocean. Technol.* **2009**, *26*, 23–32. [CrossRef]

56. Liu, C.-M.; Young, C.-Y.; Lee, Y.-C. Influence of Asian Dust Storms on Air Quality in Taiwan. *Sci. Total. Environ.* **2006**, *368*, 884–897. [CrossRef] [PubMed]

57. Chuang, M.-T.; Chou, C.-C.-K.; Lin, N.-H.; Takami, A.; Hsiao, T.-C.; Lin, T.-H.; Fu, J.-S.; Pani, S.K.; Lu, Y.R.; Yang, T.Y. A simulation study on PM$_{2.5}$ sources and meteorological characteristics at the northern tip of Taiwan in the early stage of the Asian haze period. *Aerosol Air Qual. Res.* **2017**, *17*, 3166–3178. [CrossRef]

58. Pani, S.K.; Ou-Yang, C.F.; Wang, S.H.; Ogren, J.A.; Sheridan, P.J.; Sheu, G.R.; Lin, N.H. Relationship between long-range transported atmospheric black carbon and carbon monoxide at a high-altitude background station in East Asia. *Atmos. Environ.* **2019**, *210*, 86–99. [CrossRef]

59. Wang, S.H.; Hung, R.-Y.; Lin, N.-H.; Gómez-Losada, A.; Pires, J.C.M.; Shimada, K.; Hatakeyama, S.; Takami, A. Estimation of background PM$_{2.5}$ concentrations for an air-polluted environment. *Atmos. Res.* **2020**, *231*, 104636. [CrossRef]

60. Campbell, J.R.; Hlavka, D.L.; Welton, E.J.; Flynn, C.J.; Turner, D.D.; Spinhirne, J.D.; Scott, V.S.; Hwang, I.H. Full-time, Eye-Safe Cloud and Aerosol Lidar Observation at Atmospheric Radiation Measurement Program Sites: Instrument and Data Processing. *J. Atmos. Ocean. Technol.* **2002**, *19*, 431–442. [CrossRef]

61. Welton, E.J.; Campbell, J.R.; Spinhirne, J.D.; Scott, V.S. Global monitoring of clouds and aerosols using a network of micro-pulse lidar systems. In *Lidar Remote Sensing for Industry and Environmental Monitoring*; Singh, U.N., Itabe, T., Sugimoto, N., Eds.; SPIE: Bellingham, WA, USA, 2001; Volume 4153, pp. 151–158.

62. Holben, B.N.; Eck, T.F.; Slutsker, I.A.; Tanre, D.; Buis, J.P.; Setzer, A.; Vermote, E.; Reagan, J.A.; Kaufman, Y.J.; Nakajima, T.; et al. AERONET: A federated instrument network and data archive for aerosol characterization. *Remote Sens. Environ.* **1998**, *66*, 1–16. [CrossRef]

63. Fernald, F.G. Analysis of atmospheric lidar observations: Some comments. *App. Opt.* **1984**, *23*, 652–653. [CrossRef] [PubMed]

64. Ansmann, A.; Riebesell, M.; Wandinger, U.; Weitkamp, C.; Voss, E.; Lahmann, W.; Michaelis, W. Combined Raman elastic backscatter lidar for vertical profiling of moisture, aerosol extinction, backscatter, and lidar ratio. *Appl. Phys.* **1992**, *B55*, 18–28. [CrossRef]

65. Barnaba, F.; De Tomasi, F.; Gobbi, G.P.; Perrone, M.R.; Tafuro, A. Extinction versus backscatter relationship for lidar applications at 351 nm: Maritime and desert aerosol simulations and comparison with observations. *Atmos. Res.* **2004**, *70*, 229–259. [CrossRef]

66. Draxler, R.R.; Hess, G.D. An overview of the HYSPLIT-4modeling system for trajectories, dispersion and deposition. *Aust. Meteorol. Mag.* **1998**, *47*, 295–308.

67. Stein, A.F.; Draxler, R.R.; Rolph, G.D.; Stunder, B.J.B.; Cohen, M.D.; Ngan, F. NOAA's HYSPLIT Atmospheric Transport and Dispersion Modeling System. *Bull. Amer. Meteor. Soc.* **2015**, *96*, 2059–2077. [CrossRef]

68. Kishcha, P.; Wang, S.H.; Lin, N.H.; da Silva, A.; Lin, T.H.; Lin, P.H.; Liu, G.R.; Starobinets, B.; Alpert, P. Differentiating between Local and Remote Pollution over Taiwan. *Aerosol Air Qual. Res.* **2018**, *18*, 1788–1798. [CrossRef]

69. Pani, S.K.; Wang, S.H.; Lin, N.H.; Tsay, S.C.; Lolli, S.; Chuang, M.T.; Lee, C.T.; Chantara, S.; Yu, J.Y. Assessment of aerosol optical property and radiative effect for the layer decoupling cases over the northern South China Sea during the 7-SEAS/Dongsha experiment. *J. Geophys. Res. Atmos.* **2016**, *121*, 4894–4906. [CrossRef]

70. Dubovik, O.; Sinyuk, A.; Lapyonok, T.; Holben, B.N.; Mishchenko, M.; Yang, P.; Eck, T.F.; Volten, H.; Munoz, O.; Veihelmann, B.; et al. Application of spheroid models to account for aerosol particle nonsphericity in remote sensing of desert dust. *J. Geophys. Res.* **2006**, *111*, D11208. [CrossRef]

71. Huang, K.; Fu, J.S.; Lin, N.-H.; Wang, S.-H.; Dong, X.; Wang, G. Superposition of Gobi dust and Southeast Asian biomass burning: The effect of multisource long-range transport on aerosol optical properties and regional meteorology modification. *J. Geophys. Res. Atmos.* **2019**, *124*, 9464–9483. [CrossRef]

72. Müller, D.; Ansmann, A.; Mattis, I.; Tesche, M.; Wandinger, U.; Althausen, D.; Pisani, G. Aerosol-type-dependent lidar ratios observed with Raman lidar. *J. Geophys. Res.* **2007**, *112*, D16202. [CrossRef]
73. Voss, K.J.; Welton, E.J.; Quinn, P.K.; Frouin, R.; Miller, M.; Reynolds, R.M. Aerosol optical depth measurements during the Aerosols99 experiment. *J. Geophys. Res.* **2001**, *106*, 20811–20819. [CrossRef]
74. Powell, D.M.; Reagan, J.; Rubio, M.A.; Erxleben, W.H.; Spinhirne, J.D. ACE-2 multiple angle micro-pulse lidar observations from Las Galletas, Tenerife, Canary Islands. *Tellus B Chem. Phys. Meteo.* **2000**, *52*, 652–661. [CrossRef]
75. Lau, K.M.; Kim, M.K.; Kim, K.M. Asian monsoon anomalies induced by aerosol direct effects. *Clim. Dyn.* **2006**, *26*, 855–864. [CrossRef]
76. Kuntz, L.B.; Schrag, D.P. Impact of Asian aerosol forcing on tropical Pacific circulation and the relationship to global temperature trends. *J. Geophys. Res. Atmos.* **2016**, *121*, 14403–14413. [CrossRef]
77. Cheng, F.-Y.; Hsu, C.-H. Long-term variations in $PM_{2.5}$ concentrations under changing meteorological conditions in Taiwan. *Sci. Rep.* **2019**, *9*, 6635. [CrossRef]
78. Chang, F.-J.; Chang, L.-C.; Kang, C.-C.; Wang, Y.-S.; Huang, A. Explore spatio-temporal $PM_{2.5}$ features in northern Taiwan using machine learning techniques. *Sci. Total Environ.* **2020**, *736*, 139656. [CrossRef]
79. Shin, S.-K.; Tesche, M.; Kim, K.; Kezoi, M.; Tatarov, B.; Müller, D.; Noh, Y. On the spectral depolarisation and lidar ratio mineral dust provided in AERONET version 3 inversion product. *Atmos. Chem. Phys.* **2018**, *18*, 12735–12746. [CrossRef]

Article

Observation of Turbulent Mixing Characteristics in the Typical Daytime Cloud-Topped Boundary Layer over Hong Kong in 2019

Tao Huang [1], Steve Hung-lam Yim [1,2,3,*], Yuanjian Yang [2,4], Olivia Shuk-ming Lee [5], David Hok-yin Lam [5], Jack Chin-ho Cheng [1] and Jianping Guo [6]

[1] Department of Geography and Resource Management, The Chinese University of Hong Kong, Hong Kong 999077, China; taohuang@link.cuhk.edu.hk (T.H.); jackcch@cuhk.edu.hk (J.C.-h.C.)

[2] Institute of Environment, Energy and Sustainability, The Chinese University of Hong Kong, Hong Kong 999077, China; yyj1985@nuist.edu.cn

[3] Stanley Ho Big Data Decision Analytics Research Centre, The Chinese University of Hong Kong, Hong Kong 999077, China

[4] School of Atmospheric Physics, Nanjing University of Information Science and Technology, Nanjing 210000, China

[5] Hong Kong Observatory, Hong Kong 999077, China; olee@hko.gov.hk (O.S.-m.L.); dlam@hko.gov.hk (D.H.-y.L.)

[6] State Key Laboratory of Severe Weather, Chinese Academy of Meteorological Sciences, Beijing 100081, China; jpguo@cma.gov.cn

* Correspondence: steveyim@cuhk.edu.hk

Received: 24 April 2020; Accepted: 7 May 2020; Published: 11 May 2020

Abstract: Turbulent mixing is critical in affecting urban climate and air pollution. Nevertheless, our understanding of it, especially in a cloud-topped boundary layer (CTBL), remains limited. High-temporal resolution observations provide sufficient information of vertical velocity profiles, which is essential for turbulence studies in the atmospheric boundary layer (ABL). We conducted Doppler Light Detection and Ranging (LiDAR) measurements in 2019 using the 3-Dimensional Real-time Atmospheric Monitoring System (3DREAMS) to reveal the characteristics of typical daytime turbulent mixing processes in CTBL over Hong Kong. We assessed the contribution of cloud radiative cooling on turbulent mixing and determined the altitudinal dependence of the contribution of surface heating and vertical wind shear to turbulent mixing. Our results show that more downdrafts and updrafts in spring and autumn were observed and positively associated with seasonal cloud fraction. These results reveal that cloud radiative cooling was the main source of downdraft, which was also confirmed by our detailed case study of vertical velocity. Compared to winter and autumn, cloud base heights were lower in spring and summer. Cloud radiative cooling contributed ~32% to turbulent mixing even near the surface, although the contribution was relatively weaker compared to surface heating and vertical wind shear. Surface heating and vertical wind shear together contributed to ~45% of turbulent mixing near the surface, but wind shear can affect up to ~1100 m while surface heating can only reach ~450 m. Despite the fact that more research is still needed to further understand the processes, our findings provide useful references for local weather forecast and air quality studies.

Keywords: turbulent mixing; cloud; LiDAR; Hong Kong

---

1. Introduction

Turbulent mixing is a crucial part of the atmospheric boundary layer (ABL), which modulates the variation in temperature, flow velocity, moisture, and atmospheric composition and thus acts as a bridge between the top of the ABL and the surface [1,2]. Vertical velocity, which is the key parameter reflecting the characteristics of turbulent mixing, is generally driven by sources such as heat transfer from a warm ground surface (surface heating), vertical wind shear, or a combination of both processes in a cloud-free boundary layer [3]. When reaching the capping inversion area at the top of the ABL, updrafts due to surface heating (thermal) can penetrate into the stable layer above, inducing dry air intrusion which can even reach the heights close to the surface in form of downdrafts [4]. Studies concentrating on convectively driven turbulent mixing have been well documented both in observation [5,6] and numerical modeling of boundary layer meteorology [7,8].

However, the frequent occurrence of boundary-layer clouds reduces buoyancy by suppressing direct radiative forcing during the day and also reduces thermal loss during the night [9,10]. Cloud-topped-cooling in stratocumulus layers was highlighted as it results in top-down mixing from the cloud layer toward the surface during day and night while less of a cooling effect was found in cloudless or cumulus-topped boundary layers [11]. This kind of cooling effect has been investigated both in observation [12,13] and numerical modeling [14]. Meanwhile, in terms of airflow, it has been well established that low-level vertical wind shear induced by surface friction helps to organize and maintain convective systems through exchanging moist thermals [15] and regulating aerosol effects on deep convective clouds [16]. Nevertheless, the detailed contribution of low-level vertical wind shear to the turbulent mixing generated in a cloud-topped boundary layer (CTBL) is still vague, especially in the daytime. Although cloud-topped radiative cooling effect and vertical wind shear process are critical to control the structure of turbulence and aerosol distribution, which is essential for human health, long-term observations of the high-time-resolution vertical profile of updrafts and downdrafts still remain relatively limited [3].

Except for the in situ observations by radiosondes and aircrafts [17–19], many studies have been carried out based on the ground-based remote sensing observations such as ceilometer, radar wind profiler, or different types of Light Detection and Ranging (LiDAR) systems [20–24]. For example, observation projects such as Cloudnet, European Aerosol Research Lidar Network (EARLINET), EUMETNET Profiling Programme (E-PROFILE), and Aerosol, Clouds and Trace Gases Research Infrastructure (ACTRIS) in European countries combined radar, ceilometer, and LiDAR together to obtain aerosol, clouds, trace gases, and wind profiles [25–28]. Among all the LiDAR techniques such as micro-pulse LiDAR (MPL), elastic/Raman LiDAR, and depolarization LiDAR [29,30], Doppler LiDARs can simultaneously provide wind profile data as well as aerosol attenuated backscatter at a high vertical and temporal resolution [31–33]. Doppler LiDAR performance has been proved in several previous studies on the mixing layer [34]. For example, Pearson et al. (2010) directly observed the mixing process using the Doppler LiDAR and argued that this was the most appropriate methodology to analyze the dispersion in the lower atmosphere [31]. In addition, the turbulence measured by Doppler LiDARs was used to derive a mixing layer [35], whereas other studies combined several techniques such as multi-wavelength LiDAR and microwave radiometers [36,37]. Overall, Doppler LiDAR measurements can provide data at high time and vertical resolutions, allowing for detailed analyses of turbulent mixing.

Hong Kong is a typical coastal city with a significant ABL variation which is frequently affected by boundary layer clouds. Most previous observational studies based on ceilometer and aerosol LiDARs mainly focused on the retrieval method and diurnal variation of the height of ABL (ABLH) using backscatter coefficients [20,38]. Few studies applied high time- and vertical-resolution observations for vertical wind profile in Hong Kong. Recently, the 3-Dimensional Real-time Atmospheric Monitoring System (3DREAMS) was established to measure and analyze the vertical profiles of horizontal wind speed and direction, vertical wind velocity as well as aerosol attenuated backscatter in Hong Kong [33]. Three 1.5-μm Doppler LiDAR units (Halo Photonics Stream Line XR Scanning Doppler LiDAR system) were installed at Hong Kong Observatory weather station: King's Park (LiDAR KP), the Physical Geography Experimental Station of the Chinese University of Hong Kong (LiDAR CUHK), and Hung Shui Kiu Church at Yuen Long (LiDAR HSK) to better observe the atmospheric boundary layer conditions at the south, east, and northwest of Hong Kong, respectively. LiDARs located at King's Park and the Chinese University of Hong Kong were operated regularly since late 2017, while the one at Yuen Long was established recently. For sufficient data availability, this study used the data of the LiDAR KP from 3DREAMS to determine the turbulent mixing characteristics in the typical daytime CTBL over Hong Kong in 2019.

Using vertical profiles of vertical velocity and horizontal wind at high time- and vertical-resolutions from a ground-based Doppler LiDAR combined with surface and upper-air meteorological data, we characterized the typical diurnal variation of ground-level meteorological parameters and vertical wind profile with a focus on clouds over Hong Kong in 2019. Then four cloud-topped cases in different seasons in 2019 were selected to determine the relative contribution of cloud radiative cooling, surface heating, and vertical wind shear to turbulent mixing during daytime in CTBL. In Section 2, the operating specification of Doppler LiDAR employed in this study, together with all the other meteorological data, is described. Major results are shown and discussed in Sections 3 and 4, respectively. Finally, a conclusion is given in Section 5.

2. Materials and Methods

2.1. Ground-Based Doppler LiDAR Measurement Using 3DREAMS

The Doppler LiDAR (LiDAR KP) used in this paper was located at the King's Park Meteorological Station (KPMS, 22.311 N°, 114.173 E°, Figure 1) of the Hong Kong Observatory (HKO). KPMS is the only upper-air sounding station in Hong Kong. This Doppler LiDAR is a part of 3DREAMS [33]. The instrument was fixed on a concrete foundation in a flat grass field, with an altitude of 65 m above the mean sea level. The observation site was located on a small hill in the urban area with lots of buildings in the surroundings. The east, west, and south sides are close to the sea, with the north close to the mountains.

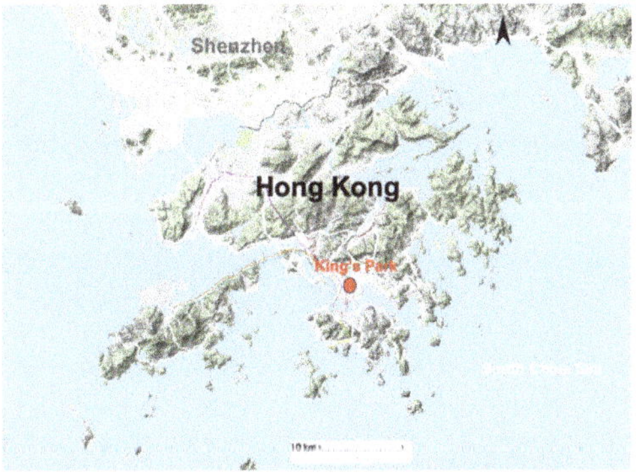

Figure 1. Location (red dot) of the LiDAR at King's Park (LiDAR KP). The figure was built on a map obtained from Google map, https://www.google.com/maps.

The Doppler LiDAR is a Halo Photonics 1.5-μm pulsed Doppler instrument from Halo Photonics Company. The instrument had been used in studies exploring the characteristics of the planetary boundary layer (PBL) in the tropical and mid-latitude environments [33,39]. The Doppler LiDAR employed in this study operated round the clock and has been set up for an optimized vertical resolution of 30 m in the boundary layer up to around 3 km altitude. It was operated in velocity azimuth display (VAD) scanning mode for obtaining one horizontal wind profiles every 10 min using 6 beams around a circle at an elevation angle of 75°. The Doppler LiDAR was also operated in vertical stare mode to collect attenuated backscatter coefficient measurements for the rest of the time, with vertical spatial and temporal resolutions of 30 m and one second, respectively. Erroneous data, which might be caused by measurement or instrument errors, were removed based on signal intensity

(SI). A threshold value of 1.01 was defined for SI, which refers to a signal-noise-ratio (SNR) of −20 dB. All data below this threshold were removed. The quality of Doppler wind data was checked before using it in our analysis. Results show that the percentage difference between the averaged horizontal wind speeds provided by LiDAR and upper air sounding data for heights less than 1.00 km was less than 10%, which indicates a sufficient agreement between both data sources [33].

2.2. Meteorological Data from Hong Kong Observatory

The ground surface meteorological data and upper-air sounding data were provided by the Hong Kong Observatory (HKO). The regular wind profile observations at KPMS at Hong Kong Time (HKT) 08:00 (UTC 00:00) and HKT 20:00 (UTC 12:00) were used to validate the wind profiles from our Doppler LiDAR at KPMS. Daily total rainfall recorded at the KPMS station in 2019 can be found at HKO websites (https://www.weather.gov.hk/en/cis/dailyExtract.htm?y=2019). Hourly surface meteorological data recorded at KPMS, including relative humidity, solar radiation, and cloud fraction, were also used.

2.3. Information on the Sampling Days

Clear and cloudy days were defined according to the averaged cloud fraction between 07:00 and 17:00 HKT. Firstly, we removed the days when rainfall was recorded according to the meteorological data from HKO. 172 no-rainfall days were selected totally, accounting for 47% of all days in 2019. Of which, the average cloud fraction less than 20% were defined as clear days, whereas the remaining days were defined as cloudy days. Table 1 shows the information of the samples used in our research.

Table 1. Numbers of no rain days and cloudy days in each season, and the percentage of no rain days with a cloudy sky.

Season	Number of No Rain Days	Number of Cloudy Days	Percentage of No Rain Days with A Cloudy Sky
DJF	51	40	78%
MAM	38	32	84%
JJA	25	20	80%
SON	58	35	61%

2.4. Validation of LiDAR and Definitions for Turbulent Mixing

Wind profile of our LiDAR has been well validated in 3DREAMS. Moreover, ABLH is also an important parameter for the LiDAR's validation. Currently, there are many mature algorithms such as potential temperature gradient or Richardson number methods for radiosondes, attenuated backscatter coefficient methods for ceilometers, and particle extinction and backscatter coefficient methods for Raman and elastic LiDARS, respectively to retrieve the ABLH [40,41]. Some newly developed algorithms also obtained finer results for this purpose [42–44]. In order to verify the accuracy of the retrieval ABL from our LiDAR, here we applied the gradient method in the profile of water vapor mixing ratio and virtual potential temperature from regular sounding data to compare with the gradient of our attenuated backscatter coefficient (i.e., maximum $-\nabla\beta$, as $\nabla\beta$ should always be negative) on a clear case: 08:00 HKT (UTC 00) 16 October (surface wind speed: 4 m/s; wind direction: Northeast; temperature: 24.2 °C; Cloud fraction: 0). An attenuated backscatter coefficient profile should be moving averaged using a time window of 3 min and vertical window of 3 layers [22,32]. The ABLHs retrieved based on LiDAR's attenuated backscatter coefficients and upper air sounding profiles of water vapor mixing ratio and virtual potential temperature show good consistency (Figure 2), confirming the accuracy of our LiDAR's capacity in clear-sky ABLH retrieval.

The turbulent mixing was typically defined as the vertical velocity variance $\sigma_w^2(z)$ within 1 minute higher than a threshold of 0.1 m^2s^{-2} at a certain layer [31]. Therefore, here the definitions of variance and skewness are given:

$$\sigma_w^2(z) = \overline{w'(z)^2}, \tag{1}$$

$$s_w(z) = \overline{\left(\frac{w'(z)}{\sigma_w(z)}\right)^3},$$ (2)

where z is the height (m), w is the vertical wind velocity (m/s). w' is the deviation between instant vertical velocity and the mean. As $s_w(z)$ is a noisy profile, a 10-min moving average was applied for the visualization in Figure 7. Correspondingly, the 1-min skewness $s_w(z)$ of vertical velocity was also calculated to represent the sources of turbulent mixing [4,32]. If $s_w(z) > 0$, the mixing is induced by surface heating and vertical wind shear processes. If $s_w(z) < 0$, the mixing is induced by cloud-driven ones.

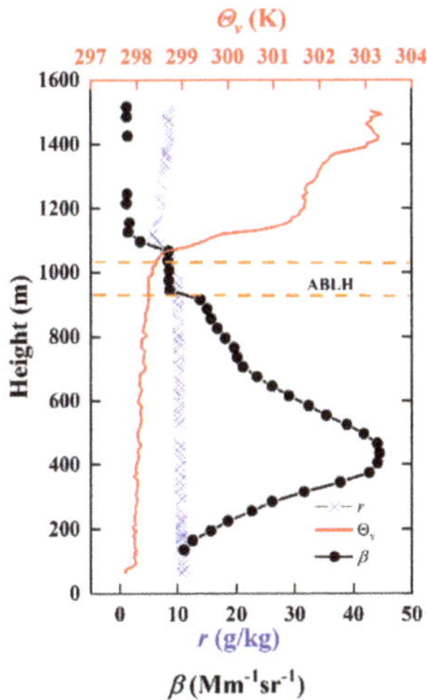

Figure 2. Vertical profiles of water vapor mixing ratio r (blue cross), virtual potential temperature θ_v (red line), and attenuated backscatter coefficient (black dot) at 08:00 Hong Kong Time (HKT) (UTC 00) on 16 October. The range of atmospheric boundary layer height (ABLH) has been marked with orange dash lines.

Wind shear was found to play an important role in a radiative-convective equilibrium system as well as in the regulating of the aerosol distribution [16,45]. It has significant effects on the momentum transport under different weather conditions, especially in a CTBL in which thermals are relatively lower than that in a clear one. Bulk vertical wind shear (VWS) was used to represent the intensity of vertical wind shear between two layers and was defined as:

$$\text{VWS} = \frac{\left(\Delta u^2 + \Delta v^2\right)^{1/2}}{\Delta z},$$ (3)

where u and v are the two components of horizontal wind, whereas Δz is the layer thickness which was 30 m in this study. We also detected the cloud base height according to a threshold of $\log(\beta) > -4$ and combined the method documented in Manninen's study [32].

3. Results

3.1. Vertical Wind Profiles and Ground-Level Meteorological Parameters on Cloudy Days in 2019

This section provides a description of meteorological conditions on cloudy days. Notably, among all the 172 no rain days in 2019, 74% of them were cloudy days. Figure 3a shows that the largest cloud fraction occurred in spring followed by summer, winter, and autumn. A similar seasonal variation was also observed in relative humidity (RH). The maximum of RH (90%) was observed at

07:00 in spring, while the minimum (60%) occurred at 13:00 in autumn. Nevertheless, the difference between maximum and minimum in each season was within 15%. In terms of solar radiation, a negative association with cloud fraction was presented. Although the temperature in summer was the highest, the higher cloud fraction in summer blocked part of solar radiation and made the surface solar radiation lower than that in autumn. Overall, cloud fraction was larger than 40% during the daytime of cloudy days in 2019, indicating that the effects associated with clouds on CTBL are ineluctable.

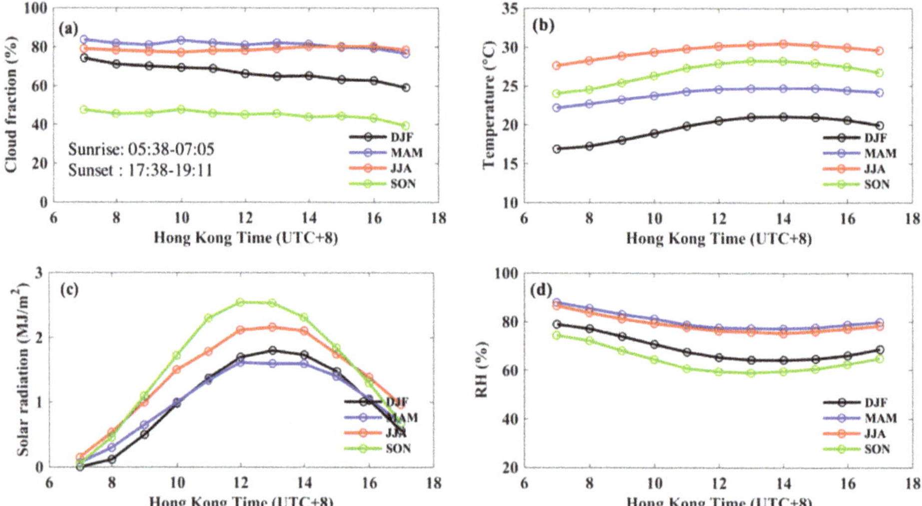

Figure 3. Diurnal variation of meteorological parameters of (**a**) cloud fraction, (**b**) temperature, (**c**) solar radiation, and (**d**) relative humidity (RH) from ground-level observation data at King's Park Meteorological Station (KPMS) on cloudy days in 2019. Sunrise and sunset marked in the figure were extracted from the Hong Kong Observatory website (https://www.hko.gov.hk/tc/gts/astron2019/almanac2019_index.htm). A t-test for the differences between seasons was conducted. The t-test results show that all the seasonal differences can pass a 95% test except the solar radiation difference between winter and spring ($p < 0.2$) and RH difference between spring and summer ($p < 0.3$).

Figure 4 shows the diurnal variation of a cloudy-day wind profile within 1500 m in different seasons. For horizontal wind speed, the maximum occurred at 1035 m in summer with a value of 12.22 m/s at 13:00 HKT. Similarly, an obvious increase in wind speed appeared above 700 m in each season. On cloudy days, the prevailing wind direction was southeast in summer, while that was northwest in autumn and winter. Figure 4f shows that the prevailing wind direction varied significantly in spring. For vertical velocity, downdrafts were more common in spring while more updrafts were observed in autumn.

As revealed by Figure 3a, the highest cloud fraction was observed in spring while the lowest was obtained in autumn. The lower cloud fraction allowed more solar radiation reaching the ground that enhanced surface heating and thus updrafts, which were more common in the afternoon in autumn as shown in Figure 4l. Figure 5 shows the probability density function curve of cloud base height on cloudy days in different seasons. Cloud base height with the highest probability was 1395 m and 1425 m in winter and autumn, respectively, whereas that was 1095 m in spring and 1185 m in summer. The lower cloud base height in spring and summer reflected the more synoptic systems in the two seasons such as trough and typhoon, which were associated with low and thick clouds. It should also be noted that the distribution of cloud base height in winter was flatter than that in other seasons, showing the largest variations in cloud base height in winter.

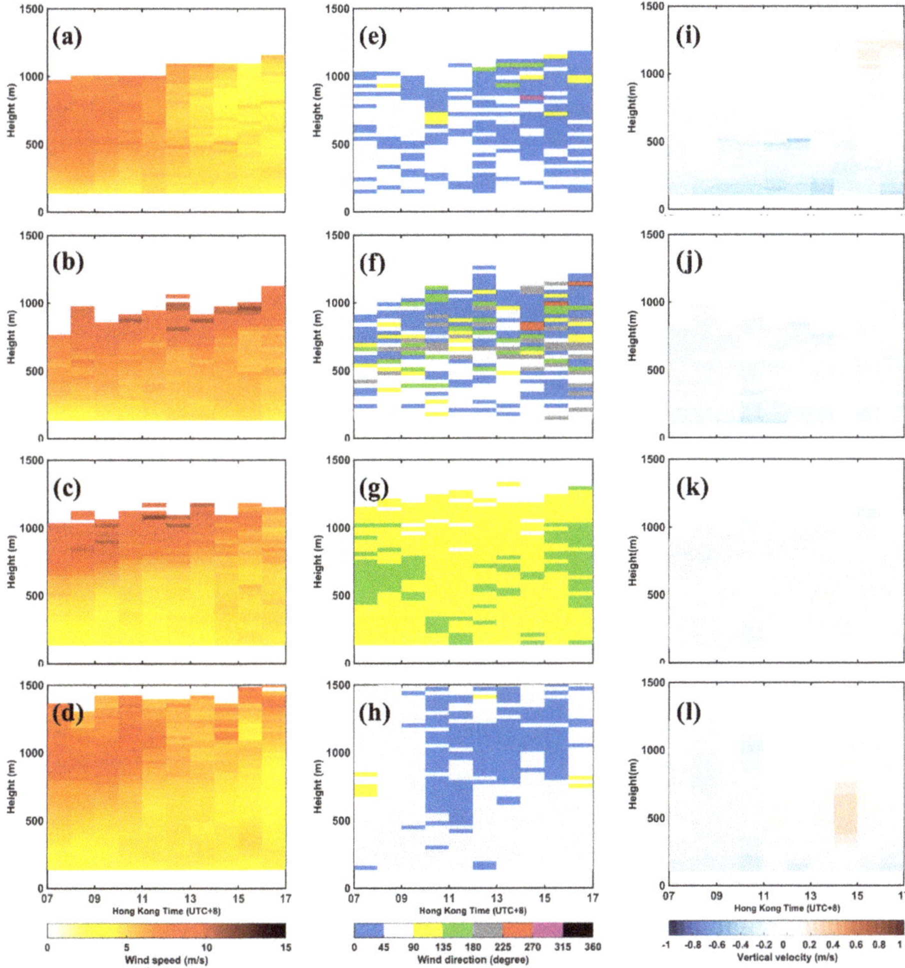

Figure 4. Diurnal variation of wind profiles of horizontal wind speed (left), horizontal wind direction (middle), and vertical velocity (right) on cloudy days in 2019. The rows from top to bottom represent winter, spring, summer and autumn, respectively.

3.2. Case Study of Vertical Velocity and Turbulent Mixing Characteristics

To assess vertical velocity and turbulent mixing characteristics on cloudy days, four cases with high cloud fraction and thick clouds in different seasons were selected for a detailed analysis. The four cases included 1 Feb, 1 May, 22 Aug and 23 Nov. Figure 6 shows the vertical velocity profiles from 07:00 to 17:00 HKT in the four cases. 10-min cloud base height derived from attenuated backscatter coefficient was marked as black dots. As shown in the results, evident downdrafts were observed under the cloud base especially during 10:00 to 10:30 on 01 Feb, 10:00 to 12:00 on 01 May, 09:30 to 12:00 on 22 Aug and 16:00 to 17:00 on 23 Nov. Some evident red patches occurred in each case might be driven by relatively strong surface heating at that time. These characteristics further confirm that cloud-induced radiative cooling is one of the main sources of downdrafts during the daytime on cloudy days.

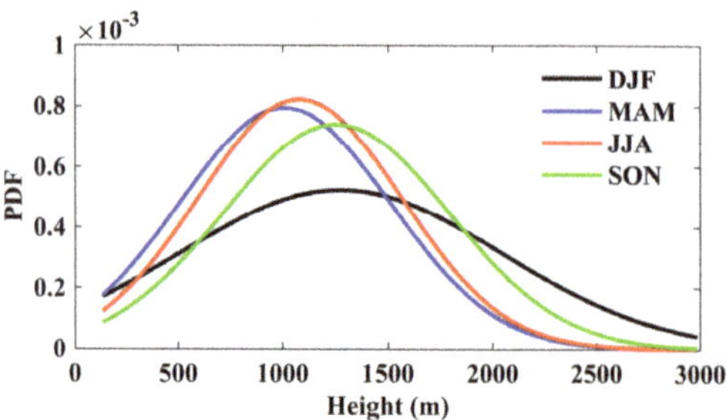

Figure 5. The probability distribution function of cloud base height within the atmospheric boundary layer during 07:00 to 17:00 HKT on cloudy days in 2019. It is noted that the curves were built based on data at a 10-min resolution.

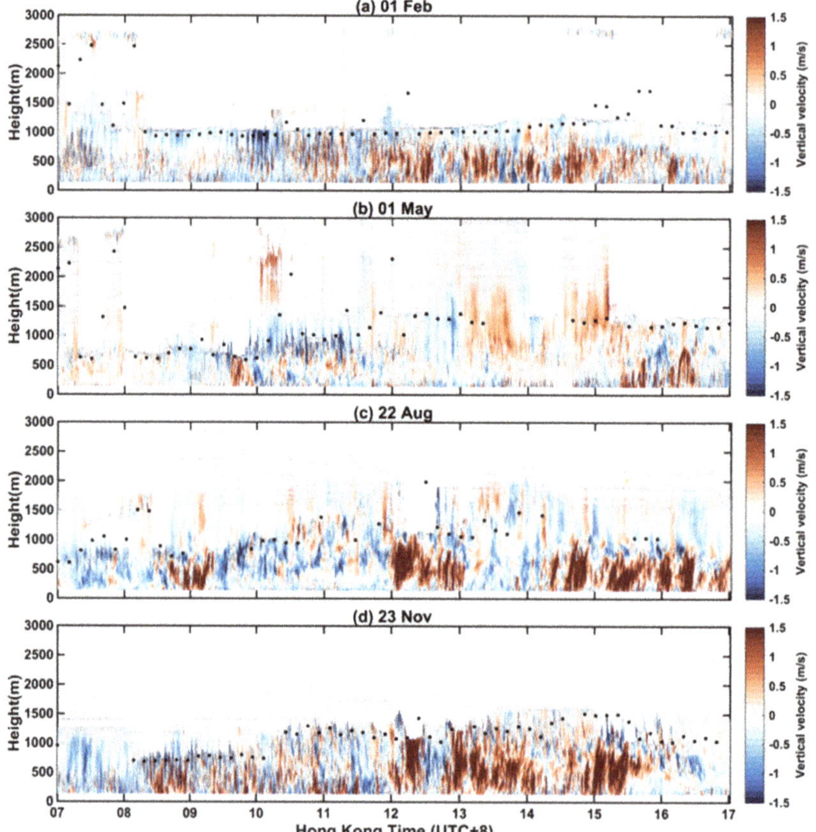

Figure 6. Vertical velocity profiles from 07:00 to 17:00 HKT on (**a**) 01 Feb, (**b**) 01 May, (**c**) 22 Aug, (**d**) 23 Nov. 10-min cloud base height has been marked as black dots in each profile. Positive vertical velocity indicates updraft.

To further understand the characteristics of turbulent mixing in the CTBL, the variance and skewness of the 1-min interval vertical velocity were calculated. Figure 7 (left panels) shows significant vertical velocity variances ($\sigma_w^2 > 0.1$), implying the significant turbulent mixing generated from the near-surface layer to the bottom of clouds. Figure 7 (right panels) shows the skewness in the four cases. Three sources (i.e., cloud radiative cooling, surface heating and vertical wind shear) were considered as the main driving forces of turbulent mixing in the CTBL. It is noted that negative skewness implies the turbulent mixing due to cloud radiative cooling whereas the positive one implies that the turbulent mixing was induced by surface heating, vertical wind shear, or a combination of them [32]. As expected, turbulent mixing near the ground surface was dominantly controlled by positive skewness. Nevertheless, the negative skewness was also evidently observed especially in Figure 7f,g. In the variance of these two cases, a clear weaker variance was derived where there was negative skewness compared to the place where skewness was positive. This may imply that surface heating and vertical wind shear are relatively larger than cloud radiative cooling effect in terms of contribution to turbulent mixing.

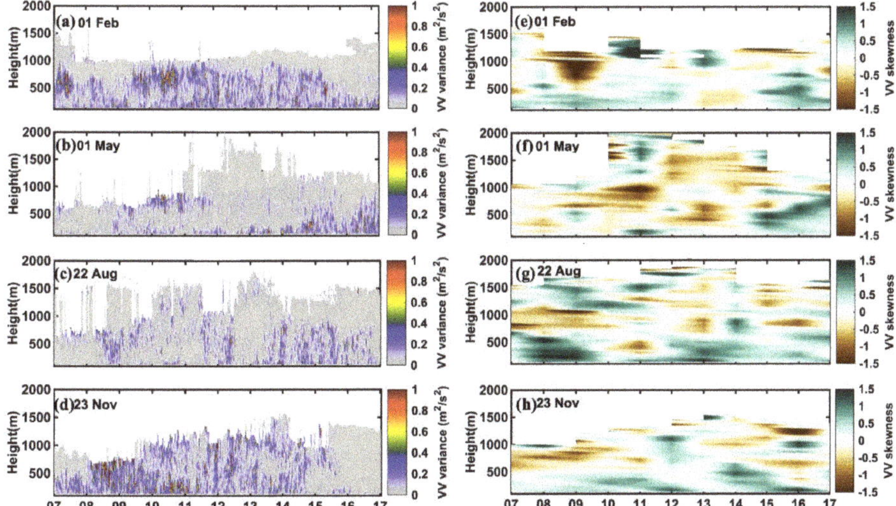

Figure 7. Vertical profile of vertical velocity variance σ_w^2 (left panel) and skewness s_w (right panel). Both σ_w^2 and s_w are 1-min time interval. For visualization, s_w has been moving averaged in a 10-min window size.

3.3. Cloud Contribution to the Turbulent Mixing

To further reveal the contribution of cloud radiative cooling to the intensity of turbulent mixing in different layers in the CTBL, we calculated the proportion of variance with a negative skewness to represent cloud contribution in Figure 8a and with a positive skewness to represent surface heating and vertical wind shear in Figure 8b, respectively. Note that the sum of the two contributions was not equal to 1 as there were many blank zones in the skewness profile where the value was 0. This part was also reflected by Manninen et al. who called this phenomenon as intermittent [32]. Figure 8a shows that cloud radiative cooling contributed ~32% to the turbulent mixing near the surface. It was comparable with the one from surface heating and vertical wind shear. The contribution of cloud-topped cooling increased with the height, but the peak emerged at ~800 m and then the contribution started to decrease again. One explanation is that most clouds in these four cases appeared at around 1000 m or even lower (see Figure 6a–c), and the cloud-topped region can be affected by the entrainment of warm and dry air from above the cloud and the turbulent instability, which has recently been studied by Mellado et al. [46]. Figure 8b explains that the overall contribution of surface heating and vertical wind shear to the turbulent mixing in each layer decreased with height, with the highest mean value of 45% near the surface. Meanwhile, the standard deviation tended to be larger especially above 800 m, indicating that the significant contribution was up to that height.

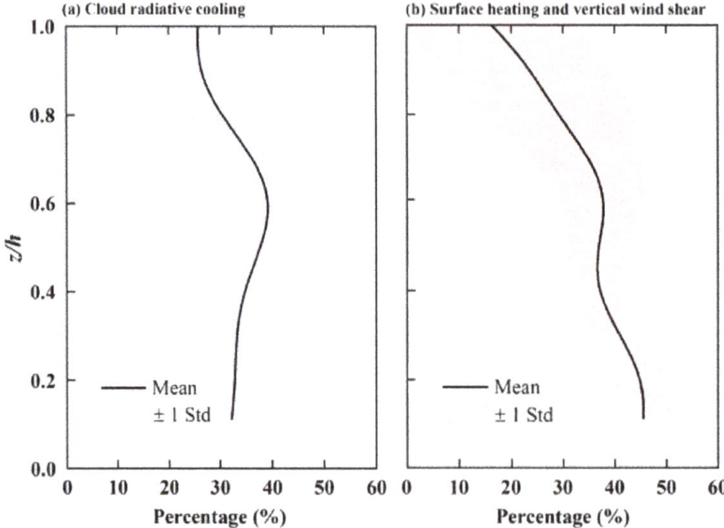

Figure 8. Contribution of (**a**) cloud radiative cooling and (**b**) surface heating and vertical wind shear to the turbulent mixing within the daytime cloud-topped boundary layer (CTBL) in all the cases. Y axis is the altitude (z) normalized by cloud base height (h).

Except for cloud radiative cooling, both surface heating and vertical wind shear have significant impacts on the turbulent mixing. Hence, we further distinguished the effects of these two factors. Figure 9 shows the daytime mean intensity of vertical wind shear between different levels in each case. The highest magnitude of wind shear always occurred between two nearest layers (i.e., 30 m in our study). A significant increase was observed in each profile at ~600 m to 1000 m. The magnitude above this layer height stayed higher than that near-surface. This result reveals that, on cloudy days, the contribution of vertical wind shear on turbulent mixing was more significant aloft than near the surface.

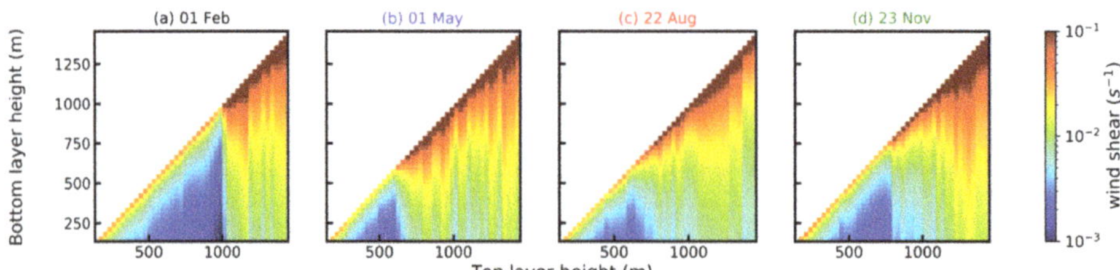

Figure 9. Vertical profile of the averaged daytime vertical wind shear between different heights for (**a**) 01 Feb, (**b**) 01 May, (**c**), 22 Aug, (**d**) 23 Nov. Shading denotes the intensity of wind shear at least two consecutive layers where x-axis is the top layer height and y-axis is the bottom layer height. Note that the colormap is presented at a log scale.

Figure 10 depicts the correlation between turbulent mixing and surface heating due to solar radiation and vertical wind shear. In Figure 10a, surface radiation was used to represent the level of surface heat flux. Significant correlation coefficients were derived from 135 m to 435 m and from 855 m to 975 m, respectively. The appearance of the high-level significant correlation may be still induced by the entrainment discussed above, which was associated with the temperature inversion, resulting in the adverse direction of the reducing effect from surface heating. Figure 10a also shows that, in a typical daytime CTBL, surface heating significantly contributed to the turbulent mixing up to 435 m from ground level. On the other hand, in Figure 10b, vertical wind shear was calculated between every two consecutive layers (30 m). A much higher correlation coefficient can be derived in Figure 10b compared to Figure 10a. Moreover, significant results can reach up to 1095 m in a typical daytime CTBL, indicating that vertical wind shear (mechanical process) played a more important role than surface heating in a CTBL.

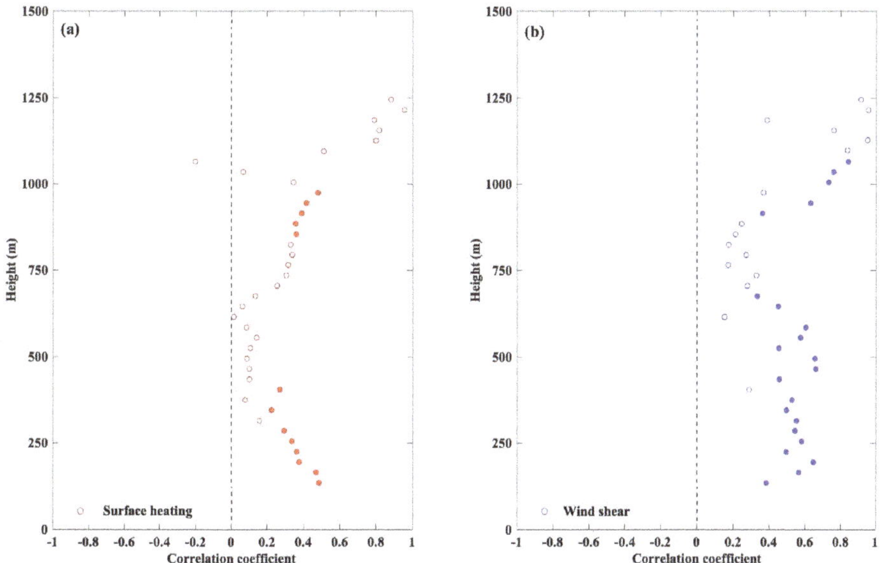

Figure 10. The correlation between turbulent mixing and the two factors: (**a**) surface heating due to solar radiation and (**b**) vertical wind shear. Solid dots represent data with high statistical significance ($p < 0.05$).

4. Discussion

Since Hong Kong is a typical coastal high-density city in the subtropics, the variations of the ABL there are complex. Due to its coastal location and complex terrain conditions, the formation and movement of clouds within the boundary layer occur often throughout a year, and its radiative cooling effect on the ground in the form of downdrafts is considerable. The findings that most clouds appeared at around 1000 m to 1500 m allowed us to figure out the regular depth of the typical CTBL in Hong Kong. Within this range, the surface heating may be important, but radiative fluxes induced by clouds produced local sources of cooling within the CTBL and can greatly influence its turbulent structure and dynamics [1]. Although we took the altitude into account, we did not categorize cloud types in detail and did not consider the situation when clouds are overlapped in the vertical direction, which may have an impact on the ABL characteristics [47].

Until now, the existing research based on observations with a high-time-resolution of the vertical wind profile gave us a general understanding of the characteristics of the turbulent mixing in a CTBL [3,11]. We can see in our results that even in the lowest layer near the surface, the cloud radiative cooling effects can still contribute ~32% of the turbulent mixing, indicating the importance of cloud-induced radiative cooling. This finding provides useful reference for turbulence and aerosol-radiation-cloud interaction research [48–50]. While the influence of meteorology on air quality is critical [51–54], the framework of 3DREAMS provides useful wind and aerosol backscatter data to understand air pollution, especially in the transboundary air pollution [55,56]. For example, during a transboundary air pollution (TAP) episode [56–60], the 3DREAMS can help us to understand the atmospheric stability, and also the contributions of cloud, buoyancy and wind shear in the local-scale vertical mixing procedure. In addition, as revealed by our results that the vertical wind shear can affect up to over 1000 m while the surface heating can only reach around 400 m in the CTBL, we should also consider the influence of wind structure on the regulating of aerosol, especially during heavy pollutant episodes. Pioneering work has been conducted by Yang et al. in 2019 [39].

The lack of direct long-time and high vertical resolution observations has brought great difficulties to atmospheric turbulence research, mesoscale weather forecasting, and thus air pollution studies. More intuitive observations, such as continuous observation of vertical wind profiles, are relatively rare. Once there is a direct observation of vertical wind profile, we can obtain the various boundary layer characteristics under different types of weather conditions, thus the research of boundary layer will have a significant improvement, which has been confirmed by Hogan et al. (2009) that long-term Doppler LiDAR observations would be useful for diagnosing the source of turbulence [11].

In general, our intention was to systematically observe and diagnose the boundary layer characteristics in Hong Kong and clarify the transfer of atmospheric energy in the boundary layer. The recently developed 3DREAMS can monitor the aerosol distribution and 3-dimension wind profile simultaneously. The monitoring of wind profiles can not only clarify the characteristics of atmospheric energy transfer but also monitor aerosols to help attribute local and transboundary air pollutions. Furthermore, our observations can also provide reliable validation data for mesoscale weather models and turbulence models. Our future monitoring will still be conducted under the framework of 3DREAMS to explore the influence of terrain on the boundary layer structure.

5. Conclusions

We employed high temporal and spatial (vertical) observations from a Doppler LiDAR to explore the turbulent mixing characteristics in the daytime cloud-topped boundary layer over Hong Kong in 2019 using 3DREAMS. Ground-level meteorological parameters and typical diurnal variation of vertical wind profile associated with clouds over Hong Kong in 2019 were derived to illustrate the cloud characteristics. Four typical cases from each season in 2019 were selected to illustrate the turbulent mixing characteristics based on the variance and skewness profiles of vertical velocity. Finally, the contribution of cloud radiative cooling, surface heating, and vertical wind shear on turbulent mixing were analyzed.

On cloudy days in 2019, the highest cloud fraction was observed in spring while the lowest was obtained in autumn. Meanwhile, downdrafts were more common in spring while more updrafts were observed in autumn, revealing that cloud radiative cooling is the main source of downdraft. On cloudy days, low-level clouds occurred generally within the range of 1000 to 1500 m. Compared to winter and autumn ones, cloud base heights were lower in spring and summer. Case studies of vertical velocity confirmed that in a typical daytime CTBL in Hong Kong, boundary layer clouds always act as the sink of heat near the bottom of the cloud layer. Although cloud radiative cooling effect on turbulent mixing was relatively weaker compared to surface heating and vertical wind shear, it still contributed ~32% even near the surface. Surface heating and vertical wind shear contributed ~45% together near the surface, but the effect of wind shear can be up to ~1100 m while that of surface heating can only reach ~450 m, revealing that vertical wind shear (mechanical process) plays a more important role than surface heating (thermal) in CTBL.

In general, our findings improved our knowledge of the turbulent mixing layer and provided useful references for weather forecast and air quality studies.

Author Contributions: Conceptualization, S.H.-l.Y.; Formal analysis, T.H., S.H.-l.Y., Y.Y., O.S.-m.L. and D.H.-y.L.; Funding acquisition, S.H.-l.Y.; Methodology, T.H.; Resources, S.H.-l.Y.; Software, J.C.-h.C.; Supervision, S.H.-l.Y.; Visualization, T.H. and Y.Y.; Writing–original draft, T.H.; Writing–review editing, S.H.-l.Y., Y.Y., O.S.-m.L., D.H.-y.L., J.C.-h.C., and J.G. All authors have read and agreed to the published version of the manuscript.

Funding: This work was jointly funded by The Vice-Chancellor's Discretionary Fund of The Chinese University of Hong Kong (grant no. 4930744) the Early Career Scheme of Research Grants Council of Hong Kong (grant no. ECS-24301415), and the project from Environmental Sustainability and Resilience (ENSURE) partnership between CUHK and UoE (http://www.exeter.ac.uk/cuhkpartnership/).

Acknowledgments: We would like to thank the Hong Kong Observatory for providing meteorological data.

Conflicts of Interest: The authors declare no conflict of interest.

References

1. Stull, R.B. *An Introduction to Boundary Layer Meteorology*; Atmospheric and Oceanographic Sciences Library; Springer: Dordrecht, The Netherlands, 1988; ISBN 978-90-277-2768-8.
2. Seibert, P.; Beyrich, F.; Gryning, S.-E.; Joffre, S.; Rasmussen, A.; Tercier, P. Review and intercomparison of operational methods for the determination of the mixing height. *Atmos. Environ.* **2000**, *34*, 1001–1027. [CrossRef]
3. Berg, L.K.; Newsom, R.K.; Turner, D.D. Year-Long Vertical Velocity Statistics Derived from Doppler Lidar Data for the Continental Convective Boundary Layer. *J. Appl. Meteor. Climatol.* **2017**, *56*, 2441–2454. [CrossRef]
4. Ansmann, A.; Fruntke, J.; Engelmann, R. Updraft and downdraft characterization with Doppler lidar: Cloud-free versus cumuli-topped mixed layer. *Atmos. Chem. Phys.* **2010**, *10*, 7845–7858. [CrossRef]
5. Greenhut, G.K.; Singh Khalsa, S.J. Convective Elements in the Marine Atmospheric Boundary Layer. Part I: Conditional Sampling Statistics. *J. Climate Appl. Meteor.* **1987**, *26*, 813–822. [CrossRef]
6. Williams, A.G.; Hacker, J.M. The composite shape and structure of coherent eddies in the convective boundary layer. *Bound.-Layer Meteorol.* **1992**, *61*, 213–245. [CrossRef]
7. Bretherton, C.S.; McCaa, J.R.; Grenier, H. A New Parameterization for Shallow Cumulus Convection and Its Application to Marine Subtropical Cloud-Topped Boundary Layers. Part I: Description and 1D Results. *Mon. Wea. Rev.* **2004**, *132*, 864–882. [CrossRef]
8. Larson, V.E.; Schanen, D.P.; Wang, M.; Ovchinnikov, M.; Ghan, S. PDF Parameterization of Boundary Layer Clouds in Models with Horizontal Grid Spacings from 2 to 16 km. *Mon. Wea. Rev.* **2011**, *140*, 285–306. [CrossRef]
9. Sicart, J.E. Atmospheric controls of the heat balance of Zongo Glacier (16°S, Bolivia). *J. Geophys. Res.* **2005**, *110*, D12106. [CrossRef]
10. De Szoeke, S.P.; Yuter, S.; Mechem, D.; Fairall, C.W.; Burleyson, C.D.; Zuidema, P. Observations of Stratocumulus Clouds and Their Effect on the Eastern Pacific Surface Heat Budget along 20°S. *J. Clim.* **2012**, *25*, 8542–8567. [CrossRef]
11. Hogan, R.J.; Grant, A.L.M.; Illingworth, A.J.; Pearson, G.N.; O'Connor, E.J. Vertical velocity variance and skewness in clear and cloud-topped boundary layers as revealed by Doppler lidar. *Q. J. R. Meteorol. Soc.* **2009**, *135*, 635–643. [CrossRef]
12. Boers, R.; Spinhirne, J.D.; Hart, W.D. Lidar Observations of the Fine-Scale Variability of Marine Stratocumulus Clouds. *J. Appl. Meteor.* **1988**, *27*, 797–810. [CrossRef]
13. Kollias, P.; Albrecht, B. The Turbulence Structure in a Continental Stratocumulus Cloud from Millimeter-Wavelength Radar Observations. *J. Atmos. Sci.* **2000**, *57*, 2417–2434. [CrossRef]
14. Pelly, J.L.; Belcher, S.E. A mixed-layer model of the well-mixed stratocumulus-topped boundary layer. *Bound. Layer Meteorol.* **2001**, *100*, 171–187. [CrossRef]
15. Chen, Q.; Fan, J.; Hagos, S.; Gustafson, W.I.; Berg, L.K. Roles of wind shear at different vertical levels: Cloud system organization and properties. *J. Geophys. Res. Atmos.* **2015**, *120*, 6551–6574. [CrossRef]
16. Fan, J.; Yuan, T.; Comstock, J.M.; Ghan, S.; Khain, A.; Leung, L.R.; Li, Z.; Martins, V.J.; Ovchinnikov, M. Dominant role by vertical wind shear in regulating aerosol effects on deep convective clouds. *J. Geophys. Res. Atmos.* **2009**, *114*. [CrossRef]
17. Guo, J.; Li, Y.; Cohen, J.B.; Li, J.; Chen, D.; Xu, H.; Liu, L.; Yin, J.; Hu, K.; Zhai, P. Shift in the Temporal Trend of Boundary Layer Height in China Using Long-Term (1979–2016) Radiosonde Data. *Geophys. Res. Lett.* **2019**, *46*, 6080–6089. [CrossRef]
18. Lenschow, D.H.; Sun, J. The spectral composition of fluxes and variances over land and sea out to the mesoscale. *Bound. Layer Meteorol.* **2007**, *125*, 63–84. [CrossRef]
19. Martin, S.; Beyrich, F.; Bange, J. Observing Entrainment Processes Using a Small Unmanned Aerial Vehicle: A Feasibility Study. *Bound. Layer Meteorol.* **2014**, *150*, 449–467. [CrossRef]

20. Su, T.; Li, J.; Li, C.; Xiang, P.; Lau, A.K.-H.; Guo, J.; Yang, D.; Miao, Y. An intercomparison of long-term planetary boundary layer heights retrieved from CALIPSO, ground-based lidar, and radiosonde measurements over Hong Kong. *J. Geophys. Res. Atmos.* **2017**, *122*, 3929–3943. [CrossRef]

21. Liu, B.; Ma, Y.; Guo, J.; Gong, W.; Zhang, Y.; Mao, F.; Li, J.; Guo, X.; Shi, Y. Boundary Layer Heights as Derived From Ground-Based Radar Wind Profiler in Beijing. *IEEE Trans. Geosci. Remote Sens.* **2019**, *57*, 8095–8104. [CrossRef]

22. Tang, G.; Zhang, J.; Zhu, X.; Song, T.; Münkel, C.; Hu, B.; Schäfer, K.; Liu, Z.; Zhang, J.; Wang, L.; et al. Mixing layer height and its implications for air pollution over Beijing, China. *Atmos. Chem. Phys.* **2016**, *16*, 2459–2475. [CrossRef]

23. Lolli, S.; Delaval, A.; Loth, C.; Garnier, A.; Flamant, P.H. 0.355-micrometer direct detection wind lidar under testing during a field campaign in consideration of ESA's ADM-Aeolus mission. *Atmos. Meas. Tech.* **2013**, *6*, 3349–3358. [CrossRef]

24. Lolli, S.; Khor, W.Y.; Matjafri, M.Z.; Lim, H.S. Monsoon Season Quantitative Assessment of Biomass Burning Clear-Sky Aerosol Radiative Effect at Surface by Ground-Based Lidar Observations in Pulau Pinang, Malaysia in 2014. *Remote Sens.* **2019**, *11*, 2660. [CrossRef]

25. Illingworth, A.J.; Hogan, R.J.; O'Connor, E.J.; Bouniol, D.; Brooks, M.E.; Delanoé, J.; Donovan, D.P.; Eastment, J.D.; Gaussiat, N.; Goddard, J.W.F.; et al. Cloudnet: Continuous Evaluation of Cloud Profiles in Seven Operational Models Using Ground-Based Observations. *Bull. Am. Meteorol. Soc.* **2007**, *88*, 883–898. [CrossRef]

26. Boesenberg, J.; Matthias, V.; Amodeo, A.; Amoiridis, V.; Ansmann, A.; Baldasano, J.M.; Balin, I.; Balis, D.; Bockmann, C.; Boselli, A.; et al. *EARLINET: A European Aerosol Research Lidar Network to Establish an Aerosol Climatology*; Report/Max-Planck-Institut für Meteorologie, No. 348: Hamburg, Germany, 2003. [CrossRef]

27. Weisshaupt, N.; Arizaga, J.; Maruri, M. The role of radar wind profilers in ornithology. *Ibis* **2018**, *160*, 516–527. [CrossRef]

28. Crenn, V.; Sciare, J.; Croteau, P.L.; Verlhac, S.; Fröhlich, R.; Belis, C.A.; Aas, W.; Äijälä, M.; Alastuey, A.; Artiñano, B.; et al. ACTRIS ACSM intercomparison—Part 1: Reproducibility of concentration and fragment results from 13 individual Quadrupole Aerosol Chemical Speciation Monitors (Q-ACSM) and consistency with co-located instruments. *Atmos. Meas. Tech.* **2015**, *8*, 5063–5087. [CrossRef]

29. Behrendt, A.; Nakamura, T.; Onishi, M.; Baumgart, R.; Tsuda, T. Combined Raman lidar for the measurement of atmospheric temperature, water vapor, particle extinction coefficient, and particle backscatter coefficient. *Appl. Opt. AO* **2002**, *41*, 7657–7666. [CrossRef]

30. Snels, M.; Cairo, F.; Colao, F.; Donfrancesco, G.D. Calibration method for depolarization lidar measurements. *Int. J. Remote Sens.* **2009**, *30*, 5725–5736. [CrossRef]

31. Pearson, G.; Davies, F.; Collier, C. Remote sensing of the tropical rain forest boundary layer using pulsed Doppler lidar. *Atmos. Chem. Phys.* **2010**, *10*, 5891–5901. [CrossRef]

32. Manninen, A.J.; Marke, T.; Tuononen, M.; O'Connor, E.J. Atmospheric boundary layer classification with doppler lidar. *J. Geophys. Res. Atmos.* **2018**, *123*, 8172–8189. [CrossRef]

33. Yim, S.H.L. Development of a 3D Real-Time Atmospheric Monitoring System (3DREAMS) Using Doppler LiDARs and Applications for Long-Term Analysis and Hot-and-Polluted Episodes. *Remote Sens.* **2020**, *12*, 1036. [CrossRef]

34. Milroy, C.; Martucci, G.; Lolli, S.; Loaec, S.; Sauvage, L.; Xueref-Remy, I.; Lavrič, J.V.; Ciais, P.; Feist, D.G.; Biavati, G.; et al. An assessment of pseudo-operational ground-based light detection and ranging sensors to determine the boundary-layer structure in the coastal atmosphere. *Adv. Meteorol.* **2012**, *2012*, 929080. [CrossRef]

35. Barlow, J.F.; Dunbar, T.M.; Nemitz, E.G.; Wood, C.R.; Gallagher, M.W.; Davies, F.; O'Connor, E.; Harrison, R.M. Boundary layer dynamics over London, UK, as observed using Doppler lidar during REPARTEE-II. *Atmos. Chem. Phys.* **2011**, *11*, 2111–2125. [CrossRef]

36. De Arruda Moreira, G.; Da Silva Lopes, F.J.; Guerrero-Rascado, J.L.; João Da Silva, J.; Gomes, A.A.; Landulfo, E.; Alados-Arboledas, L. Analyzing the atmospheric boundary layer using high-order moments obtained from multiwavelength lidar data: Impact of wavelength choice. *Atmos. Meas. Tech.* **2019**, *12*, 4261–4276. [CrossRef]

37. De Arruda Moreira, G.; Guerrero-Rascado, J.L.; Bravo-Aranda, J.A.; Benavent-Oltra, J.A.; Ortiz-Amezcua, P.; Róman, R.; Bedoya-Velásquez, A.E.; Landulfo, E.; Alados-Arboledas, L. Study of the planetary boundary

layer by microwave radiometer, elastic lidar and Doppler lidar estimations in Southern Iberian Peninsula. *Atmos. Res.* **2018**, *213*, 185–195. [CrossRef]

38. Yang, D.; Li, C.; Lau, A.K.-H.; Li, Y. Long-term measurement of daytime atmospheric mixing layer height over Hong Kong. *J. Geophys. Res. Atmos.* **2013**, *118*, 2422–2433. [CrossRef]

39. Yang, Y.; Yim, S.H.L.; Haywood, J.; Osborne, M.; Chan, J.C.S.; Zeng, Z.; Cheng, J.C.H. Characteristics of Heavy particulate matter pollution events over hong kong and their relationships with vertical wind profiles using high-time-resolution doppler lidar measurements. *J. Geophys. Res. Atmos.* **2019**, *124*, 9609–9623. [CrossRef]

40. Melfi, S.H.; Whiteman, D.; Ferrare, R. Observation of Atmospheric Fronts Using Raman Lidar Moisture Measurements. *J. Appl. Meteorol.* **1989**, *28*, 789–806. [CrossRef]

41. Ansmann, A.; Wandinger, U.; Riebesell, M.; Weitkamp, C.; Michaelis, W. Independent measurement of extinction and backscatter profiles in cirrus clouds by using a combined Raman elastic-backscatter lidar. *Appl. Opt. AO* **1992**, *31*, 7113–7131. [CrossRef]

42. Morille, Y.; Haeffelin, M.; Drobinski, P.; Pelon, J. STRAT: An Automated Algorithm to Retrieve the Vertical Structure of the Atmosphere from Single-Channel Lidar Data. *J. Atmos. Ocean. Technol.* **2007**, *24*, 761–775. [CrossRef]

43. De Bruine, M.; Apituley, A.; Donovan, D.P.; Klein Baltink, H.; de Haij, M.J. Pathfinder: Applying graph theory to consistent tracking of daytime mixed layer height with backscatter lidar. *Atmos. Meas. Tech.* **2017**, *10*, 1893–1909. [CrossRef]

44. Geiß, A.; Wiegner, M.; Bonn, B.; Schäfer, K.; Forkel, R.; von Schneidemesser, E.; Münkel, C.; Chan, K.L.; Nothard, R. Mixing layer height as an indicator for urban air quality? *Atmos. Meas. Tech.* **2017**, *10*, 2969–2988. [CrossRef]

45. Zhang, Y.; Guo, J.; Yang, Y.; Wang, Y.; Yim, S. Vertical Wind Shear Modulates Particulate Matter Pollutions: A Perspective from Radar Wind Profiler Observations in Beijing, China. *Remote Sens.* **2020**, *12*, 546. [CrossRef]

46. Mellado, J.P. Cloud-Top Entrainment in Stratocumulus Clouds. *Annu. Rev. Fluid Mech.* **2017**, *49*, 145–169. [CrossRef]

47. Dang, R.; Yang, Y.; Li, H.; Hu, X.-M.; Wang, Z.; Huang, Z.; Zhou, T.; Zhang, T. Atmosphere Boundary Layer Height (ABLH) Determination under Multiple-Layer Conditions Using Micro-Pulse Lidar. *Remote Sens.* **2019**, *11*, 263. [CrossRef]

48. Liu, Z.; Ming, Y.; Zhao, C.; Lau, N.C.; Guo, J.; Bollasina, M.; Yim, S.H.L. Contribution of local and remote anthropogenic aerosols to a record-breaking torrential rainfall event in Guangdong Province, China. *Atmos. Chem. Phys.* **2020**, *20*, 223–241. [CrossRef]

49. Liu, Z.; Ming, Y.; Wang, L.; Bollasina, M.; Luo, M.; Lau, N.-C.; Yim, S.H.-L. A Model Investigation of Aerosol-Induced Changes in the East Asian Winter Monsoon. *Geophys. Res. Lett.* **2019**, *46*, 10186–10195. [CrossRef]

50. Liu, Z.; Yim, S.H.L.; Wang, C.; Lau, N.C. The Impact of the Aerosol Direct Radiative Forcing on Deep Convection and Air Quality in the Pearl River Delta Region. *Geophys. Res. Lett.* **2018**, *45*, 4410–4418. [CrossRef]

51. Yim, S.H.L.; Wang, M.; Gu, Y.; Yang, Y.; Dong, G.; Li, Q. Effect of urbanization on ozone and resultant health effects in the pearl river delta region of China. *J. Geophys. Res. Atmos.* **2019**, *124*, 11568–11579. [CrossRef]

52. Tong, C.H.M.; Yim, S.H.L.; Rothenberg, D.; Wang, C.; Lin, C.-Y.; Chen, Y.D.; Lau, N.C. Projecting the impacts of atmospheric conditions under climate change on air quality over the Pearl River Delta region. *Atmos. Environ.* **2018**, *193*, 79–87. [CrossRef]

53. Tong, C.H.M.; Yim, S.H.L.; Rothenberg, D.; Wang, C.; Lin, C.-Y.; Chen, Y.; Lau, N.C. Assessing the impacts of seasonal and vertical atmospheric conditions on air quality over the Pearl River Delta region. *Atmos. Environ.* **2018**, *180*, 69–78. [CrossRef]

54. Yim, S.H.L.; Fung, J.C.H.; Ng, E.Y.Y. An assessment indicator for air ventilation and pollutant dispersion potential in an urban canopy with complex natural terrain and significant wind variations. *Atmos. Environ.* **2014**, *94*, 297–306. [CrossRef]

55. Shi, C.; Nduka, I.C.; Yang, Y.; Huang, Y.; Yao, R.; Zhang, H.; He, B.; Xie, C.; Wang, Z.; Yim, S.H.L. Characteristics and meteorological mechanisms of transboundary air pollution in a persistent heavy PM2.5 pollution episode in Central-East China. *Atmos. Environ.* **2020**, *223*, 117239. [CrossRef]

56. Gu, Y.; Yim, S.H.L. The air quality and health impacts of domestic trans-boundary pollution in various regions of China. *Environ. Int.* **2016**, *97*, 117–124. [CrossRef] [PubMed]

57. Hou, X.; Chan, C.K.; Dong, G.H.; Yim, S.H.L. Impacts of transboundary air pollution and local emissions on PM2.5 pollution in the Pearl River Delta region of China and the public health, and the policy implications. *Environ. Res. Lett.* **2018**. [CrossRef]

58. Yim, S.H.L.; Gu, Y.; Shapiro, M.A.; Stephens, B. Air quality and acid deposition impacts of local emissions and transboundary air pollution in Japan and South Korea. *Atmos. Chem. Phys.* **2019**, *19*, 13309–13323. [CrossRef]

59. Luo, M.; Hou, X.; Gu, Y.; Lau, N.-C.; Yim, S.H.-L. Trans-boundary air pollution in a city under various atmospheric conditions. *Sci. Total Environ.* **2018**, *618*, 132–141. [CrossRef]

60. Yim, S.H.L.; Hou, X.; Guo, J.; Yang, Y. Contribution of local emissions and transboundary air pollution to air quality in Hong Kong during El Niño-Southern Oscillation and heatwaves. *Atmos. Res.* **2019**, *218*, 50–58. [CrossRef]

Article

Development of a 3D Real-Time Atmospheric Monitoring System (3DREAMS) Using Doppler LiDARs and Applications for Long-Term Analysis and Hot-and-Polluted Episodes

Steve Hung Lam YIM [1,2,3]

[1]   Department of Geography and Resource Management, The Chinese University of Hong Kong, Hong Kong, China; steveyim@cuhk.edu.hk; Tel.: +852-3943-6534
[2]   Stanley Ho Big Data Decision Analytics Research Centre, The Chinese University of Hong Kong, Hong Kong, China
[3]   Institute of Environment, Energy and Sustainability, The Chinese University of Hong Kong, Hong Kong, China

Received: 6 February 2020; Accepted: 18 March 2020; Published: 24 March 2020

Abstract: Heatwaves and air pollution are serious environmental problems that adversely affect human health. While related studies have typically employed ground-level data, the long-term and episodic characteristics of meteorology and air quality at higher altitudes have yet to be fully understood. This study developed a 3-Dimensional Real-timE Atmospheric Monitoring System (3DREAMS) to measure and analyze the vertical profiles of horizontal wind speed and direction, vertical wind velocity as well as aerosol backscatter. The system was applied to Hong Kong, a highly dense city with complex topography, during each season and including hot-and-polluted episodes (HPEs) in 2019. The results reveal that the high spatial wind variability and wind characteristics in the lower atmosphere in Hong Kong can extend upwards by up to 0.66 km, thus highlighting the importance of mountains for the wind environment in the city. Both upslope and downslope winds were observed at one site, whereas downward air motions predominated at another site. The high temperature and high concentration of fine particulate matter during HPEs were caused by a significant reduction in both horizontal and vertical wind speeds that established conditions favorable for heat and air pollutant accumulation, and by the prevailing westerly wind promoting transboundary air pollution. The findings of this study are anticipated to provide valuable insight for weather forecasting and air quality studies. The 3DREAMS will be further developed to monitor upper atmosphere wind and air quality over the Greater Bay Area of China.

Keywords: Doppler LiDAR; spatial wind variability; air quality

---

1. Introduction

Heatwaves and air pollution are major environmental problems [1–4] that adversely affect human health [5–19]. Despite the substantial burdens on human health, the present understanding of such weather and air quality problems remains limited by insufficient data. In particular, data concerning the wind and air quality in the upper atmosphere are required for data analyses and modeling. This absence of data limits research into heatwaves and air pollution.

Extremely high temperatures and air pollution may occur simultaneously because of their shared atmospheric driving conditions. For this study, an event featuring extremely high temperatures and air pollution was defined as a hot-and-polluted episode (HPE). The synergy between high temperatures and air pollution can result in serious public health burdens. Lee et al. [20] investigated a period of abnormally high temperatures and air pollution in the United Kingdom. They determined that the regional entrainment of air from the upper atmosphere caused early morning increases in ozone during the episode and increased biogenic emissions under high temperatures. Stedman [21] reported that this increase in ozone and particulates caused more than 400 and 700 additional mortalities in England and Wales, respectively. Therefore, a more comprehensive understanding of the formation mechanism of such episodes is required.

In addition to local emissions, transboundary air pollution (TAP) is a major contributor to serious air pollution [14,22]. For example, studies have reported the severe TAP in the Greater Bay Area of China [23–27] and the resultant health impacts [10,28]. However, the understanding of upper atmosphere TAP remains limited. To mitigate this limitation and provide forecasting of TAP and HPEs, a monitoring system is required to collect extensive data of atmospheric variables from the upper atmosphere.

Conventional measurements have focused on ground-level air quality. Since 2013, China has been releasing hourly ground-level air pollutant data covering the entire country. Previous studies have intensively investigated ground-level air quality by using surface monitoring networks [29–32] or satellite-retrieval approaches [33]. One previous study attempted to retrieve vertical structures of aerosol from various satellite observations [34]. Nevertheless, air quality at higher altitudes and the meteorological driving conditions have yet to be fully understood. Some studies have employed upper air sounding data to investigate the meteorology at various altitudes [35,36]. However, such data are typically measured only two to three times per day because of the high cost of data collection. In addition, upper air sounding data are collected using a helium balloon that carries devices for meteorological measurement. Therefore, the measurement locations are dependent on the horizontal and vertical wind velocities at various altitudes. Although sounding data provide a valuable data set for upper atmosphere meteorology, methodological characteristics prevent the colocation and high time resolution required for conducting air quality studies.

To overcome data availability problems, Light Detection and Ranging (LiDAR) can be used for remote atmospheric sensing [37]. For example, an intercomparison study took place at Mace Head, Ireland [38] where one LiDAR and two co-located ceilometers were validated against the boundary-layer height derived from radiosoundings. As preparation for the European Space Agency ADM Mission [39], an intercomparison campaign that held in southern France showed the feasibility of the direct detection Dopple wind LiDAR technique to retrieve the horizontal wind speed atmospheric profile. A single Doppler LiDAR unit can perform wind profiling and air quality monitoring [40–42]. For example, Hong Kong International Airport applied Doppler Lidar to monitor wind shear near the airport [43–45]. The laser beam of their lidar points toward the departure and approach runways. Their application in Dec 2005 captured ~76% of reported wind shear events [46]. Another study applied Doppler LiDAR for evaluating offshore wind characteristics for wind energy [47]. A previous study assessed internal boundary layer structure in Hong Kong under sea-breeze conditions. The authors compared their simulated internal boundary layer with that derived from Doppler LiDAR data. This shows the importance of LiDAR for model developments [48]. Doppler LiDAR was also used for air quality research. A study applied Doppler LiDAR to investigate the characteristics of heavy particular matter pollution [49]. These studies have shown the capabilities of Doppler LiDAR for meteorological and air quality studies.

For cities with considerable spatial and temporal variations in wind, a single LiDAR unit may be insufficient to provide a complete description of complex atmospheric conditions. Previous studies have reported substantial spatial wind variability [50] and TAP [25,26] in Hong Kong. To improve understanding of the wind environment and its influence on air quality in this highly dense city, a 3-Dimensional (3D) Real-timE Atmospheric Monitoring System (3DREAMS) was developed in the present study. This system can be used for air pollution studies with high spatial wind variability or TAP, especially those in Hong Kong as well as in the Greater Bay Area to investigate the interactions between cities in the region.

The aim of this study was to develop the 3DREAMS using more than one Doppler LiDAR unit. The method is described in Section 2. Section 3 presents the results of comparison between the LiDAR data and upper air sounding data as well as those of the analyses for annual and episodic wind profiles. Conclusions are provided in Section 4.

2. Materials and Methods

Two 1.5-µm Doppler LiDAR units (Halo Photonics Stream Line Scanning Doppler LiDAR system) were employed for the development of the first stage of the 3DREAMS. The One LiDAR unit was installed at the Physical Geography Experimental Station of the Chinese University of Hong Kong (CUHK), and the other unit was installed at the Hong Kong Observatory weather station: King's Park (KP). The locations of the two LiDAR units are shown in Figure 1. The CUHK site is located in northeastern Hong Kong and has an elevation of 5 m above sea level. This site is surrounded by mountains, including Ma On Shan (≈700 m), Lion Rock (≈500 m), Kam Shan (≈370 m), and Tai Mo Shan (≈957 m), as shown in Figure 1. The KP site is located in the downtown area of Hong Kong and has an elevation of 65 m above sea level.

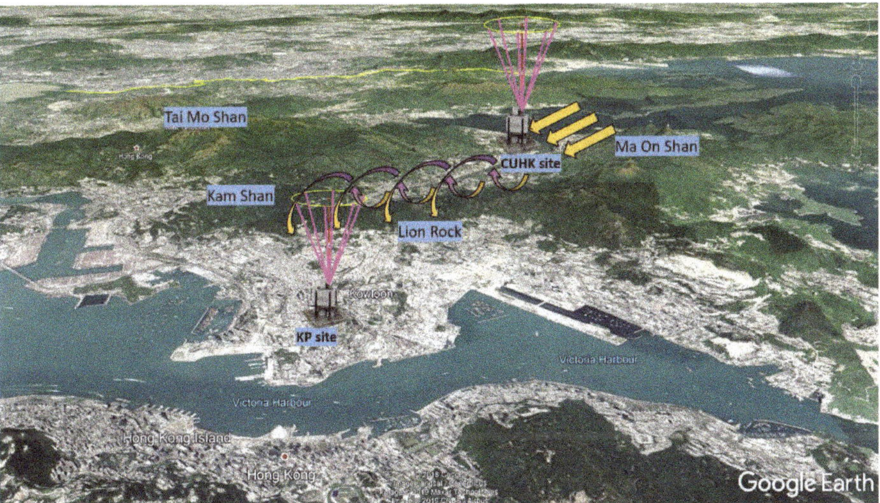

Figure 1. Locations of the two LiDAR sites (CUHK and KP), topography near the CUHK site, and schematic of the upslope and downslope winds. The arrows refer to wind direction. The arrow colors refer to altitude: orange: lower altitude; purple: higher altitude. This CUHK site is surrounded by mountains, including Ma On Shan (≈700 m), Lion Rock (≈500 m), Kam Shan (≈370 m), and Tai Mo Shan (≈957 m). The figure was built on a map obtained from Google Earth, earth.google.com/web/.

The LiDAR units were configured to retrieve aerosol backscatter and wind profiles in the boundary layer up to approximately 3 km above ground level. The principle of the Doppler LiDARs is that laser pulses are emitted by a transmitter and the reflected signals scattered by particles are received by a receiver which is built with the transmitter in the same unit. Particles, which are transported by horizontal wind, induce a Doppler shift as reflected in backscattered light signals. Through measuring the line-of-sight Doppler wind values, the optical heterodyning in the receiver determines the horizontal wind vector.

This study employed the both stare and velocity-azimuth display (VAD) scanning methods to measure horizontal wind information. Figure 2 depicts the schematic diagram of the stare and VAD scanning methods. The stare scan refers to a scanning using a continuous vertically pointing laser beam with $\Phi = 0°$ and $\theta = 90°$. The stare scan was configured to operate at height and temporal resolutions of 30 m and 1 s, respectively. For the VAD scan, the LiDAR units were set to have six azimuthal positions at a constant interval of an azimuth angle $\alpha = 60°$ with a constant evaluation angle of $\theta = 75°$ ($\Phi = 15°$ with respect to the zenith) at a 10-min interval. For quality control, the data with a signal-to-noise ratio less than -20 dB were removed.

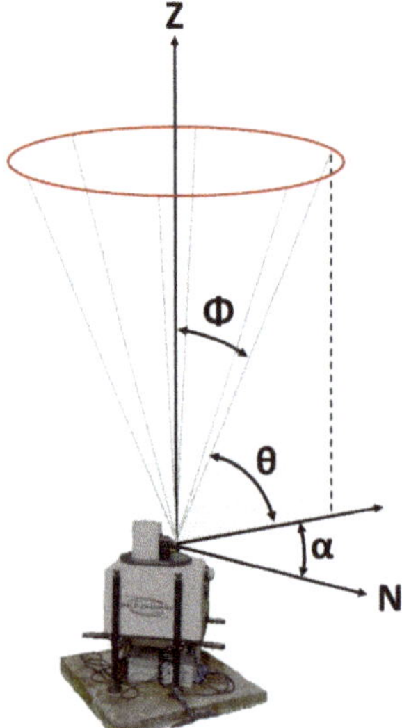

Figure 2. The schematic diagram of both stare and velocity-azimuth display (VAD) scanning methods. The number of bean directions for a VAD scan was 6. The Doppler LiDAR laser beam pointed upward with a constant elevation angle θ and a constant angle Φ with respect to the vertical Z. The laser beam rotated around the vertical Z with a constant interval of an azimuth angle α.

Before their operations, the two LiDAR units were calibrated by the LiDAR manufacturer and validated using a colocation comparison test. The colocation comparison results confirmed consistency between the LiDAR units with a 95% confidence interval, thus indicating an acceptable calibration. Regular maintenance checks were conducted during operation. The LiDAR windows were cleaned hourly through automatic wiping and weekly with an optical cleaning solution. In addition, horizontal leveling checks were conducted weekly to correct for any settling, which can partly affect beam-pointing accuracy. The precision of the leveling was within $\pm 1°$. The literature has reported the limitation of LiDAR performance in complex terrain [51–53]. The LiDARs' were configured and checked to make sure surrounding mountains would not significantly affect the LiDARs' performance, which could be further confirmed by a comparison between LiDAR data and available upper air sounding data at the same site. The comparison results are shown in Section 3.1.

Analyses in this study were conducted using data collected during the spring, summer, fall, and winter of 2019, which were defined as February–April, May–August, September–October, and November–January, respectively. The annual data availability for the CUHK and KP sites was 94% and 82%, respectively.

The annual mean vertical profiles of horizontal wind speed and direction and vertical wind velocity were investigated, and the vertical profiles during HPEs were analyzed. HPEs were defined by a temperature of ≥ 28.2 °C [54] and more than 50% of air quality stations reporting a concentration of fine particulate matter with an aerodynamic diameter ≤ 2.5 μm ($PM_{2.5}$) higher than or equal to the median of the 90th percentile of daily $PM_{2.5}$ of all stations for the year (33.2 μg/m^3). The previous study [54] reported a statistically significant increase in premature mortality risk for an average 1 °C increase in daily average temperature above 28.2 °C. Overall, nine HPEs were identified during 2019.

Information concerning these HPEs is provided in Table 1. The $PM_{2.5}$ concentration data at 16 air quality monitoring stations were obtained from the Hong Kong Environmental Protection Department (http://www.aqhi.gov.hk/en.html). The temperature and surface wind, and the upper air sounding data at the KP weather station (latitude: 22°18′43″; longitude: 114°10′22″) were obtained from the Hong Kong Observatory (https://www.hko.gov.hk/en/cis/stn.htm).

Table 1. Nine identified HPEs during 2019 with the mean daily $PM_{2.5}$ concentration ($\mu g/m^3$) at both general and roadside stations and daily mean temperature (°C).

HPE #	Month	Day	Mean of daily $PM_{2.5}$ Concentration at General Stations ($\mu g/m^3$)	Mean of Daily $PM_{2.5}$ Concentration at Roadside Stations ($\mu g/m^3$)	Daily Mean Temperature (°C)
1	7	17	31.9	38.6	30.3
2	7	18	43.5	54.8	31.0
3	8	9	32.2	38.1	31.0
4	8	24	45.4	61.9	30.7
5	9	29	44.5	51.8	28.3
6	9	30	60.5	79.3	29.9
7	10	1	48.1	52.7	29.9
8	10	2	36.5	41.8	29.0
9	10	11	31.3	38.0	28.3

3. Results

3.1. Comparison with Upper Air Sounding Data

LiDAR data were compared with the available upper air sounding data located at the same site. Figure 3 shows that the LiDAR units captured the general vertical profiles of horizontal wind speed and direction. Differences at heights less than 0.50 km were negligible, whereas those for heights greater than 0.50 km were more substantial. As discussed in the methods section, the measurement locations of the upper air sounding data varied with the wind velocities at the various measurement altitudes, possibly resulting in different measurement locations for the two data sources. The percentage difference between the averaged horizontal wind speeds of the two data sources for heights less than 1.00 km was less than 10%, which indicates sufficient agreement between LiDAR and upper air sounding data.

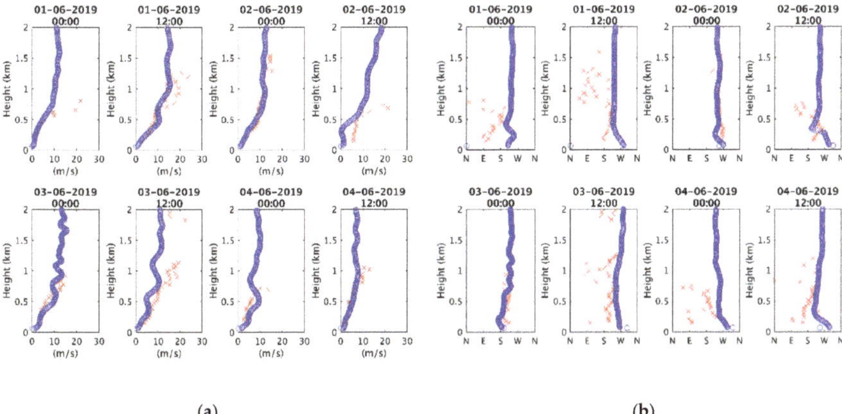

Figure 3. (a) Comparison of horizontal wind speed (m/s) according to KP LiDAR data (red cross) and upper air sounding data (blue circles) over a 4-day period (1 June 2019, 00:00 UTC to 4 June 2019, 12:00 UTC). (b) Comparison of horizontal wind direction according to LiDAR data (red cross) and upper air sounding data (blue circles) over a 4-day period (1 June 2019, 00:00 UTC to 4 June 2019, 12:00 UTC).

Figure 4c depicts the seasonal vertical profiles of horizontal wind speed based on the LiDAR measurements at the KP site, whereas Figure 4d shows the seasonal vertical profiles of horizontal wind speed averaged from the upper air sounding data throughout the entire year. The results show a high agreement between the profiles, despite the fact that the time resolution of the two data was different.

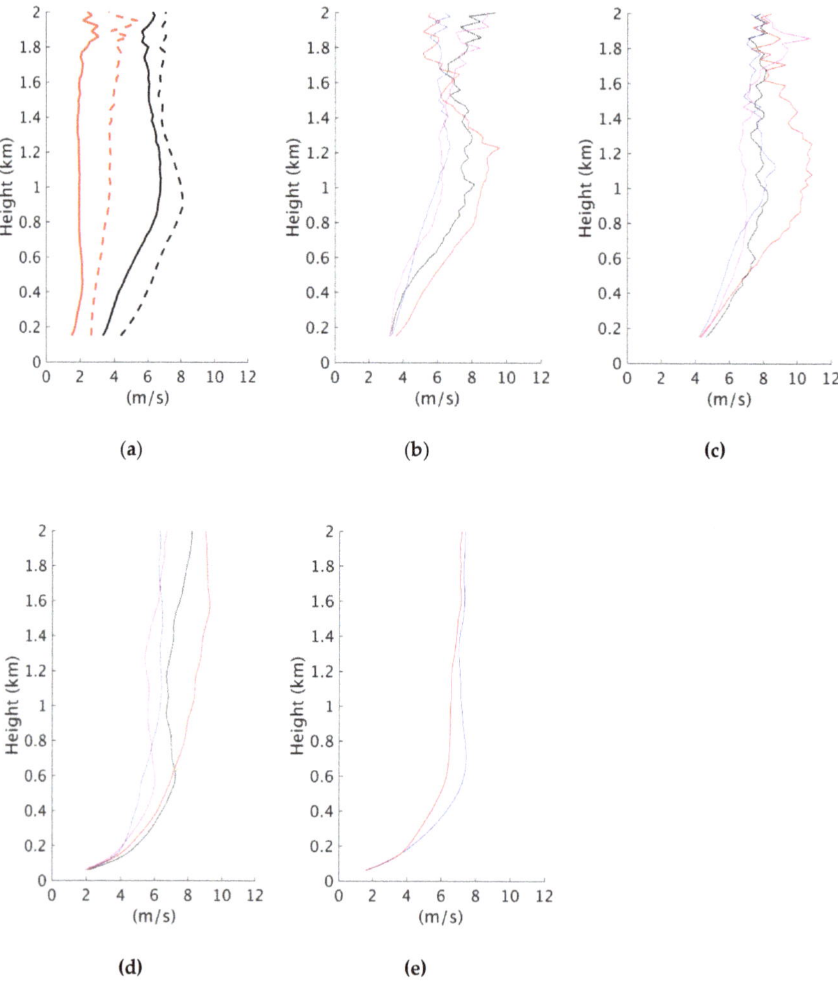

Figure 4. (**a**) Annual (black) and HPE (red) vertical profiles of horizontal wind speed (m/s) at CUHK (solid) and KP (dashed) sites. Seasonal vertical profiles at (**b**) CUHK and (**c**) KP derived from LiDAR data. (**d**) Seasonal and (**e**) annual vertical profiles of horizontal wind speed (m/s) based on the upper air sounding data collected at 08:00 (HKT) and 20:00 (HKT) every day. For (**b–d**): spring (black), summer (red), fall (blue), and winter (magenta). For (**e**): 08:00 (HKT) (blue) and 20:00 (HKT) (red).

3.2. Horizontal Wind Speed

3.2.1. Annual and Seasonal Vertical Profiles

Figure 4a presents the annual vertical profiles of horizontal wind speed at the CUHK and KP sites. The results reveal the horizontal wind speed was higher at KP than at CUHK. The KP site is located in an urban area, whereas the CUHK site is located in a suburban area. The lower horizontal wind speed at the CUHK site may be attributable to the complex topography near the site [50].

Figure 4b,c present the seasonal vertical profiles of horizontal wind speed. At the CUHK site, the mean wind speed for heights less than 1.00 km was higher in summer (6.4 m/s) than in other seasons (5.4, 4.6, and 4.6 m/s for spring, fall, and winter, respectively). For heights less than 0.42 km, the wind speeds in spring and fall were similar, whereas, for height between 0.42 km to 1.00 km, the wind speed in spring was clearly higher than that in fall. For heights of 1.60 to 2.00 km, the wind speeds in spring and winter were higher than those in summer and fall.

Similarly, at the KP site, the mean wind speed for heights less than 1.00 km was the highest in summer (7.6 m/s), followed by spring (6.8 m/s), winter (6.2 m/s) and fall (6.1 m/s). For heights less than 0.48 km, the wind speed was the highest in spring; for heights between 0.48 km and 1.74 km, the wind speed in summer was clearly higher than those of other seasons. For heights of 1.74 to 2.00 km, the wind speed was the highest in winter; nevertheless the seasonal difference at that level at KP was not significant as that at CUHK.

To understand the seasonal variation, the seasonal vertical profiles of wind speed measured by upper air sounding as well as the wind speed at 10 m above ground (wsd$_{10m}$) at three automatic weather stations were analyzed. The similar seasonal trend was also shown in the upper air sounding data, see Figure 4d. Despite the fact that the upper air sounding only provided two data points (08:00 and 20:00 HKT) in a day, the sounding vertical profiles confirm the LiDAR seasonal profiles. Table 2 lists seasonal wsd$_{10m}$ at various Hong Kong Observatory (HKO) stations. The wsd$_{10m}$ at the Waglan Island station shows an obvious seasonal wind variation with the highest wind speed in winter, followed by summer, spring and fall. Nevertheless, the wsd$_{10m}$ at Sha Tin station (near to the CUHK site) and the KP site was the highest in summer and spring, followed by fall and winter. It is noted that the Waglan Island station is located at southeast Hong Kong, which is not affected by any mountains and buildings, and thus serves as a background weather station for Hong Kong. The different seasonal wind speed clearly demonstrates the influence of topography on wind environment in Hong Kong and supports to the findings of vertical wind profiles.

Table 2. The seasonal surface horizontal wind speed (unit: m/s) at the Hong Kong Observatory automatic weather stations: Sha Tin, King's Park and Waglan Island. The data was extracted from https://www.hko.gov.hk/en/cis/climat.htm on 7 Mar. 2020.

	Sha Tin	King's Park	Waglan Island
>Latitude	22°24′09′′	22°18′43′′	22°10′56′′
>Longitude	114°12′36′′	114°10′22′′	114°18′12′′
>spring	7.37	10.40	23.27
>summer	8.40	9.88	23.60
>fall	6.30	9.65	22.35
winter	6.80	9.67	25.00

3.2.2. Diurnal Vertical Profiles

Figure 5 shows the diurnal vertical profiles of horizontal wind speed. Similar to the annual vertical profiles, the diurnal horizontal wind was stronger at the KP site than at the CUHK site. The diurnal vertical profiles exhibited clear peak wind speeds for heights of 0.84 to 1.98 km at CUHK and 0.81 to 1.98 km at KP. The results reveal a clear vertical gradient of horizontal wind speed at the two LiDAR sites from 21:00 to 12:00 HKT. The maximum hourly mean wind speed was 10.6 m/s, which occurred at 20:00 HKT and 15:00 HKT at CUHK and KP, respectively. The minimum hourly mean wind speeds were 6.6 m/s and 7.8 m/s, which occurred at 17:00 HKT and 19:00 HKT at CUHK and KP, respectively.

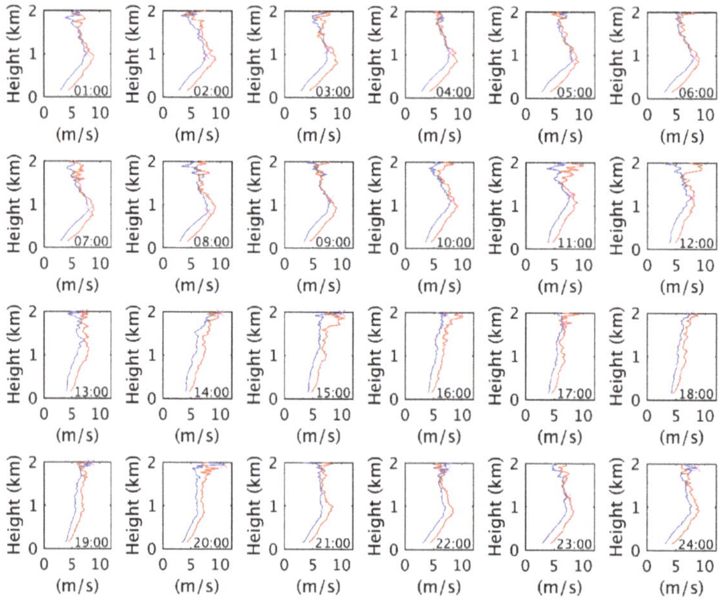

Figure 5. Diurnal vertical profiles of horizontal wind speed at the CUHK (blue) and KP (red) sites.

The vertical gradient was shown clearly in the averaged upper air sounding profiles, see Figure 4e. Similar to the LiDAR profiles at 08:00 HKT and 20:00 HKT, the averaged mean vertical profile at 08:00 HKT has a clear vertical gradient with a peak at around 0.80 km, whereas it was not shown in the 20:00 HKT upper air sounding profile. Figure 6 shows the diurnal vertical profiles of prevailing wind direction and the corresponding mean wind speed at the two LiDAR stations. The weaker vertical gradient of horizontal wind speed at the two LiDAR sites between 13:00 HKT and 20:00 HKT was due to the relatively weak northerly wind between 0.8 km and 1.4 km. It is noted that the northerly wind should be blocked by the topography at the north of the two sites.

3.3. Horizontal Wind Direction

Figure 7 shows the annual vertical profiles of horizontal wind direction frequency at CUHK and KP. At CUHK, the prevailing horizontal wind direction was north for heights less than 0.20 km. The prevailing horizontal wind direction clearly shifted to northeast for heights of 0.20 to 1.80 km and to north and northeast for heights of 1.80 to 2.00 km. At KP, the prevailing horizontal wind direction was northeast for heights less than 1.70 km. Within this range, easterly wind was observed for heights less than 1.10 km and northerly wind was observed for heights from 1.10 to 1.70 km. For heights from 1.70 to 2.00 km, the prevailing horizontal wind directions were north and northeast. These results demonstrated the high spatial wind variability for heights less than 0.30 km and consistent prevailing horizontal wind direction for heights greater than 0.39 km. The high spatial wind variability was mainly due to the complex topography of Hong Kong [50], as discussed in Section 3.2.1.

Figure 8 shows the seasonal vertical profiles of horizontal wind direction frequency at the CUHK and KP sites. In spring, the prevailing horizontal wind directions at CUHK were north and northeast for heights less than 0.60 km, whereas those at KP were northeast and east. For heights from 0.60 to 1.70 km, the prevailing horizontal wind directions at these sites were more consistent (northeast and east). For heights of 1.70 to 2.00 km, the prevailing horizontal wind directions shifted west and southwest at these two sites.

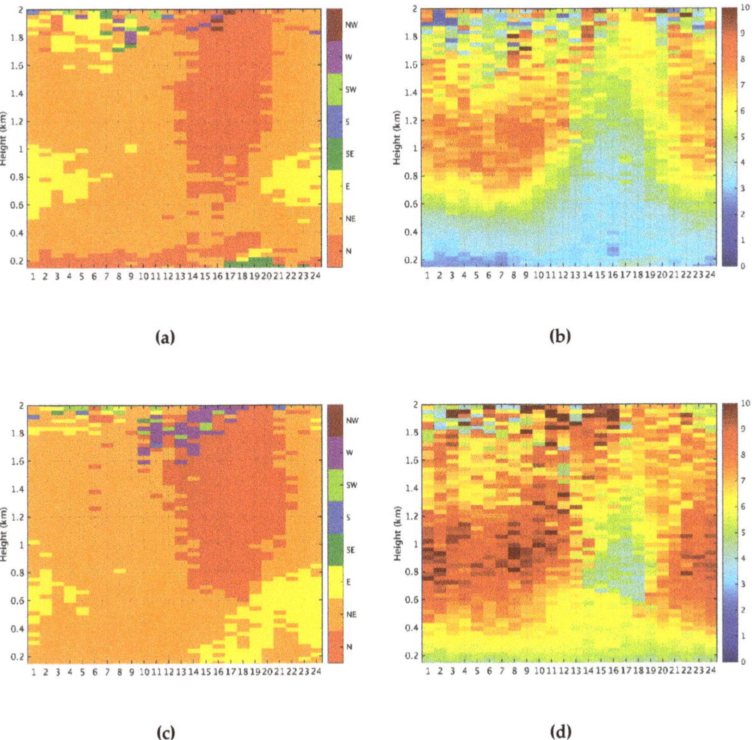

Figure 6. The diurnal vertical profiles of prevailing wind direction (**a**,**c**) and the mean horizontal wind speed (m/s) of the prevailing horizontal wind direction (**b**,**d**) at the two LiDAR stations, CUHK (upper panel) and KP (lower panel).

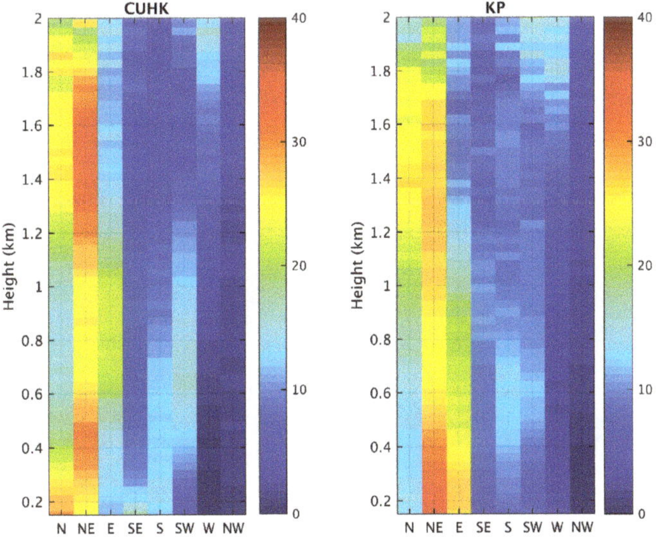

Figure 7. Annual vertical profiles of horizontal wind direction frequency at CUHK (**left**) and KP (**right**).

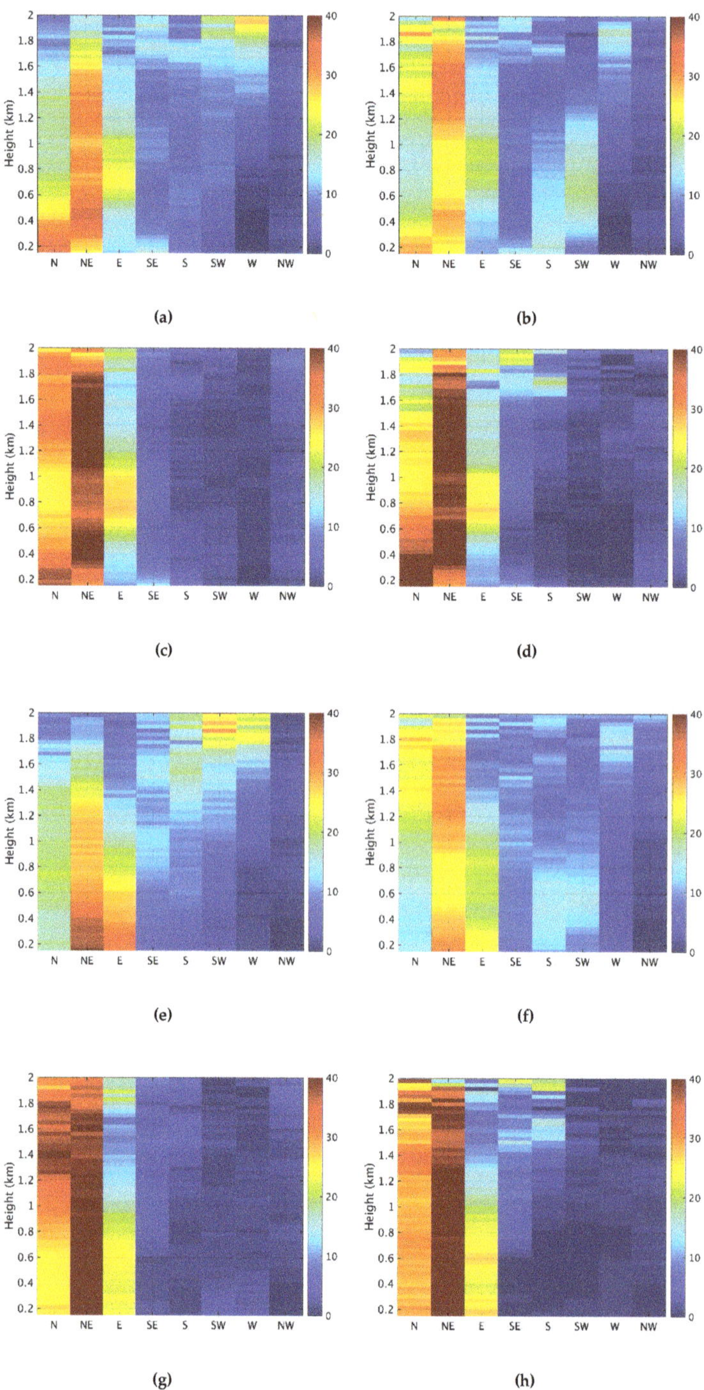

Figure 8. Seasonal vertical profiles of horizontal wind direction frequency in spring (**a**,**e**), summer (**b**,**f**), fall (**c**,**g**), and winter (**d**,**h**) at CUHK and KP, respectively. The upper panel (**a–d**) is CUHK, whereas the lower panel (**e–h**) is KP.

In summer, the prevailing horizontal wind directions at ground level were similar to those in spring except that southerly and southwesterly winds were more frequently noted for heights less than 1.00 km. At heights from 1.00 to 2.00 km, the prevailing horizontal wind directions at the two sites consistently shifted to north and northeast. In fall and winter, the prevailing horizontal wind directions at the two LiDAR sites were relatively stable, mainly northeast, followed by north and east.

3.4. Vertical Wind Velocity

Figure 9 displays the annual and seasonal vertical profiles of vertical wind speed (cm/s) at the two LiDAR sites. At CUHK, the average vertical wind velocity for heights less than 1.00 km was −0.26 cm/s. At this site, upward and downward air motions were observed. The annual profile indicates that upward air motions dominated for heights less than 0.60 km, whereas downward air motions dominated for heights from 0.60 to 2.00 km. The upward air motions at heights less than 0.60 km may be induced by the surrounding topography. As shown in Figure 7, the prevailing horizontal wind directions at CUHK were north and northeast. The northerly and northeasterly air flows were lifted up by the surrounding topography, thus inducing upward air motions. Seasonal variations were also observed. Positive mean vertical velocities were observed in spring, fall, and winter (2.56, 1.34, and 0.59 cm/s, respectively), whereas the mean vertical velocity observed in summer (−1.34 cm/s) was consistent with the annual mean.

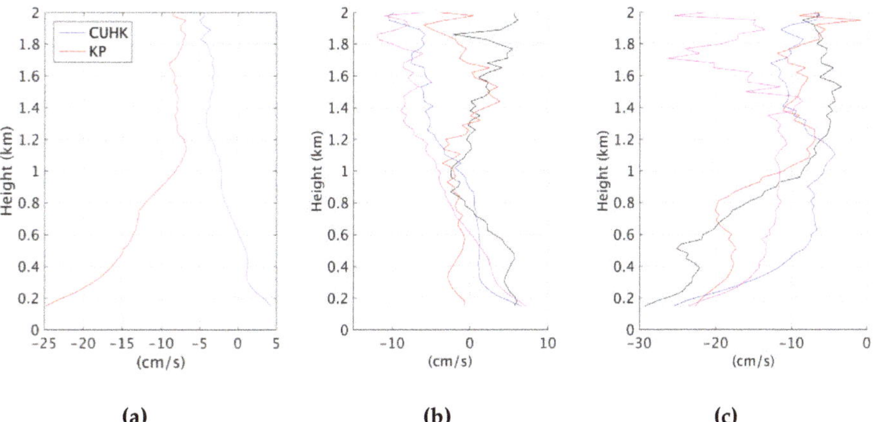

Figure 9. (a) Annual vertical profiles of vertical wind velocity (cm/s) at CUHK (blue) and KP (red) and seasonal vertical profiles at CUHK (**b**) and KP (**c**) during spring (black), summer (red), fall (blue), and winter (magenta). Positive values refer to upward motions; whereas negative values refer to download motions.

At KP, downward air motions dominated for heights less than 2.00 km. The annual mean vertical wind velocity for heights less than 1.00 km was −14.92 cm/s, highlighting that downward motions at KP were stronger than those at CUHK. Negative vertical wind velocities were consistent for all seasons. The strongest downward air motion was observed during spring (−20.07 cm/s), followed by summer (−18.42 cm/s), winter (−14.18 cm/s), and fall (10.90 cm/s).

Notably, when the prevailing northerly or northeasterly wind approaches Hong Kong, CUHK is located upwind, whereas KP is located downwind. For heights less than 1.00 km, the upslope wind at CUHK was clearly weaker than the downslope wind at KP. However, this difference was less apparent in summer (Figure 9b). This may be because of the unstable atmosphere that occurs during summer as a result of stronger solar radiation in that season. Relatively strong buoyancy force could be generated from the warmer ground surface. The unstable atmosphere is favorable for wind going over mountains, enhancing upslope flow [55]. On the other hand, the induced vertical rising air motion may suppress downslope flow. These two effects may reduce the difference between upslope and downslope winds in summer.

3.5. Wind Profiles in HPEs

By examining the temperature and $PM_{2.5}$ data for the thresholds described in the methods section, nine HPEs were identified. Detailed information regarding each HPE is provided in Table 1 This section details wind and backscatter analyses for the HPEs.

3.5.1. Horizontal Wind Speed and Direction

Figure 4 displays the annual and HPE vertical profiles of horizontal wind speed at the two LiDAR sites, and Figure 10 depicts the vertical profiles of horizontal wind direction frequency during HPEs. Compared with annual means, the horizontal wind speeds were clearly lower during HPEs. At CUHK, the horizontal wind speed for heights less than 1.00 km during HPEs was 61.4% lower than the annual mean; the horizontal wind speed over these heights was 51.7% lower at KP. The relatively low horizontal wind speed was unfavorable for air pollutant dispersion.

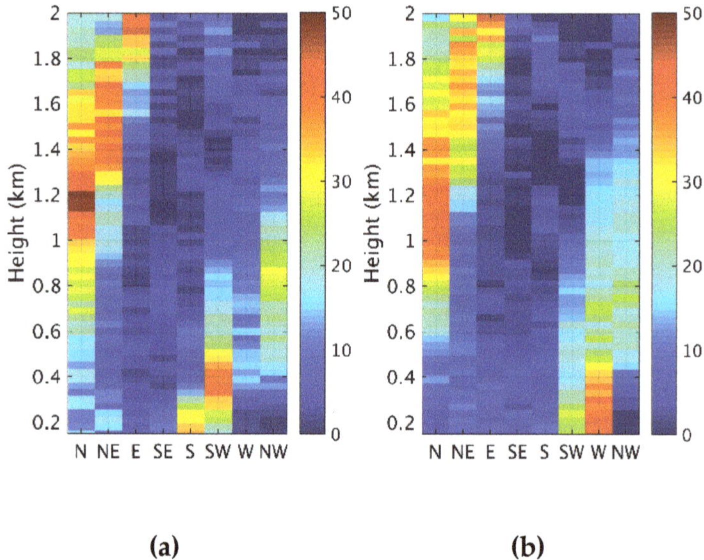

(a)   (b)

Figure 10. Vertical profiles of horizontal wind direction frequency during HPEs at CUHK (**a**) and KP (**b**).

The prevailing horizontal wind directions for heights less than 0.60 km were south and southwest at the CUHK site and west and southwest at the KP site. In addition to lower horizontal wind speed, the prevailing westerly wind during HPEs introduced regional TAP from the Greater Bay Area to Hong Kong. Figure 11 depicts the weather chart of each HPE. The weather charts show that, in each HPE, a typhoon was located in South China sea (east of Hong Kong). The westerly wind was induced by the typhoon-associated counterclockwise wind flow, providing a wind environment for transboundary air pollution within the region [25] and thus, the formation of a HPE. For heights from 0.60 to 1.20 km, northerly and northwesterly wind predominated. For heights from 1.20 to 1.90 km, the prevailing horizontal wind directions was north and northeast. For heights greater than 1.90 km, the easterly winds predominated.

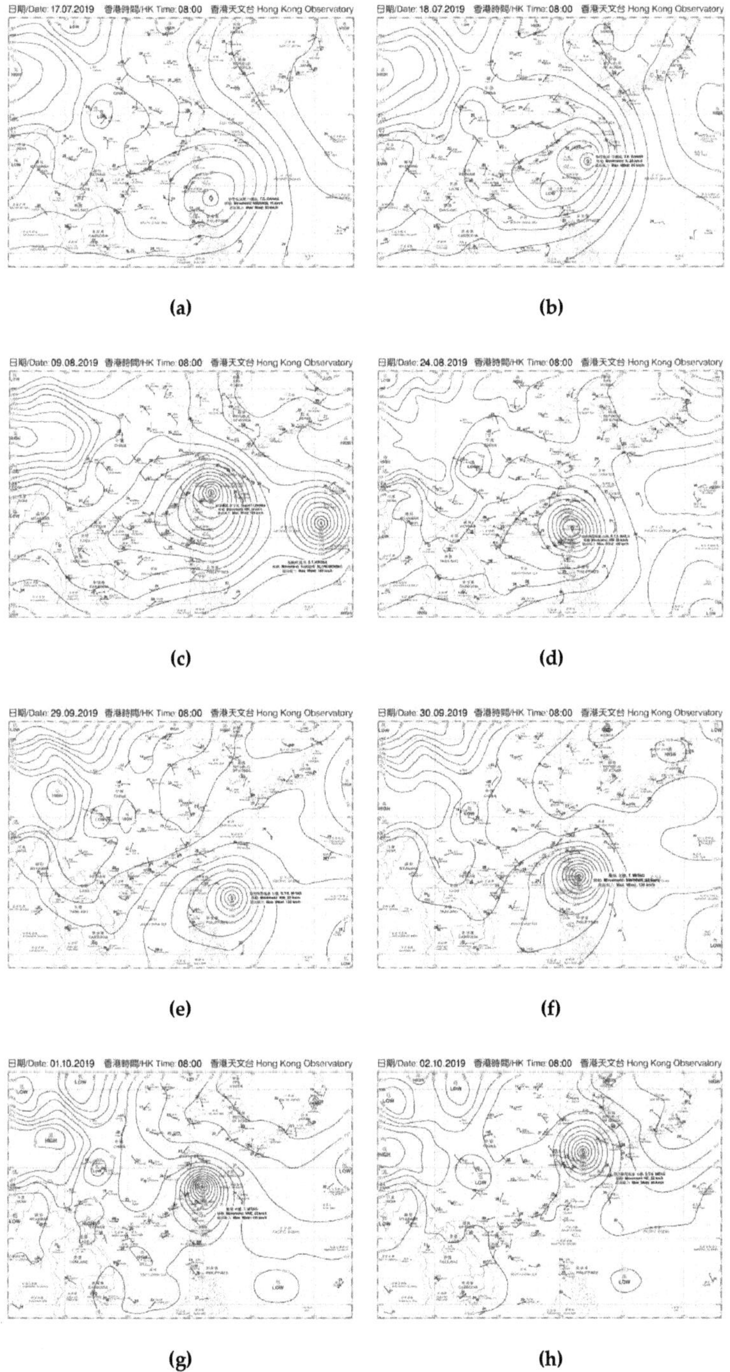

Figure 11. *Cont.*

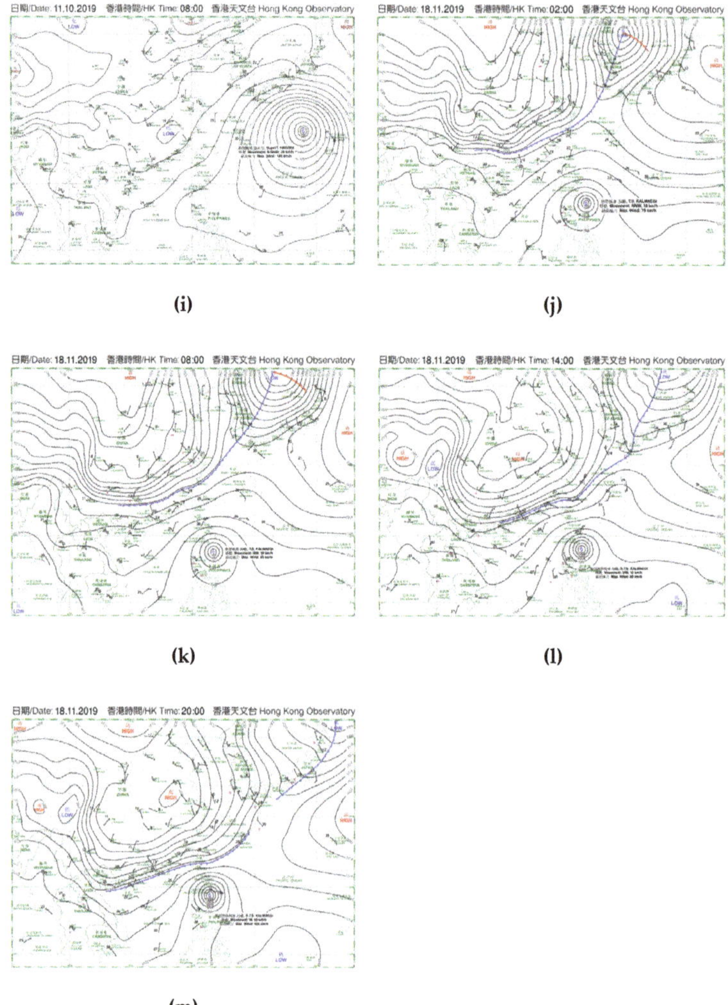

Figure 11. The weather charts (08:00 HKT) of each HPE (**a–i**) and of a cold front case occurred on 18 Nov. 2019: (**j**) 02:00 HKT; (**k**) 08:00 HKT; (**l**) 14:00 HKT and (**m**) 20:00 HKT. The weather charts were obtained from the Hong Kong Observatory web site (https://www.hko.gov.hk/en/wxinfo/currwx/wxcht.htm) on 2 Mar. 2020.

3.5.2. Vertical Wind Velocity

Figure 12 shows the vertical profiles of vertical wind velocity at the two LiDAR sites. At CUHK, the annual vertical wind velocity for heights less than 0.66 km exhibited upward air motion (positive), whereas the HPE profiles exhibited downward air motion. Although upward air motions were observed for heights from 0.66 to 1.00 km, these motions were weak. For heights less than 1.00 km at KP, the strong downward air motions in the annual profile weakened during HPEs. The overall vertical air motions caused the accumulation of heat energy and air pollutants close to ground level.

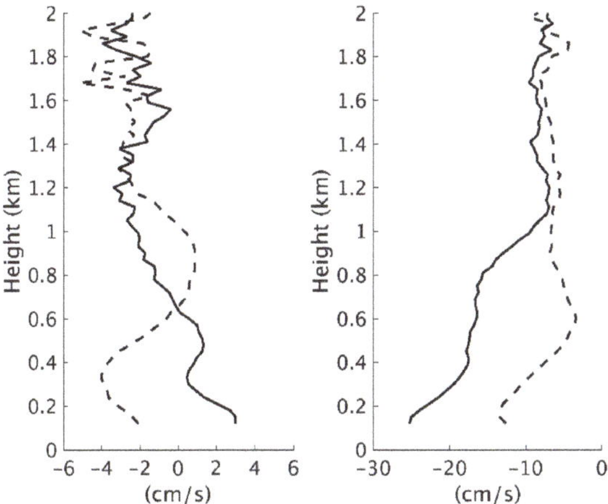

Figure 12. Vertical profiles of annual (solid) and HPE (dashed) vertical wind velocity at CUHK (**left**) and KP (**right**). Positive values refer to upward air motions, whereas negative values represent downward air motions.

Figure 13 shows the mean aerosol backscatter profiles at the two LiDAR sites during the nine identified HPEs. Higher aerosol backscatter was observed for heights less than 1.00 km at nighttime and heights less than 1.20 km in the afternoon. The increased height of the top of the aerosol layer in the afternoon may be attributable to the higher mixing height at this time due to the peak in solar radiation. The aerosol backscatter profiles also reveal that that aerosol backscatter was typically the highest for heights less than 0.60 km. This demonstrates an association with the vertical wind velocity profiles in Figure 12. For heights less than 0.66 km, downward air motions were observed at the CUHK site; the air motions for heights from 0.66 to 1.00 km were directed upward. At KP, the magnitude of downward wind velocity decreased with heights up to approximately 0.60 km and then increased up to 0.90 km. Compared with CUHK, KP had stronger downward air motions that resulted in higher aerosol backscatter.

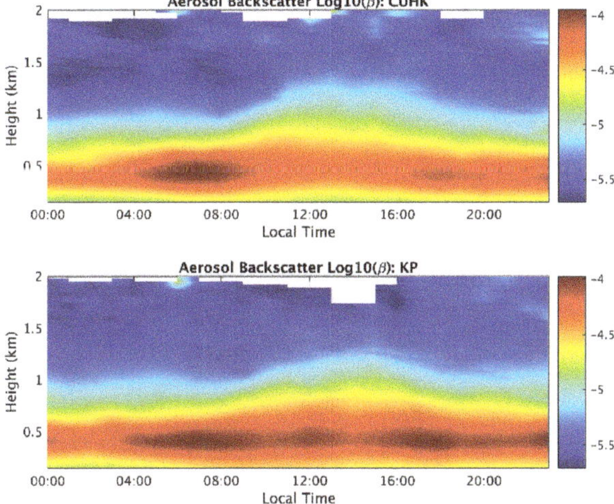

Figure 13. Mean aerosol backscatter log10(β) at CUHK (upper) and KP (lower) during the nine identified HPEs.

3.6. A demonstration Case: Cold Front

To demonstrate the real-time monitoring capability of 3DREAMS, a cold front case occurred on 18 Nov. 2019 was discussed. Figure 11j–m depicts the weather charts of the cold front case at 02:00, 08:00, 14:00 and 20:00 HKT on 18 Nov. 2019, respectively. The cold front arrived at HK in the afternoon on 18 Nov. 2019. Figure 14 shows the high-temporal-resolution vertical profiles of horizontal wind speed at the two sites on 18 and 19 Nov. 2019. The results show that 3DREAMS captured the first arrival of the cold front at the CUHK LiDAR site at between 16:30 and 17:00 HKT. The earlier arrival of the cold front at the CUHK LiDAR site was due to the fact that the site is located at the northeastern Hong Kong. When approaching from northwest, the cold front first hit the CUHK site. In addition, the CUHK LiDAR shows a clear increase in wind speed at the altitudes between 0.50 km and 1.00 km, which lasted until 08:00 HKT on the next day. It is noted that the increase in wind speed near the ground level at KP LiDAR site was not as high as that at the CUHK LiDAR site. This difference was due to the topographical effect that blocked the northerly flow to the KP LiDAR site.

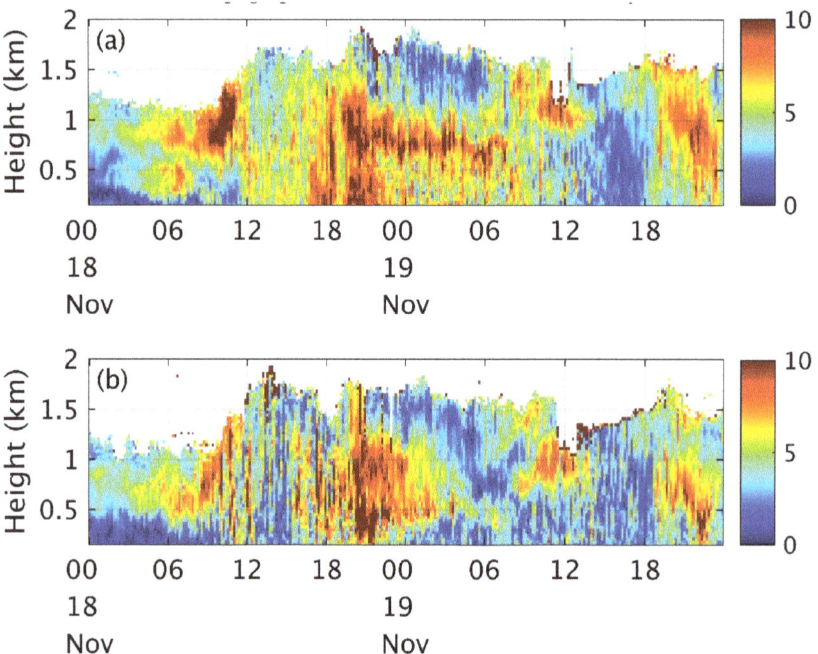

Figure 14. The high-temporal-resolution (10 minutes) vertical profiles of horizontal wind speed (m/s) at the (**a**) CUHK and (**b**) KP LiDAR sites.

4. Discussion

This study developed the 3DREAMS using two Doppler LiDAR units for instantaneously measuring wind and aerosol backscatter profiles, thus providing valuable datasets for wind and aerosol backscatter that can be employed in weather and air quality studies. Previous studies have relied heavily on data measured at the ground level. However, Tong et al. [24] identified significant influences of upper atmosphere meteorology on air quality. Although some studies have measured the vertical profiles of wind and air quality, the typically short measurement periods have resulted in limited understanding of the long-term characteristics of and relationships between weather and air quality. Other studies have relied on upper air sounding data that can be measured only two to three times per day, whereas the present study developed the 3DREAMS to collect real-time upper atmosphere wind and aerosol backscatter information critical for long-term meteorological and air quality studies. It should be highlighted that, while upper air sounding measurements are useful, LiDARs can fill the missing data gaps, providing higher time-resolution measurements.

Spatial wind variability at various altitudes is critical for weather forecasting and air quality studies. For example, local air pollutant emissions may be mixed or transported by lower-level wind; whereas transboundary air pollution may be transported by a higher-level wind. Significant spatial wind variability may affect local weather and thus, air quality at different altitudes and locations. Yim et al. [50] identified high spatial wind variability in Hong Kong due to its complex topography. Nevertheless, their study was limited to ground-level spatial wind variability. The current study reveals that this high spatial wind variability extends to heights of 0.30 km for the horizontal wind direction and 0.60 km for the vertical wind velocity. These findings suggest that further studies should investigate wind shear at various altitudes in Hong Kong. Improved understanding of wind shear is particularly critical for aviation safety at Hong Kong International Airport, which is located in an area with complex topography. The demonstrated cold front case provided a useful example of how the 3DREAMS can be used to study spatial and vertical variations of various horizontal wind speed in weather events.

Previous studies have mainly focused on the effects of extremely hot weather or air pollution episodes. The present study introduced and investigated HPEs, which can adversely affect human health as a result of the synergistic effects of high temperature and high $PM_{2.5}$ concentration. The results reveal that a prevailing horizontal wind direction introducing TAP from the Greater Bay Area and significant reductions in horizontal wind speed at all altitudes and vertical wind velocity for heights less than 0.66 km enhanced the accumulation of heat and air pollutants in the lower atmosphere.

5. Conclusions

This study developed the 3DREAMS to measure the long-term vertical profiles of horizontal wind speed and direction and vertical wind velocity and aerosol backscatter in a highly dense city with complex topography. In addition, the vertical profiles of nine identified HPEs were analyzed to assess the influences of horizontal wind speed and direction and vertical wind velocity on heat and aerosol accumulation. The results reveal high spatial wind variability for heights less than approximately 0.60 km in Hong Kong, highlighting the influence of mountainous topography on the wind environment in the city. Both upslope and downslope winds were observed at CUHK site, whereas downward air motions predominated at KP site. The different air vertical motions resulted in different vertical profiles of aerosol backscatter at the two sites during HPEs. Combining the analyses of horizontal wind speed and direction and vertical wind velocity reveal that high temperatures and $PM_{2.5}$ concentrations were due to a prevailing westerly wind that introduced TAP from the Greater Bay Area. Moreover, a substantial reduction in horizontal wind speed and vertical wind velocity resulted in heat and air pollutant accumulation during HPEs. The findings of this study can provide critical insight for weather forecast and future air quality research. The 3DREAMS will be further developed to integrate new and existing LiDARs into the system and to include more sites to monitor wind and air quality for the Greater Bay Area.

Author Contributions: The research was solely done by S.H.L.Y. including conceptualization, methodology, software, validation, formal analysis, investigation, resources, data curation, writing—original draft preparation, writing—review and editing, visualization, funding acquisition. The author has read and agreed to the published version of the manuscript.

Funding: This work was jointly funded by The Vice-Chancellor's Discretionary Fund of The Chinese University of Hong Kong (grant no. 4930744) and Dr. Stanley Ho Medicine Development Foundation (grant no. 8305509).

Acknowledgments: The author would like to thank the Hong Kong Observatory (HKO) for providing not only weather data but also the measurement site at King's Park weather station. Without the HKO full support, the 3DREAMS may not be able to be built smoothly. In addition, the author also appreciates the Hong Kong Environmental Protection Department for providing air quality data.

Conflicts of Interest: The authors declare no conflict of interest.

References

1. Akimoto, H. Global Air Quality and Pollution. *Science* **2003**, *302*, 1716–1719. [CrossRef]
2. Mage, D.; Ozolins, G.; Peterson, P.; Webster, A.; Orthofer, R.; Vandeweerd, V.; Gwynne, M. Urban air pollution in megacities of the world. *Atmos. Environ.* **1996**, *30*, 681–686. [CrossRef]
3. Miralles, D.G.; Teuling, A.J.; van Heerwaarden, C.C.; Vilà-Guerau de Arellano, J. Mega-heatwave temperatures due to combined soil desiccation and atmospheric heat accumulation. *Nat. Geosci.* **2014**, *7*, 345–349. [CrossRef]
4. Perkins, S.E.; Alexander, L.V.; Nairn, J.R. Increasing frequency, intensity and duration of observed global heatwaves and warm spells. *Geophys. Res. Lett.* **2012**, *39*. [CrossRef]
5. Brauer, M.; Freedman, G.; Frostad, J.; van Donkelaar, A.; Martin, R.V.; Dentener, F.; van Dingenen, R.; Estep, K.; Amini, H.; Apte, J.S.; et al. Ambient Air Pollution Exposure Estimation for the Global Burden of Disease 2013. *Environ. Sci. Technol.* **2016**, *50*, 79–88. [CrossRef] [PubMed]
6. Cohen, A.J.; Brauer, M.; Burnett, R.; Anderson, H.R.; Frostad, J.; Estep, K.; Balakrishnan, K.; Brunekreef, B.; Dandona, L.; Dandona, R.; et al. Estimates and 25-year trends of the global burden of disease attributable to ambient air pollution: An analysis of data from the Global Burden of Diseases Study 2015. *Lancet* **2017**, *389*, 1907–1918. [CrossRef]
7. Grynszpan, D. Lessons from the French heatwave. *Lancet* **2003**, *362*, 1169–1170. [CrossRef]
8. Gu, Y.; Yim, S.H.L. The air quality and health impacts of domestic trans-boundary pollution in various regions of China. *Environ. Int.* **2016**, *97*, 117–124. [CrossRef]
9. Gu, Y.; Wong, T.W.; Law, S.C.; Dong, G.H.; Ho, K.F.; Yang, Y.; Yim, S.H.L. Impacts of sectoral emissions in China and the implications: Air quality, public health, crop production, and economic costs. *Environ. Res. Lett.* **2018**, *13*, 084008. [CrossRef]
10. Hou, X.; Chan, C.K.; Dong, G.H.; Yim, S.H.L. Impacts of transboundary air pollution and local emissions on PM2.5 pollution in the Pearl River Delta region of China and the public health, and the policy implications. *Environ. Res. Lett.* **2018**, *14*, 034005. [CrossRef]
11. Lee, H.-H.; Iraqui, O.; Gu, Y.; Yim, S.H.-L.; Chulakadabba, A.; Tonks, A.Y.-M.; Yang, Z.; Wang, C. Impacts of air pollutants from fire and non-fire emissions on the regional air quality in Southeast Asia. *Atmos. Chem. Phys.* **2018**, *18*, 6141–6156. [CrossRef]
12. Lye, M.; Kamal, A. Effects of a Heatwave on mortality-rates in elderly inpatients. *Lancet* **1977**, *309*, 529–531. [CrossRef]
13. Semenza, J.C.; Rubin, C.H.; Falter, K.H.; Selanikio, J.D.; Flanders, W.D.; Howe, H.L.; Wilhelm, J.L. Heat-Related Deaths during the July 1995 Heat Wave in Chicago. *N. Engl. J. Med.* **1996**, *335*, 84–90. [CrossRef] [PubMed]
14. Wang, M.Y.; Yim, S.H.L.; Wong, D.C.; Ho, K.F. Source contributions of surface ozone in China using an adjoint sensitivity analysis. *S. Total Environ.* **2019**, *662*, 385–392. [CrossRef]
15. Wang, Y.; Chan, A.; Lau, G.N.-C.; Li, Q.; Yang, Y.; Yim, S.H.L. Effects of urbanization and global climate change on regional climate in the Pearl River Delta and thermal comfort implications. *Int. J. Climatol.* **2019**, in press. [CrossRef]
16. Yim, H.L.S.; Wang, M.Y.; Gu, Y.; Yang, Y.; Dong, G.H.; Li, Q. Effect of Urbanization on Ozone and Resultant Health Effects in the Pearl River Delta Region of China. *J. Geophys. Res. Atmos.* **2019**, *124*, 11568–11579. [CrossRef]
17. Yim, S.H.L.; Barrett, S.R.H. Public Health Impacts of Combustion Emissions in the United Kingdom. *Environ. Sci. Technol.* **2012**, *46*, 4291–4296. [CrossRef]
18. Yim, S.H.L.; Lee, G.L.; Lee, I.H.; Allroggen, F.; Ashok, A.; Caiazzo, F.; Eastham, S.D.; Malina, R.; Barrett, S.R.H. Global, regional and local health impacts of civil aviation emissions. *Environ. Res. Lett.* **2015**, *10*, 034001. [CrossRef]
19. Yim, S.H.L.; Stettler, M.E.J.; Barrett, S.R.H. Air quality and public health impacts of UK airports. Part II: Impacts and policy assessment. *Atmos. Environ.* **2013**, *67*, 184–192. [CrossRef]
20. Lee, J.D.; Lewis, A.C.; Monks, P.S.; Jacob, M.; Hamilton, J.F.; Hopkins, J.R.; Watson, N.M.; Saxton, J.E.; Ennis, C.; Carpenter, L.J.; et al. Ozone photochemistry and elevated isoprene during the UK heatwave of august 2003. *Atmos. Environ.* **2006**, *40*, 7598–7613. [CrossRef]
21. Stedman, J.R. The predicted number of air pollution related deaths in the UK during the August 2003 heatwave. *Atmos. Environ.* **2004**, *38*, 1087–1090. [CrossRef]

22. Yim, S.H.L.; Gu, Y.; Shapiro, M.; Stephens, B. Air quality and acid deposition impacts of local emissions and transboundary air pollution in Japan and South Korea. *Atmos. Chem. Phys.* **2019**, *19*, 13309–13323. [CrossRef]

23. Liu, Z.; Ming, Y.; Zhao, C.; Lau, N.C.; Guo, J.; Yim, S.H.L. Contribution of local and remote anthropogenic aerosols to intensification of a record-breaking torrential rainfall event in Guangdong province. *Atmos. Chem. Phys. Discuss.* **2018**, 1–35. [CrossRef]

24. Tong, C.H.M.; Yim, S.H.L.; Rothenberg, D.; Wang, C.; Lin, C.-Y.; Chen, Y.; Lau, N.C. Assessing the impacts of seasonal and vertical atmospheric conditions on air quality over the Pearl River Delta region. *Atmos. Environ.* **2018**, *180*, 69–78. [CrossRef]

25. Luo, M.; Hou, X.; Gu, Y.; Lau, N.-C.; Yim, S.H.-L. Trans-boundary air pollution in a city under various atmospheric conditions. *Sci. Total Environ.* **2018**, *618*, 132–141. [CrossRef] [PubMed]

26. Yim, S.H.L.; Hou, X.; Guo, J.; Yang, Y. Contribution of local emissions and transboundary air pollution to air quality in Hong Kong during El Niño-Southern Oscillation and heatwaves. *Atmos. Res.* **2019**, *218*, 50–58. [CrossRef]

27. Tong, C.H.M.; Yim, S.H.L.; Rothenberg, D.; Wang, C.; Lin, C.-Y.; Chen, Y.D.; Lau, N.C. Projecting the impacts of atmospheric conditions under climate change on air quality over the Pearl River Delta region. *Atmos. Environ.* **2018**, *193*, 79–87. [CrossRef]

28. Wang, M.Y.; Yim, S.H.L.; Dong, G.H.; Ho, K.F.; Wong, D.C. Mapping ozone source-receptor relationship and apportioning the health impact in the Pearl River Delta region using adjoint sensitivity analysis. *Atmos. Environ.* **2020**, *222*, 117026. [CrossRef]

29. Rohde, R.A.; Muller, R.A. Air pollution in China: Mapping of concentrations and sources. *PLoS ONE* **2015**, *10*, e0135749. [CrossRef]

30. Wang, Y.; Ying, Q.; Hu, J.; Zhang, H. Spatial and temporal variations of six criteria air pollutants in 31 provincial capital cities in China during 2013–2014. *Environ. Int.* **2014**, *73*, 413–422. [CrossRef]

31. Guo, J.; Su, T.; Li, Z.; Miao, Y.; Li, J.; Liu, H.; Xu, H.; Cribb, M.; Zhai, P. Declining frequency of summertime local-scale precipitation over eastern China from 1970 to 2010 and its potential link to aerosols. *Geophys. Res. Lett.* **2017**, *44*, 5700–5708. [CrossRef]

32. Guo, J.; Deng, M.; Lee, S.S.; Wang, F.; Li, Z.; Zhai, P.; Liu, H.; Lv, W.; Yao, W.; Li, X. Delaying precipitation and lightning by air pollution over the Pearl River Delta. Part I: Observational analyses. *J. Geophys. Res. Atmos.* **2016**, *121*, 6472–6488. [CrossRef]

33. Guo, J.-P.; Zhang, X.-Y.; Wu, Y.-R.; Zhaxi, Y.; Che, H.-Z.; La, B.; Wang, W.; Li, X.-W. Spatio-temporal variation trends of satellite-based aerosol optical depth in China during 1980–2008. *Atmos. Environ.* **2011**, *45*, 6802–6811. [CrossRef]

34. Guo, J.; Liu, H.; Wang, F.; Huang, J.; Xia, F.; Lou, M.; Wu, Y.; Jiang, J.H.; Xie, T.; Zhaxi, Y.; et al. Three-dimensional structure of aerosol in China: A perspective from multi-satellite observations. *Atmos. Res.* **2016**, *178*, 580–589. [CrossRef]

35. Guo, J.; Miao, Y.; Zhang, Y.; Liu, H.; Li, Z.; Zhang, W.; He, J.; Lou, M.; Yan, Y.; Bian, L.; et al. The climatology of planetary boundary layer height in China derived from radiosonde and reanalysis data. *Atmos. Chem. Phys.* **2016**, *16*, 13309–13319. [CrossRef]

36. Guo, J.; Li, Y.; Cohen, J.B.; Li, J.; Chen, D.; Xu, H.; Liu, L.; Yin, J.; Hu, K.; Zhai, P. Shift in the Temporal Trend of Boundary Layer Height in China Using Long-Term (1979–2016) Radiosonde Data. *Geophys. Res. Lett.* **2019**, *46*, 6080–6089. [CrossRef]

37. Collis, R.T.H. Lidar: A new atmospheric probe. *Q. J. R. Meteorol. Soc.* **1966**, *92*, 220–230. [CrossRef]

38. Milroy, C.; Martucci, G.; Lolli, S.; Loaec, S.; Sauvage, L.; Xueref-Remy, I.; Lavrič, J.V.; Ciais, P.; Feist, D.G.; Biavati, G.; et al. An Assessment of Pseudo-Operational Ground-Based Light Detection and Ranging Sensors to Determine the Boundary-Layer Structure in the Coastal Atmosphere. Available online: https://www.hindawi.com/journals/amete/2012/929080/ (accessed on 7 February 2020).

39. Lolli, S.; Delaval, A.; Loth, C.; Garnier, A.; Flamant, P.H. 0.355-micrometer direct detection wind lidar under testing during a field campaign in consideration of ESA's ADM-Aeolus mission. *Atmos. Meas. Tech.* **2013**, *6*, 3349–3358. [CrossRef]

40. Bozier, K.E.; Pearson, G.N.; Collier, C.G. Doppler lidar observations of Russian forest fire plumes over Helsinki. *Weather* **2007**, *62*, 203–208. [CrossRef]

41. Pearson, G.; Davies, F.; Collier, C. Remote sensing of the tropical rain forest boundary layer using pulsed Doppler lidar. *Atmos. Chem. Phys.* **2010**, *10*, 5891–5901. [CrossRef]

42. Pearson, G.; Davies, F.; Collier, C. An Analysis of the Performance of the UFAM Pulsed Doppler Lidar for Observing the Boundary Layer. *J. Atmos. Ocean. Technol.* **2009**, *26*, 240–250. [CrossRef]

43. Chan, P.W. Generation of an Eddy Dissipation Rate Map at the Hong Kong International Airport Based on Doppler Lidar Data. *J. Atmos. Ocean. Technol.* **2010**, *28*, 37–49. [CrossRef]

44. Chan, P.W.; Lee, Y.F. Application of Short-Range Lidar in Wind Shear Alerting. *J. Atmos. Ocean. Technol.* **2011**, *29*, 207–220. [CrossRef]

45. Hon, K.K.; Chan, P.W. Application of LIDAR-derived eddy dissipation rate profiles in low-level wind shear and turbulence alerts at Hong Kong International Airport. *Meteorol. Appl.* **2014**, *21*, 74–85. [CrossRef]

46. Shun, C.M.; Chan, P.W. Applications of an Infrared Doppler Lidar in Detection of Wind Shear. *J. Atmos. Ocean. Technol.* **2008**, *25*, 637–655. [CrossRef]

47. Shu, Z.R.; Li, Q.S.; He, Y.C.; Chan, P.W. Observations of offshore wind characteristics by Doppler-LiDAR for wind energy applications. *Appl. Energy* **2016**, *169*, 150–163. [CrossRef]

48. Liu, H.; Chan, J.C.L.; Cheng, A.Y.S. Internal boundary layer structure under sea-breeze conditions in Hong Kong. *Atmos. Environ.* **2001**, *35*, 683–692. [CrossRef]

49. Yang, Y.; Yim, S.H.L.; Haywood, J.; Osborne, M.; Chan, J.C.S.; Zeng, Z.; Cheng, J.C.H. Characteristics of Heavy Particulate Matter Pollution Events over Hong Kong and Their Relationships with Vertical Wind Profiles Using High-Time-Resolution Doppler Lidar Measurements. *J. Geophys. Res. Atmos.* **2019**, *124*, 9609–9623. [CrossRef]

50. Yim, S.H.L.; Fung, J.C.H.; Lau, A.K.H.; Kot, S.C. Developing a high-resolution wind map for a complex terrain with a coupled MM5/CALMET system. *J. Geophys. Res.* **2007**, *112*, 1–15. [CrossRef]

51. Hofsäß, M.; Clifton, A.; Cheng, P.W. Reducing the Uncertainty of Lidar Measurements in Complex Terrain Using a Linear Model Approach. *Remote Sens.* **2018**, *10*, 1465. [CrossRef]

52. Pauscher, L.; Vasiljevic, N.; Callies, D.; Lea, G.; Mann, J.; Klaas, T.; Hieronimus, J.; Gottschall, J.; Schwesig, A.; Kühn, M.; et al. An Inter-Comparison Study of Multi-and DBS Lidar Measurements in Complex Terrain. *Remote Sens.* **2016**, *8*, 782. [CrossRef]

53. Bingöl, F.; Mann, J.; Foussekis, D. Conically scanning lidar error in complex terrain. *Meteorol. Z.* **2009**, *18*, 189–195. [CrossRef]

54. Chan, E.Y.Y.; Goggins, W.B.; Kim, J.J.; Griffiths, S.M. A study of intracity variation of temperature-related mortality and socioeconomic status among the Chinese population in Hong Kong. *J. Epidemiol. Community Health* **2012**, *66*, 322–327. [CrossRef] [PubMed]

55. Pal Arya, S. *Air Pollution Meteorology and Dispersion*; Oxford University Press: Oxford, UK, 1999; ISBN 0-19-507398-3.

 *remote sensing*

Article

The Determination of Aerosol Distribution by a No-Blind-Zone Scanning Lidar

Jie Wang [1,2], Wenqing Liu [1], Cheng Liu [3], Tianshu Zhang [1,4,*], Jianguo Liu [1], Zhenyi Chen [1], Yan Xiang [1,4] and Xiaoyan Meng [5]

[1] Key Laboratory of Environmental Optics and Technology, Anhui Institutes of Optics and Fine Mechanics, Chinese Academy of Sciences, Hefei 230031, China; jiewang@aiofm.ac.cn (J.W.); wqliu@aiofm.ac.cn (W.L.); jgliu@aiofm.ac.cn (J.L.); zychen@aiofm.ac.cn (Z.C.); yxiang@ahu.edu.cn (Y.X.)
[2] Science Island Branch of Graduate School, University of Science and Technology of China, Hefei 230026, China
[3] Department of Precision Machinery and Precision Instrumentation, University of Science and Technology of China, Hefei 230026, China; chliu81@ustc.edu.cn
[4] Institutes of Physical Science and Information Technology, Anhui University, Hefei 230601, China
[5] China National Environmental Monitoring Centre, Beijing 100012, China; mengxy@cnemc.cn
* Correspondence: tszhang@aiofm.ac.cn

Received: 1 January 2020; Accepted: 11 February 2020; Published: 13 February 2020

Abstract: A homemade portable no-blind zone laser detection and ranging (lidar) system was designed to map the three-dimensional (3D) distribution of aerosols based on a dual-field-of-view (FOV) receiver system. This innovative lidar prototype has a space resolution of 7.5 m and a time resolution of 30 s. A blind zone of zero meters, and a transition zone of approximately 60 m were realized with careful optical alignments, and were rather meaningful to the lower atmosphere observation. With a scanning platform, the lidar system was used to locate the industrial pollution sources at ground level. The primary parameters of the transmitter, receivers, and detectors are described in this paper. Acquiring a whole return signal of this lidar system represents the key step to the retrieval of aerosol distribution with applying a linear joining method to the two FOV signals. The vertical profiles of aerosols were retrieved by the traditional Fernald method and verified by real-time observations. To effectively and reliably retrieve the horizontal distributions of aerosols, a composition of the Fernald method and the slope method were applied. In this way, a priori assumptions of even atmospheric conditions and the already-known reference point in the lidar equation were avoided. No-blind-zone vertical in-situ observation of aerosol illustrated a detailed evolution from almost 0 m to higher altitudes. No-blind-zone detection provided tiny structures of pollution distribution in lower atmosphere, which is closely related to human health. Horizontal field scanning experiments were also conducted in the Shandong Province. The results showed a high accuracy of aerosol mass movement by this lidar system. An effective quantitative way to locate pollution sources distribution was paved with the portable lidar system after validation by the mass concentration of suspended particulate matter from a ground air quality station.

Keywords: lidar; dual-field-of-view (FOV); geometric overlap factor (GOF); blind zone; transition zone; aerosol; mass concentration; stereo-monitoring networks

---

1. Introduction

The gradient of energy distribution affects the evolution of the atmospheric boundary layer (ABL), and mainly induces the uneven distribution of aerosol and the formation and deterioration of air quality [1–5]. The transport of dust storms and smoke has caused severe regional air pollution on several occasions [6–8]. Laser detection and ranging (lidar), as an active remote-sensing technique, is often utilized to detect and identify the physical properties, shapes and particle-size distribution of aerosols with a high spatial and temporal resolution from ground stationary lidar networks [9,10], mobile platforms [11], aircrafts [12] or satellites [13]. However, due to a monostatic receiving field and "geometric overlap factor (GOF)", the hundreds of meters of blind zone and the transition zone in traditional Mie-lidars [14–20] always lead to a difficulty in probing aerosols, especially in the lower troposphere [21–23]. A few groups have developed side-scattering imaging technologies [24–27] or multi-field-of-view (mFOV) techniques [28,29] to overcome such difficulties. However, the former methods are limited as the detection range ignores important information from the distant atmosphere, and the latter methods are laboratory instruments that are difficult to move. In this study, a compact portable dual-field-of-view (FOV) lidar prototype aimed at narrowing the transition zone is demonstrated. A reliable inversion method for determining the vertical and horizontal distributions of aerosols within 5 km is provided. Field experiments were also conducted to verify the applicability of the dual-FOV lidar for studying lower-atmosphere air pollution and quantitatively assessing the distribution of emission sources. With this scanning lidar, the pollution distribution near ground can be identified with no-blind-zone and quantitatively. The established lidar network with this kind of lidars can be used to study the pollution transport pathways.

2. Materials and Methods

2.1. Dual-FOV Lidar System

The system framework of the proposed dual-FOV lidar is shown in Figure 1. A homemade laser source provided a center wavelength of 532 nm and a fluctuation of ±0.3 nm. The single-pulse energy was 400 μJ. The repetition rate was 2–7 kHz. The linear-polarized laser beam has a divergence of 0.1 mrad collimated by a 10-times beam expander and shot into the air. The laser source with a green wavelength of 532 nm and a higher single-pulse energy from sophisticated commercial lasers was chosen to produce a higher signal-to-noise of observations and a less error of the extinction coefficients of aerosol.

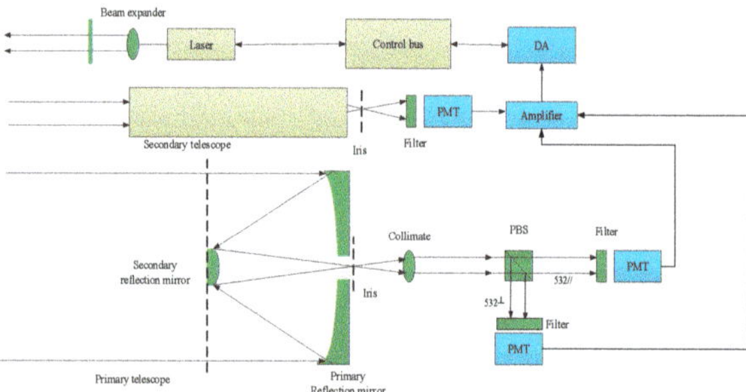

Figure 1. Systematic framework of the dual-field-of-view (FOV) laser detection and ranging (lidar).

The primary FOV's Mie backscattering signals were collected by a Cassegrain-type telescope, and the secondary FOV's signals were collected by a Newtonian-type telescope at the same time. The primary telescope was designed with a diameter of 150 mm, a focal length of 700 mm, and an FOV of 1 mrad. The secondary telescope was designed with a diameter of 30 mm, a focal length of 118 mm, and an FOV of 8 mrad. A spacing of 25 mm was maintained between the optical axes of the two telescopes. The distances of the primary FOV optical axis and secondary FOV optical axis to the laser emitting axis were 121, and 58 mm, respectively. The intersection angle of the two optical axes was 0.2–0.4°. This intersection angle was adjustable and rather important to the performance of the lidar's blind zone and transition zone. A typical GOF of the traditional dual-axis Mie-lidar is shown in Figure 2a and the calculated GOFs of the two FOVs of this lidar system are illustrated in

Figure 2b. The GOF η of the lidar in Figure 2a depends on the parameters of the laser divergence angle, receiving FOV, and the distance between laser (transmitter) and receiving telescope. Within a very short distance from the lidar, the laser beam emitted by the laser does not receive in the telescope's field of view. At this time, η = 0, and the receiving telescope cannot receive the atmospheric backscatter light. This area is called the blind zone of the lidar. Far away from the blind area, the emitted laser beam gradually enters the receiving field of view. Currently, η gradually increases from 0, but less than 1. The atmospheric backscattered light is only partially received. This area is called the transition zone of lidar. Far away from the transition region, the emitted laser beam is completely contained in the receiving field of view. At this time, there is always η = 1, and all the backscattered light in the atmosphere is received. From Figure 2b, it is demonstrated that a larger blind zone of about 75 m in primary FOV comparted with less than the 15 m of blind zone in the secondary FOV. It also illustrates that the transition zones of primary and secondary FOVs were 115.5, and 12.4 m, respectively. Here, we ignored the effects of the intersection angle of the secondary FOV in the GOF calculation.

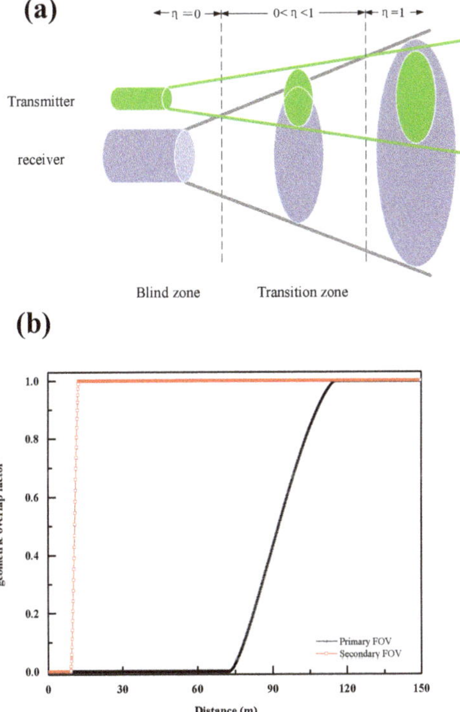

Figure 2. (**a**) The typical geometric overlap factor (GOF) of the traditional dual-axis Mie-lidar and (**b**) the calculated GOFs of primary and secondary FOVs. The effects of the intersection angle were ignored in the calculation of the secondary FOV.

The polarized backscattered signals, collected by the primary FOV, traveled through an iris, a collimator, and a polarized beam splitter (PBS). After passing through the PBS, parallel and perpendicular signals went through two narrow-band filters with a bandwidth of 1.6 nm and an optical density of 5 to suppress the scattered light. The depolarization ratio was retrieved followed by the reference [14]. The signal in the secondary FOV passed through an iris and a narrow-band interference filter. All these three signals were detected by three photo multiplier tubes (PMTs). An amplifier and a 16-bit analog-to-digital converter were used to amplify and digitize the three-channel electrical signals.

The data-acquisition rate was 20 MHz, and the transit time of PMTs was 2.7 ns. An industrial computer was embedded to save the raw data files. An outside view of the lidar, with detailed module identifications, is shown in Figure 3. The total weight of the lidar, including a scanning platform, which supports the three-dimensional (3-D) scanning of aerosols, was 65 kg. This portable lidar system is flexible for mobile applications and station observations.

Figure 3. Outside-view of dual-FOV lidar. Front view of dual FOV lidar (**left**) and field-experiment view of dual FOV lidar (**right**).

2.2. Evaluation and Joining of the Signals from the Two FOVs

The backscattering range corrected signals (PRR), received by the two FOVs, are shown in Figure 4. The primary FOV's signal (indicated by the black-square line in Figure 4) illustrated a higher signal-to-noise ratio (SNR) over 10 above 2 km. The strong intensity at 4.5 km referred to clouds. Many tiny peaks in the range of 0.75 km to approximately 4.5 km suggested more than 5 layers of aerosol distribution at the lower atmosphere. For the returned secondary FOV signal (red-dot line in Figure 4), a much higher SNR over 8 was shown below 2 km. Compared with the primary signal, sophisticated structures were clearly shown under 1 km.

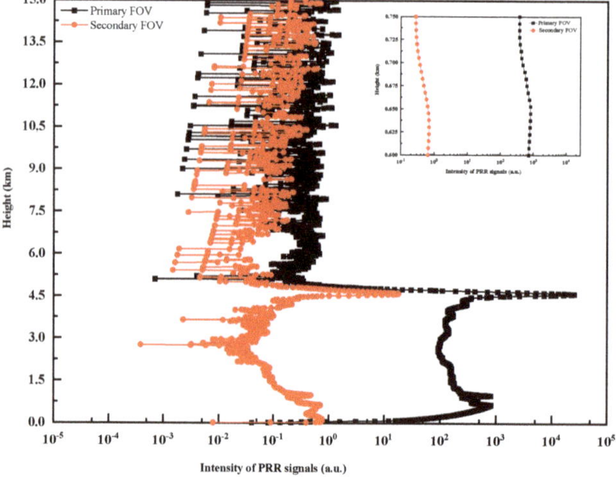

Figure 4. Intensities of backscattering range-corrected signals (PRR) from primary FOV (black square line) and secondary FOV (red dot line). A case of the primary (black square line) and secondary FOV PRR signal (red dot line) at the combining region of 0.6~0.75 km is shown in the sub-chart.

To acquire a complete signal of dual FOV lidar, a careful joining evaluation of primary and secondary FOV signals was conducted thoroughly along the laser-beam path. An optimum combination interval with the most relevant linearity between the signals of the two fields was determined by the following Equation (1)

$$O(r) = \sum_{i=m}^{m+\Delta n} \{[aP_s(r) + b] - P_P(r)\}^2 \tag{1}$$

where $O(r)$ is the square loss function of $\{P_s(r), P_p(r)\}$ at the range of r. m is the joining start-point of the combination interval, Δn is the combination length, $P_s(r)$ is the intensity of the PRR signal of the secondary FOV, and $P_p(r)$ is the intensity of the PRR signal of the primary FOV. The intensity ratio of a and the slop of b will be determined by conducting a non-linear least square method.

A case is shown in Figure 4. Here, the combination interval was determined at 0.6~0.75 km (sub-chart in Figure 4), with a mean intensity ratio of 1212.7 and a minimum standard deviation of 29.2. The signal from the primary field over 0.75 km and the signal under 0.75 km from the secondary field corrected by the intensity ratio were composed to a complete return backscattering signal of the dual-FOV lidar. The original primary-field signal, the corrected secondary-field signal (multiplied by the intensity ratio), and the combined signal are illustrated in Figure 5. It is obvious that the combined signal retained excellent characteristics of the primary-field and secondary-field signals. In this way, the SNR of the entire signal exceeded 8, which was advantageous to inverse the extinction coefficients of aerosol. Although, different combination intervals will be different for every raw lidar signal, the combination length of 150 m was fixed in inversion procedure with a resolution of 7.5 m. In this lidar system, the signals' joining start-points are always located below 1 km.

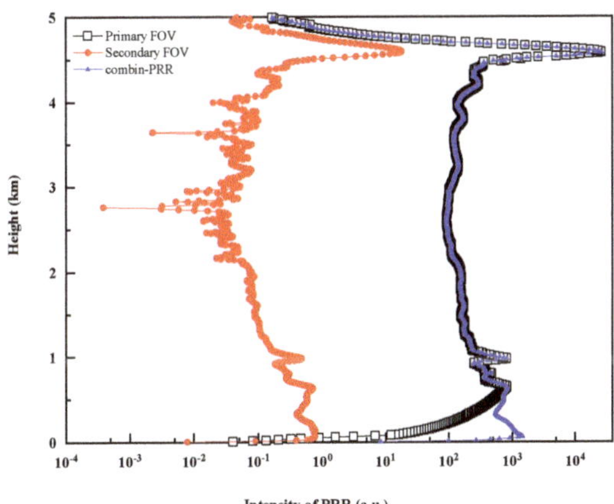

Figure 5. Intensities of combined PRR signal (blue triangle line), primary FOV PRR signal (black square line), and secondary FOV PRR signal (red dot line) of dual-FOV lidar.

2.3. Evaluation of the Blind Zone and the Transition Zone of Dual-FOV Lidar

The blind zone and transition zone of this dual-FOV lidar system can be checked from the PRR of the secondary FOV. We aligned the laser beam parallel to the ground surface to measure the PRR signal, which is shown in Figure 6. This figure shows that the blind zone decreased to 0 m, and the transition zone was 60 m. This measured no-blind-zone was mainly due to the alignment of the intersection angle of the secondary FOV axis and the transmitter axis.

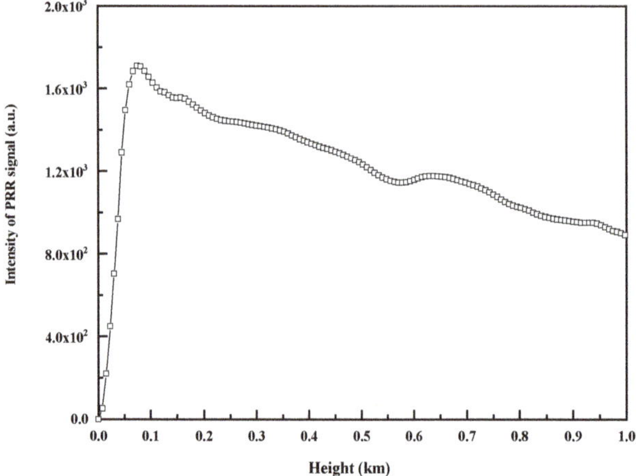

Figure 6. Evaluation of the blind zone and transition zone in dual FOV lidar with the parallel returned PRR signal of secondary FOV.

2.4. Retrieval of the Vertical Profile of Aerosols

The Fernald method [30,31], using the lidar equation given in Equation (2), was used to retrieve the vertical profile of aerosol extinction coefficients from the returned signals, which had been created by joining together the signals from the primary and secondary FOVs:

$$P(z) = \frac{C \cdot E \cdot [\beta_m(z) + \beta_a(z)] \cdot \exp\{-2 \int_0^r [\alpha_m(z') + \alpha_a(z')]dz' >\}}{z^2}. \tag{2}$$

Here, $P(z)$ is the received power from altitude z. C is lidar system constant, E is the laser emitting pulse energy, and $\beta_m(x)$ and $\beta_a(z)$ are the backscattering coefficients of the atmosphere and aerosol at the range z, respectively. While, $\alpha_m(z)$ and $\alpha_a(z)$ are the atmospheric extinction coefficient and aerosol extinction coefficient at the range z, respectively.

2.5. Retrieval of Horizontal Distribution of Aerosols

To implement the mapping of the near-surface aerosol level density distribution, a combination retrieval algorithm of the Fernald method and the slope method [32] was applied in the lidar Equation (2). The inversion procedure was as follows and the flow chart is illustrated in Figure 7.

1. At an initial point Z_i, the position with an SNR that exceeds the threshold, along with each combined return signal, was determined;
2. Then, the combined returned signal was divided into a series of parts from the origin point Z_0 to initial point Z_i as follows: part 1, Z_0~[$Z_0 + \Delta z$]; part 2, [$Z_0 + \delta z$]~[$Z_0 + \Delta z + \delta z$]; ... part n, [$Z_0 + (n-1) \times \delta z$]~[$Z_0 + \Delta z + (n-1) \times \delta z$], etc., where $n = 1, 2, \ldots$; Δz is the step of the interval and δz is the minimum spatial resolution of this lidar system;
3. The slope method [32] was applied to each part. Pairs of both, aerosol extinction coefficients α_n and the Pearson coefficient γ_n were returned from each part;
4. Finally, the pair of {α_f, Z_f} with the optimum γ_f was applied as boundary conditions in the Fernald solutions. The forward-integral and backward-integral results provide the entire profile of aerosol extinction coefficients in the horizontal distribution.

Here, the step of the interval Δz is closely related to the spatial resolution δz of the lidar system. In this study, an interval step of 75 m, which is 10 times δz, was chosen in the inversion procedure. The

SNR threshold was chosen to be 5. The extinction-to-backscatter lidar ratio is an important factor for solving lidar Equation (2). A lidar ratio of 50, which is the representative value for continental aerosol detection at 532 nm, especially in eastern Asia [33,34], was applied in the lidar Equation (2).

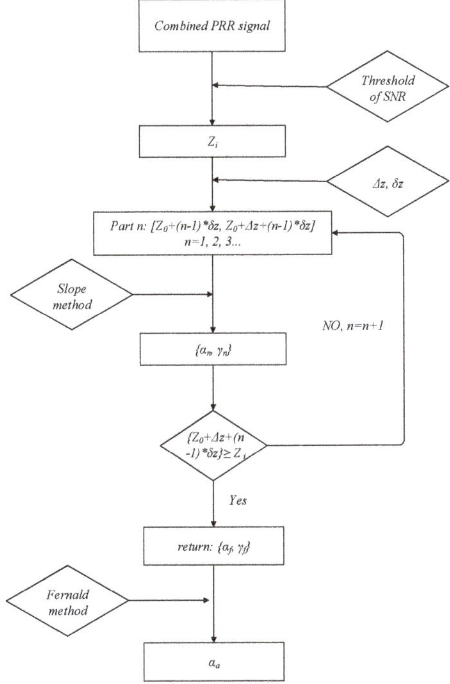

Figure 7. Flow chart of inversion procedure of dual-FOV lidar. Z_i is the initial position, whose SNR exceeds the threshold. Δz is the step of the interval and δz is the minimum spatial resolution of this lidar system. $\{\alpha_n, \gamma_n\}$ are the aerosol extinction coefficients inversed by the slope method and the Pearson coefficient at part n. $\{\alpha_f, \gamma_f\}$ are the optimum aerosol extinction coefficients with the maximum Pearson coefficient at part f. α_a is the extinction coefficient of the aerosol profile inversed by the Fernald method.

3. Results and Discussion

3.1. Vertical Distribution Observation of Aerosols

The vertical observation of aerosols with this dual-FOV lidar was performed on 14 April 2019, at Wuxi, Jiangsu, China. In the data analysis, the time coordinate used was local standard time (LST), which equates to coordinated universal time (UTC) +8 h. The lidar was set on the roof of an office building, which was approximately 30 m above ground level. A typical diurnal variation of the aerosol extinction coefficients is shown in Figure 8a. Aerosols were mainly distributed below 2 km in the early morning and night. The maximum extinction coefficient was close to 2 km^{-1} and was observed after 22:00. During the daytime, the atmospheric boundary layer (ABL) increased from 1 km at 8:00 to 2.3 km at 13:00, and the ABL decreased to 800 m at 20:00. From 7:00. Transporting aerosol layers at 1.5~2.5 km were also detected, and these layers mixed with the boundary layer at noon, which increased the extinction coefficient to 1 km^{-1}. The maximum extinction coefficient reached 1.5 km^{-1} at approximately 11:00. These tiny structures of aerosol distribution below 3 km are shown in Figure 8b. Cloud layers appeared from dawn to 18:00, with a spread range of 3~5 km, and the extinction coefficients are larger than 5 km^{-1}. Then, the extinction coefficients of clouds at noon decreased to approximately 2 km^{-1}. High-altitude clouds were also observed at 19:00 at 12 km. The depolarization observation from this lidar system is shown in Figure 8c. The depolarization ratio is significant to identify the aerosol types that include haze, dust, smoke, volcanic emissions, and particles released through pollution (e.g., carbon-based) or by the surface of the ocean, and those created by gas-to-particle conversions [14]. The aerosol distribution below 2.5 km had a depolarization ratio less than 0.15 mainly dominated by fine particles, and a depolarization ratio of more than 0.6 referred to the cloud distributions at 3~5 km and 12 km.

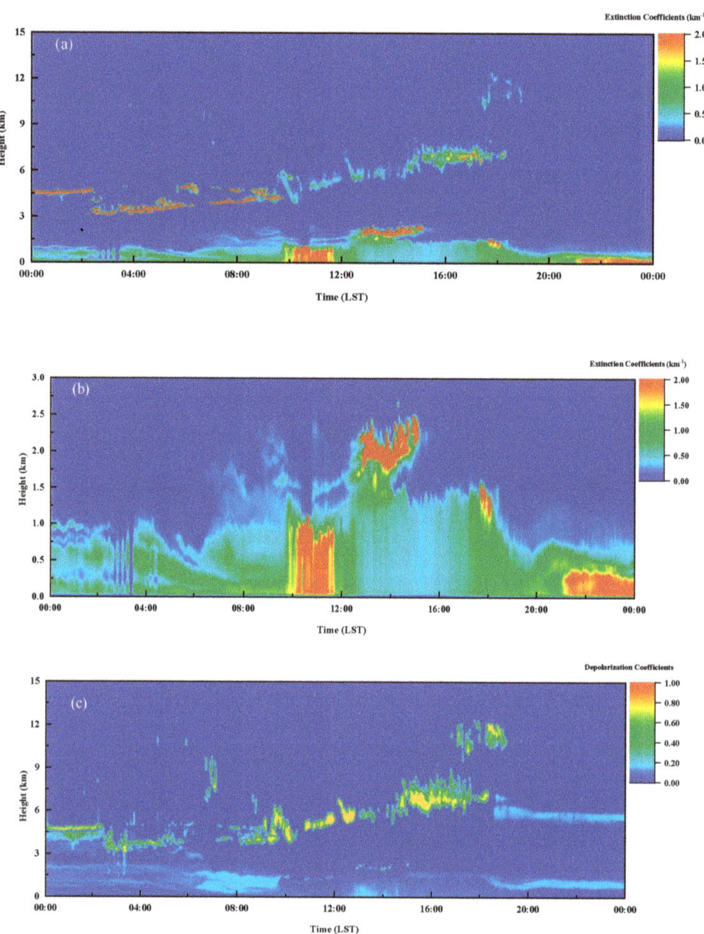

Figure 8. Vertical profile of aerosol detected by dual-FOV lidar. (**a**) Aerosol extinction coefficients in 15 km. (**b**) Aerosol extinction coefficients in 3 km. (**c**) The depolarization coefficients at 532 nm on 14 April 2019.

3.2. Horizontal Distribution Mapping of Aerosols

Field horizontal scanning experiments were conducted in the Shandong Province, China. The lidar station (LS) was set up on the roof of a primary school, which was 22 m above ground level in Hanzhuang Town, Jining City. Weishan Lake is located to the west of the LS. Some industrial plants are scattered on the south of the LS. Farmland is located at both, the north and the east of the LS. A national air quality station (AQS), Hanzhuang Station, is located 272 m northwest of the LS. This station provides in situ data of O_3, NO_2, SO_2, CO, PM_{10} and $PM_{2.5}$. Meteorological parameters, i.e., the wind speed and wind direction, were also measured. The mass concentration of particulate matters $\rho(PM_{10})$ and $\rho(PM_{2.5})$ at the AQS were measured by TEOM™ (Thermo Scientific, Waltham, MA USA). The layout of the LS and AQS is shown in Figure 9.

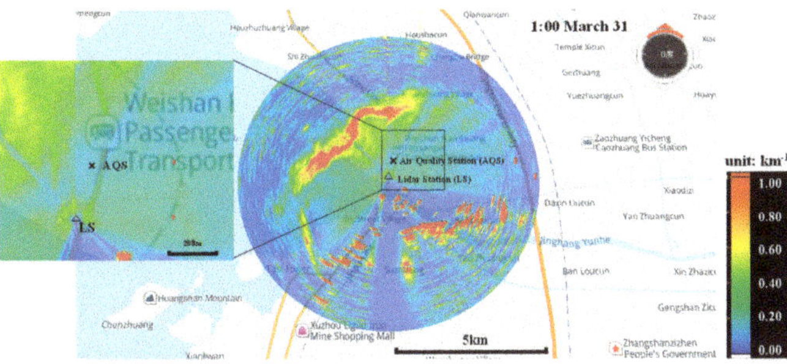

Figure 9. Horizontal extinction coefficient distribution of aerosols at 1:00, 31 March. Sub-chart illustrated the aerosol distribution in the region between lidar station (LS) and air quality station (AQS) on 31 March. (△, the position of LS. ×, the position of AQS).

The scanning parameters of the lidar system were set as follows. The elevation angle was fixed at 5°, and the azimuth angle was 0~360°. The repetition rate of the laser pulses was selected to be 7 kHz. The scanning angle resolution was 2°. Each returned signal was acquired with 150,000 times of the accumulation of the laser pulses. A total of 180 profiles were obtained for each cycle-scanning every 1.5 h. The scanned extinction coefficients of the aerosols were overlaid on the maps using a geo-information system (GIS).

One sample of aerosol distribution scanning on the early morning of 31 March is shown in Figure 9. It is noted that the aerosols were steadily suspended in the range of 2 km to the southwest, south and southeast of the LS. These areas are occupied by many large steel and iron factories and chemical plants. These pollution sources strongly contribute to the $\rho(PM)$ at the AQS with a southeast wind direction. In Figure 9, a pollution mass with an approximate length of 5 km was observed 2~2.5 km north of the LS. The right-wing of this belt significantly influenced the $\rho(PM)$ at the AQS. The corresponding $\rho(PM_{10})$ increased to 135 µg/m³ (Figure 10). The overhead extinction coefficient from the lidar system was 0.53 km⁻¹. This pollution mass led to air quality deterioration at AQS in the next few hours.

Lidar scanning began at 17:53, 29 March and ended at 07:11, 31 March 2018. In total, 30 cycle-scanning maps were acquired from this field experiment. Time series of $\rho(PM)$ at the AQS and four lidar scanning results of aerosol distribution are shown in Figure 10. $\rho(PM_{10})$ and $\rho(PM_{2.5})$ decreased from 128 µg/m³, and 27 µg/m³, respectively, at the beginning of the observation (upper panel of Figure 10). The corresponding lidar scanning illustrated few particulate matters around the AQS, and the overhead extinction coefficient of the AQS was 0.12 km⁻¹ (lower panel Figure 10a). The $\rho(PM_{10})$ and $\rho(PM_{2.5})$ climbed up to 114 µg/m³ and 47.4 µg/m³ at 08:00, 30 March (upper panel of Figure 10). The lidar observation at 07:23–08:50 confirmed the accumulation of aerosols over the AQS, with an overhead extinction coefficient of 0.21 km⁻¹ (lower panel Figure 10b). From 23:00, 30 March, the $\rho(PM_{10})$ and $\rho(PM_{2.5})$ climbed steadily from 122 µg/m³ and 41 µg/m³ to 160 µg/m³, and 109 µg/m³, respectively, at 07:00, 31 March, the end of lidar scanning (upper panel of Figure 10). At this stage, the lidar maps show that the aerosols were enriched toward the AQS and increased the PM mass concentrations. The corresponding overhead extinction coefficients were 0.43 km⁻¹ at 23:00, 30 March and 0.63 km⁻¹ at 06:00, 31 March (lower panel Figure 10c,d).

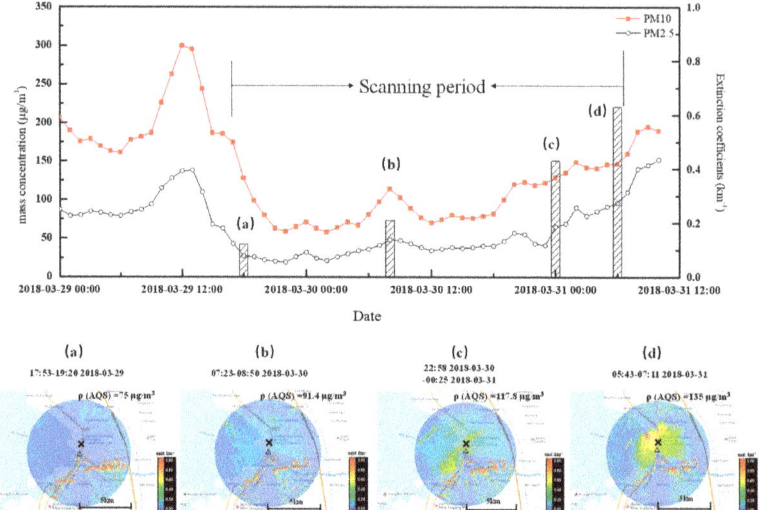

Figure 10. Time series of $\rho(PM_{10})$, $\rho(PM_{2.5})$ of the air quality station (AQS), and the overhead extinction coefficients of AQS from lidar (upper panel) and the aerosol mappings from lidar at Hanzhuang at 17:53–19:20, 29 March (lower pane (**a**)), 07:23–08:50, 30 March (lower pane (**b**)), 22:58, 30 March–00:25, 31 March (lower pane (**c**)), and 05:43–07:11, 31 March (lower pane (**d**)) (Δ, the position of lidar, LS. ×, the position of AQS).

3.3. Quantitative Evaluation of Aerosol Distribution

Mass concentration mapping of aerosols from the lidar system is an ideal choice for supporting environmental management, especially for air quality modeling simulations that rely on the following three advantages: (1) A higher spatial and temporal resolution of mass concentrations; (2) a greater observation range in both the vertical and horizontal directions; and (3) easy calibration, validation by an array of standard point instruments at ground level, and the establishment of a data set with data quality assurance.

Here, a nonlinear relationship can be fitted with the $\rho(PM_{10})$ mass concentration at the AQS and the overhead extinction coefficients α_a from lidar scanning. This nonlinear relationship is shown as follows,

$$\rho(PM_{10}) = \kappa \cdot \alpha_a{}^{\zeta} + C \tag{3}$$

where κ, ζ, and C are the fitted parameters.

Using 30 sets of $\rho(PM_{10})$ and α_a values, the fitting relationship of Equation (3) is determined and is illustrated in Figure 11. The parameters are $\kappa = 148.85\ \mu g/m^3$, $\zeta = 0.4$, and $C = 11.62\ \mu g/m^3$, and the Pearson coefficient is 0.91.

The lidar scanning results, the extinction coefficients of the aerosol distributions, can be transformed to aerosol mass concentration distributions with this fitted relationship. The early morning case on 31 March is shown again but with mass concentrations in Figure 12. The mass concentration of the point sources 2.5 km southeast of the LS was close to 250 μg/m³. The mass concentration of the "aerosol-belt" northwest of the LS exceeded 250 μg/m³. Figure 12 also ignored the low SNR regions of extinction coefficients, especially over the radius of 3 km. The calculated overhead mass concentrations of AQS were also illustrated in the lower panel in Figure 10. The overhead mass concentrations were 75 μg/m³ (99 μg/m³) at 19:00, 29 March, 91.4 μg/m³ (103 μg/m³) at 08:00, 30 March, 117.8 μg/m³ (129 μg/m³) at 00:00, 31 March, and 135 μg/m³ (147 μg/m³) at 19:00, 31 March, respectively. A difference from the ground sampling measurement from AQS (values in brackets) may be because of water vapor or relative humidity in Weishan Lake.

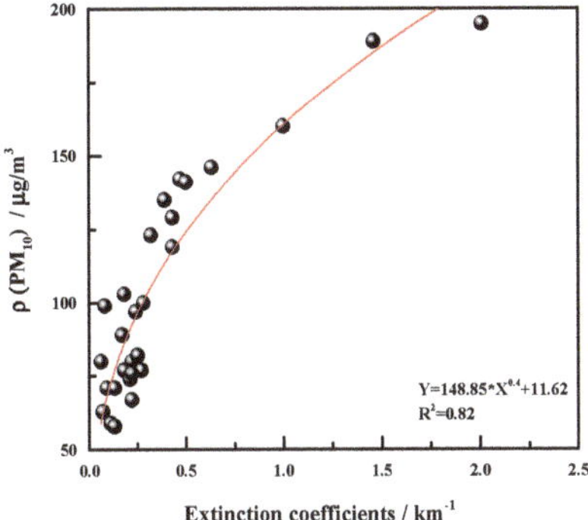

Figure 11. Fitting result with extinction coefficients and $\rho(PM_{10})$.

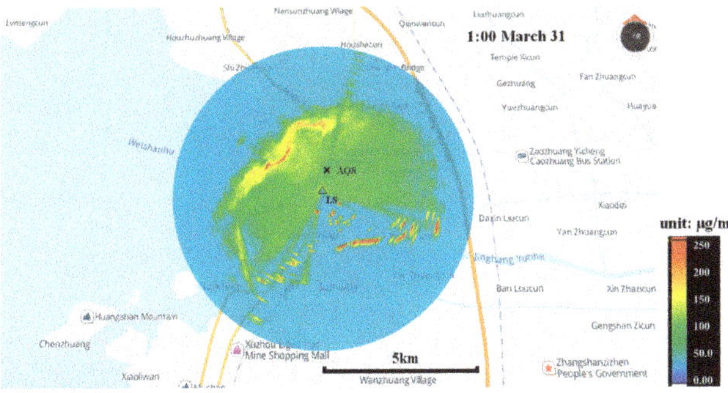

Figure 12. Level mass concentration distribution of aerosols at 1:00, 31 March, of dual-FOV lidar (at altitude of 22 m). (Δ, the position of lidar, LS. ×, the position of air quality station, AQS).

An accurate lidar ratio is important for the retrieval of extinction coefficients. In general, the lidar ratio varies with height and depends on the shape, size distribution and refractive index of the aerosol particles, as well as the lidar wavelength. Here, we applied a constant lidar ratio in the lidar equation for both the vertical and horizontal profile inversion. When we applied the "Fernald-slope" method to retrieve the horizontal distribution of aerosols, the SNR threshold was a critical definition. In the future, the effect of the SNR threshold will be discussed. It is noted that the lidar beam probed across the Weishan Lake surface, and a not negligible factor of water vapor was carefully considered when we converted the extinction coefficients to mass concentrations. The influence of the water vapor or the relative humidity on the retrieval of extinction coefficients has been studied by Zhao et al. [35], and the extinction coefficients here need to be carefully corrected in further studies. In addition, deploying a PM sampling array instead of a single lidar station and providing a few $\rho(PM_{10})$ and extinction coefficient datasets at different positions would be another reasonable measure to minimize the error of retrieving the mass concentration distribution. If possible, the chemical compositions of aerosols can be extracted to make a reasonable evaluation of the extinction coefficients.

4. Conclusions

A dual-FOV lidar was designed without a blind zone and a transition zone of 60 m to observe the lower atmosphere. Dual FOV technology was applied to overcome the traditional obstacles of Mie-lidars with hundreds of meters of blind zone. The lidar system is maneuverable and has a weight of less than 70 kg, including the scanning platform. A complete combined returned signal was acquired by carefully checking for the linear combination interval with the primary FOV signal and the secondary FOV signal. Vertical profiles of the aerosol field were retrieved with the Fernald method, while the horizontal distribution of aerosol extinction coefficients was retrieved by combining the slope method and the Fernald method. Horizontal scanning observations clearly showed the spatial-temporal distribution of aerosols. The lidar system can provide valuable and accurate pollution information with the mass concentration mapping, when validated with local quantitative PM measurements. Furthermore, the corrected mass concentration mapping data can be assimilated into 3-D air quality modeling, in order to significantly improve the accuracy of regional air quality predictions. This compact portable lidar system is also an ideal choice for the 3-D atmospheric stereo-pollution monitoring networks construction in severe polluted regions in the future, i.e., New Delhi in India, Beijing–Tianjin–Hebei (JJJ) in China, Bangkok in Southeast Asia.

Author Contributions: J.W.: original draft preparation, formal analysis, input references and drafting acknowledgement. W.L. and J.L.: conceptualization of system, investigation and administration of this project and funding. C.L.: formal analysis, data curation and input Figure 9. T.Z.: leading conceptualization of system, validation, project administration and input Figure 2, Figure 7, Figure 10. Z.C.: methodology, especially for the draft section of 2.2, validation and polishing for the draft preparation. Y.X.: Software development and algorithm optimization. X.M.: validation and data curation. All authors have read and agree to the published version of the manuscript.

Funding: This research was supported by the National Key Research and Development Program of China (2016YFC0200400, 2016YFC0200402, 2017YFC0212800, 2017YFC0212801, 2018YFC0213100, 2018YFC0213101) and the National Research Program for Key Issues in Air Pollution Control (DQGG0102).

Acknowledgments: Great thanks are owed to Jinsheng Zhu and Ling Li. It is not possible to conduct this research without their assistance in the field experiment, especially for the maintenance of lidar. The authors also gratefully acknowledge the China National Environmental Monitoring Centre (CNEMC) for providing the data set of air quality station in this research. We also thank Vivian Du for her editorial advice. We would like to thank three anonymous reviewers for their comments which greatly strengthened the manuscript. Special thanks are given to the editors of MDPI for their English editing works on this manuscript.

Conflicts of Interest: The authors declare no conflict of interest.

References

1. Liu, B.; Ma, Y.Y.; Gong, W.; Zhang, M.; Yang, J. Determination of boundary layer top on the basis of the characteristics of atmospheric particles. *Atmos. Environ.* **2018**, *178*, 140–147. [CrossRef]
2. Han, S.Q.; Liu, J.L.; Hao, T.Y.; Zhang, Y.F.; Li, P.Y.; Yang, J.B.; Wang, Q.L.; Cai, Z.Y.; Yao, Q.; Zhang, M.; et al. Boundary layer structure and scavenging effect during a typical winter haze-fog episode in a core city of BTH region, China. *Atmos. Environ.* **2018**, *179*, 187–200. [CrossRef]
3. Chen, Z.Y.; Liu, C.; Liu, W.Q.; Zhang, T.S.; Xu, J. A synchronous observation of enhanced aerosol and NO2 over Beijing, China, in winter 2015. *Sci. Total Environ.* **2017**, *575*, 429. [CrossRef] [PubMed]
4. Pal, S.; Lee, T.R.; Phelps, S.; De Wekker, S.F.J. Impact of atmospheric boundary layer depth variability and wind reversal on the diurnal variability of aerosol concentration at a valley site. *Sci. Total Environ.* **2014**, *496*, 424–434. [CrossRef]
5. Danchovski, V. Summertime Urban Mixing Layer Height over Sofia, Bulgaria. *Atmosphere* **2019**, *10*, 36. [CrossRef]
6. Bozlaker, A.; Prospero, J.M.; Fraser, M.P.; Chellam, S. Quantifying the Contribution of Long-Range Saharan Dust Transport on Particulate Matter Concentrations in Houston, Texas, Using Detailed Elemental Analysis. *Environ. Sci. Technol.* **2013**, *47*, 10179–10187. [CrossRef]
7. Salinas, S.V.; Chew, B.N.; Miettinen, J.; Campbell, J.R.; Welton, E.J.; Reid, J.S.; Yu, L.E.; Liew, S.C. Physical and optical characteristics of the October 2010 haze event over Singapore: A photometric and lidar analysis. *Atmos. Res.* **2013**, *122*, 555–570. [CrossRef]
8. Xie, C.B.; Nishizawa, T.; Sugimoto, N.; Matstui, I.; Wang, Z.F. Characteristics of aerosol optical properties in pollution and Asian dust episodes over Beijing, China. *Appl. Opt.* **2008**, *47*, 4945–4951. [CrossRef]

9.   Nishizawa, T.; Sugimoto, N.; Matsui, I.; Shimizu, A.; Hara, Y.; Itsushi, U.; Yasunaga, C.; Kudo, R.; Kim, S.W. Ground-based network observation using Mie–Raman lidars and multi-wavelength Raman lidars and algorithm to retrieve distributions of aerosol components. *J. Quant. Spectrosc. RA* **2017**, *188*, 79–93. [CrossRef]

10. Matthais, V.; Freudenthaler, V.; Amodeo, A.; Balin, I.; Balis, D.; Bösenberg, J.; Chaikovsky, A.; Chourdakis, G.; Comeron, A.; Delaval, A.; et al. Aerosol lidar intercomparison in the framework of the EARLINET project. 1. Instruments. *Appl. Opt.* **2004**, *43*, 961–976. [CrossRef]

11. Chiang, C.W.; Das, S.K.; Chiang, H.W.; Nee, J.B.; Sun, S.H.; Chen, S.W.; Lin, P.H.; Chu, J.C.; Su, C.S.; Su, L.S. A new mobile and portable scanning lidar for profiling the lower troposphere. *Geosci. Instrum. Meth.* **2015**, *4*, 35–44. [CrossRef]

12. Chazette, P.; Sanak, J.; Dulac, F. New Approach for Aerosol Profiling with a Lidar Onboard an Ultralight Aircraft: Application to the African Monsoon Multidisciplinary Analysis. *Environ. Sci. Technol.* **2007**, *41*, 8335–8341. [CrossRef] [PubMed]

13. Proestakis, E.; Amiridis, V.; Marinou, E.; Georgoulias, A.K.; Solomos, S.; Kazadzis, S.; Chimot, J.; Che, H.Z.; Alexandri, G.; Binietoglou, I.; et al. 9-year spatial and temporal evolution of desert dust aerosols over South-East Asia as revealed by CALIOP. *Atmos. Chem. Phys.* **2018**, *18*, 1–35. [CrossRef]

14. Weitkamp, C. *LIDAR: Range-Resolved Optical Remote Sensing of the Atmosphere*; Springer: New York, NY, USA, 2004.

15. Kovalev, V.A.; Eichinger, W.E. *Elastic Lidar: Theory, Practice, and Analysis Methods*; Wiley-Interscience: New York, NY, USA, 2004.

16. Sasano, Y.; Shimizu, H.; Takeuchi, N.; Okuda, M. Geometrical form factor in the laser radar equation: An experimental determination. *Appl. Opt.* **1979**, *18*, 3908–3910. [CrossRef] [PubMed]

17. Kuze, H.; Kinjo, H.; Sakurada, Y.; Takeuchi, N. Field-of-view dependence of lidar signals by use of Newtonian and Cassegrainian telescopes. *Appl. Opt.* **1998**, *37*, 3128–3132. [CrossRef]

18. Welton, E.J.; Voss, K.J.; Gordon, H.R.; Maring, H.; Smirnov, A.; Holben, B.; Schmid, B.; Livingston, J.M.; Russell, P.B.; Durkee, P.A.; et al. Ground-based lidar measurements of aerosols during ACE-2: Instrument description, results, and comparisons with other ground-based and airborne measurements. *Tellus B* **2000**, *52*, 636–651. [CrossRef]

19. Stelmaszczyk, K.; Dell'Aglio, M.; Chudzyński, S.; Stacewicz, T.; Wöste, L. Analytical function for lidar geometrical compression form-factor calculations. *Appl. Opt.* **2005**, *44*, 1323–1331. [CrossRef]

20. Hey, J.V.; Coupland, J.; Foo, M.H.; Richards, J.; Sandford, A. Determination of overlap in lidar systems. *Appl. Opt.* **2011**, *50*, 5791–5797.

21. He, T.Y.; Stanič, S.; Gao, F.; Bergant, K.; Veberič, D.; Song, X.Q.; A. Dolžan, A. Tracking of urban aerosols using combined lidar-based remote sensing and ground-based measurements. *Atmos. Meas. Tech.* **2012**, *5*, 891–900. [CrossRef]

22. Behrendt, A.; Pal, S.; Wulfmeyer, V.; Valdebenito, B.Á.M.; Lammel, G. A novel approach for the characterization of transport and optical properties of aerosol particles near sources—Part I: Measurement of particle backscatter coefficient maps with a scanning UV lidar. *Atmos. Environ.* **2011**, *45*, 2795–2802. [CrossRef]

23. Sasano, Y. Tropospheric aerosol extinction coefficient profiles derived from scanning lidar measurements over Tsukuba, Japan, from 1990 to 1993. *Appl. Opt.* **1996**, *35*, 4941–4952. [CrossRef] [PubMed]

24. Shan, H.H.; Zhang, H.; Liu, J.J.; Tao, Z.M.; Wang, S.H.; Ma, X.M.; Zhou, P.C.; Yao, L.; Liu, D.; Xie, C.B.; et al. Retrieval method of aerosol extinction coefficient profile based on backscattering, side-scattering and Raman-scattering lidar. *Opt. Commun.* **2018**, *410*, 730–732. [CrossRef]

25. Bian, Y.X.; Zhao, C.S.; Xu, W.Y.; Kuang, Y.; Tao, J.C.; Wei, W.; Ma, N.; Zhao, G.; Lian, S.P.; Tan, W.S.; et al. A novel method to retrieve the nocturnal boundary layer structure based on CCD laser aerosol detection system measurements. *Remote Sens. Environ.* **2018**, *211*, 38–47. [CrossRef]

26. Barnes, J.E.; Parikh, N.C.; Kaplan, T. Boundary layer scattering measurements with a charge-coupled device camera lidar. *Appl. Opt.* **2003**, *42*, 2647–2652. [CrossRef]

27. Meki, K.; Yamaguchi, K.; Li, X.; Saito, Y.; Kawahara, T.D.; Nomura, A. Range-resolved bistatic imaging lidar for the measurement of the lower atmosphere. *Opt. Lett.* **1996**, *21*, 1318–1320. [CrossRef] [PubMed]

28. Schmidt, J.; Wandinger, U.; Malinka, A. Dual-field-of-view Raman lidar measurements for the retrieval of cloud microphysical properties. *Appl. Opt.* **2013**, *52*, 2235–2247. [CrossRef] [PubMed]

29. Gong, W.; Li, J.; Mao, F.Y.; Zhang, J.Y. Comparison of simultaneous signals obtained from a dual-field-of-view lidar and its application to noise reduction based on empirical mode decomposition. *Chin. Opt. Lett.* **2011**, *9*, 050101. [CrossRef]

30. Fernald, F.G. Analysis of atmospheric lidar observations- Some comments. *Appl. Opt.* **1984**, *23*, 652–653. [CrossRef] [PubMed]

31. Fernald, F.G.; Herman, B.M.; Reagan, J.A. Determination of Aerosol Height Distributions by Lidar. *J. Appl. Meteorol.* **1972**, *11*, 482–489. [CrossRef]

32. Collis, R.T.H. Lidar: A new atmospheric probe. *Q. J. R. Meteor. Soc.* **1966**, *92*, 220–230. [CrossRef]

33. Chiang, C.W.; Das, S.K.; Nee, J.B. An iterative calculation to derive extinction-to-backscatter ratio based on lidar measurements. *J. Quant. Spectrosc. R.* **2008**, *109*, 1187–1195. [CrossRef]

34. Müller, D.; Ansmann, A.; Mattis, I.; Tesche, M.; Wandinger, U.; Althausen, D.; Pisani, G. Aerosol-type-dependent lidar ratios observed with Raman lidar. *J. Geophys. Res.* **2007**, *112*, D16202. [CrossRef]

35. Zhao, G.; Zhao, C.; Kuang, Y.; Tao, J.; Tan, W.; Bian, J.; Li, C. Impact of aerosol hygroscopic growth on retrieving aerosol extinction coefficient profiles from elastic-backscatter lidar signals. *Atmos. Chem. Phys.* **2017**, *17*, 12133–12143. [CrossRef]

 *remote sensing*

Article

Vertical Wind Shear Modulates Particulate Matter Pollutions: A Perspective from Radar Wind Profiler Observations in Beijing, China

Ying Zhang [1], Jianping Guo [1,*], Yuanjian Yang [2,3], Yu Wang [4] and Steve H.L. Yim [3,5,6]

[1] State Key Laboratory of Severe Weather, Chinese Academy of Meteorological Sciences, Beijing 100081, China; zhangy1@mail.ustc.edu.cn

[2] School of Atmospheric Physics, Nanjing University of Information Science and Technology, Nanjing, 210044, China; yyj1985@nuist.edu.cn

[3] Institute of Environment, Energy and Sustainability, The Chinese University of Hong Kong, Shatin, New Territories, Hong Kong 999077, China; steveyim@cuhk.edu.hk

[4] School of Earth and Space Sciences, University of Science and Technology of China, Hefei 230026, China; wangyu09@ustc.edu.cn

[5] Department of Geography and Resource Management, The Chinese University of Hong Kong, Shatin, New Territories, Hong Kong 999077, China

[6] Stanley Ho Big Data Decision Analytics Research Centre, The Chinese University of Hong Kong, Shatin, New Territories, Hong Kong 999077, China

* Correspondence: jpguo@cma.gov.cn; Tel.: +86-10-5899-3189

Received: 6 January 2020; Accepted: 4 February 2020; Published: 7 February 2020

Abstract: Vertical wind shear (VWS) is one of the key meteorological factors in modulating ground-level particulate matter with an aerodynamic diameter of 2.5 μm or less ($PM_{2.5}$). Due to the lack of high-resolution vertical wind measurements, how the VWS affects ground-level $PM_{2.5}$ remains highly debated. Here we employed the wind profiling observations from the fine-time-resolution radar wind profiler (RWP), together with hourly ground-level $PM_{2.5}$ measurements, to explore the wind features in the planetary boundary layer (PBL) and their association with aerosols in Beijing for the period from December 1, 2018, to February 28, 2019. Overall, southerly wind anomalies almost dominated throughout the whole PBL or even beyond the PBL under polluted conditions during the course of a day, as totally opposed to the northerly wind anomalies in the PBL under clean conditions. Besides, the ground-level $PM_{2.5}$ pollution exhibited a strong dependence on the VWS. A much weaker VWS was observed in the lower part of the PBL under polluted conditions, compared with that under clean conditions, which could be due to the strong ground-level $PM_{2.5}$ accumulation induced by weak vertical mixing in the PBL. Notably, weak northbound transboundary $PM_{2.5}$ pollution mainly appeared within the PBL, where relatively small VWS dominated. Above the PBL, strong northerlies winds also favored the long-range transport of aerosols, which in turn deteriorated the air quality in Beijing as well. This was well corroborated by the synoptic-scale circulation and backward trajectory analysis. Therefore, we argued here that not only the wind speed in the vertical but the VWS were important for the investigation of aerosol pollution formation mechanism in Beijing. Also, our findings offer wider insights into the role of VWS from RWP in modulating the variation of $PM_{2.5}$, which deserves explicit consideration in the forecast of air quality in the future.

Keywords: $PM_{2.5}$; radar wind profiler; Beijing; wind shear

---

1. Introduction

Particulate particle with an aerodynamic diameter of 2.5 μm or less ($PM_{2.5}$), mainly originated from industrial emissions and vehicle exhaust pollutants, and secondary aerosols forming through a series of photochemical reactions [1,2] have been shown to significantly affect the atmospheric environment [3–5], weather and climate system [6–15], and human health [16–19]. Therefore, the $PM_{2.5}$ pollution and its causes have been increasingly receiving attention in recent years [20,21].

The major drivers for deteriorating or improving $PM_{2.5}$ pollution are roughly twofold—aerosol emissions and meteorology—both of which are highly versatile and uncertain. In addition to high emissions accompanied with the rapid development of urbanization and industrialization, the roles of meteorological conditions, including large-scale synoptic patterns [22–26], and local meteorological conditions in the planetary boundary layer (PBL) [27–32] have been well recognized able to modulate the $PM_{2.5}$ concentration. For instance, Tai et al. [27] revealed that the local meteorological conditions could explain up to 50% of the daily variability of $PM_{2.5}$ in the USA from 1998 to 2008. In 68 major cities of China, ground-level $PM_{2.5}$ were found to be broadly associated with local meteorological factors at seasonal, yearly, and regional scales [33]. Among various meteorological factors, the surface wind speed was one of the variables modulating ground-level $PM_{2.5}$ over the Yangtze River Delta region of China, which showed that $PM_{2.5}$ decreased approximately by -2.42 μg m^{-3} for a 1 m s^{-1} increase in wind speed [34]. In Beijing, the heavy pollution events frequently occurred under the calm wind conditions, which was generally associated with stable atmospheric stratification and shallow PBL [35–38]. The presence of high pressure in northwest parts of Beijing, linked to strong northwesterly winds, was closely associated with a significant drop in $PM_{2.5}$ concentrations in Beijing [4]. Besides, the changes in circulation induced by local mountain-valley and urban heat island setting in Beijing and its surrounding areas were found to be able to modulate the diurnal variations of $PM_{2.5}$ in Beijing [32].

Among others, reanalysis data and model simulations were one of the most used approaches to analyze the variation of wind with different height in the lower troposphere and its impacts on air pollution, revealing a significant role of vertical wind shear (VWS, an important indicator of dynamically vertical mixing) in modulating particulate matter pollution [37,39,40]. In recent years, there has been a surge of interest in observational investigation of VWS in connection with atmospheric pollution, most of which are based on Doppler wind lidar [26,39]. However, the Doppler wind lidar has limited capability to offer vertically resolved wind observations under pretty clean or foggy conditions. To date, the associations between vertical wind profile and surface particulate matter concentrations have yet to be fully understood in Beijing.

Fortunately, the new-generation radar wind profiler (RWP) deployed by the China Meteorological Administration (CMA) in Beijing [41] offers us the best opportunity to quantify the long-term effect of local wind vector profiles and VWS on ground-level $PM_{2.5}$ pollution in Beijing. Thus, this study aims to explore the impacts of wind profiles and VWS on the wintertime $PM_{2.5}$ in Beijing based on high-resolution RWP observation along with ground-level $PM_{2.5}$ monitoring. The remaining contents of this work proceed as follows. In Section 2, the measurements of RWP, other related weather data, and ground-level $PM_{2.5}$ are described in detail. In Section 3, the impacts of wind and VWS on $PM_{2.5}$ pollution in Beijing are analyzed and discussed. Finally, the main findings are summarized in Section 4.

2. Data and Methodology

2.1. Study Area

Beijing, the capital of the People's Republic of China, is located in the north part of the North China Plain (NCP) of China and covers an area of around 16,410 square kilometers. As shown in Figure 1a, there exists a large amount of aerosol emission sources surrounding Beijing with the recent rapid economic development throughout China, especially in eastern China. Beijing is surrounded by the Yanshan Mountains to the north, and by Taihang Mountains to the west and northwestern (Figure 1b). In terms of the climate in Beijing, it typically belongs to a semi-humid continental climate in the north temperate zone, characterized by hot and humid summers due to the subsidence caused by the subtropical high, and cold, windy and dry winters which is mainly under the influence of the vast Siberian anticyclone [41–43]. Due to the huge amount of anthropogenic emission in the NCP (see Figure 1a), the atmospheric pollution (especially the $PM_{2.5}$) in Beijing has been intensively analyzed in recent years [38,39,41,44].

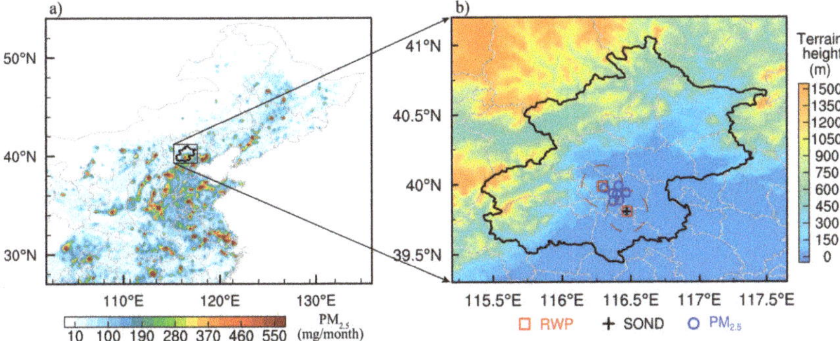

Figure 1. The spatial distribution of (**a**) particulate particle with an aerodynamic diameter of 2.5 μm or less ($PM_{2.5}$) emissions (mainly in the transportation, agriculture, industry, power and residual sectors) during winter in 2016, as acquired from the multi-resolution emission inventory for China (http://www.meicmodel.org/), and (**b**) monitoring stations in Beijing overlaid with topography (color shading). The red squares, blue circles, and black plus refer to the locations of radar wind profiler (RWP), $PM_{2.5}$, and radiosonde (SOND) sites, respectively. The administrative boundary of Beijing is denoted by the black solid lines.

2.2. Radar Wind Profiler Measurements

The RWP is a type of remote sensing instrument that detects and processes vertical-resolved wind field information by transmitting and receiving electromagnetic beams in different directions. This instrument can provide a variety of data products, including the profiles of horizontal wind speed and direction, and vertical velocity. Specifically, the RWP data used here were collected from two sites (marked by the red squares in Figure 1b), including the Tongzhou site (116.29° E; 39.99° N) and Chaoyang site (116.47° E; 39.81° N). Both RWPs deployed in Beijing are the CFL-16 profiler, which provides 25 levels of wind speed and direction below ~3 km above ground level (AGL) with a vertical resolution of 120 m, beginning at 150 m (AGL). The measurements of wind are taken at 6 min intervals, and the detailed specifications of the RWP are given in Table 1. Prior to further analysis, the raw data have to undergo strict quality control for data consistency, continuity, and deviation [41,45,46]. To match the hourly $PM_{2.5}$, the original 6 min RWP measurements were aggregated into hourly data. Our study period covers the whole boreal winter of 2018 (December of 2018 through February of 2019). To exclude the effect of wet deposition, all the measurements mentioned in this study refer to those taken on non-precipitation (i.e., rain, hail, or snow) hours, unless otherwise noted. The hourly precipitation events with precipitation amounts larger than 0.1 mm are generally defined as precipitation hours [44].

2.3. Ground-level $PM_{2.5}$ Concentration Measurements

In this study, the aerosol pollution in Beijing is denoted by hourly ground-level $PM_{2.5}$ concentration measurements, collected from seven air quality sites (marked by the blue circles in Figure 1b; Table 2) of the Ministry of Ecology and Environment of China. At each monitoring site, the hourly $PM_{2.5}$ is measured using the tapered element oscillating microbalance method and the beta absorption method. The systematic uncertainty of ground-level $PM_{2.5}$ mass concentration at these air quality monitoring stations was controlled within 15% [47]. To avoid the uncertainties caused by aerosol heterogenous distribution, averages were taken on the $PM_{2.5}$ measurements from all these seven sites.

Table 1. Summary of the operating and sampling characteristics of the CFL-16 Radar Wind Profiler (RWP) deployed in Beijing.

Parameter	Range of Respective Values
Direction accuracy	≤ 10
Speed accuracy	$1\ \mathrm{m\ s^{-1}}$
Vertical resolution	120 m
Lowest level	150 m AGL
Maximum height	16 km AGL
Operating frequency	445 MHz
Aperture	$100\ \mathrm{m^2}$
Gain	33 dB
Peak power	23 kW
Pulse width	0.8 us
Averaging time	6~60 min

Table 2. Basic information for observation stations.

Station Type	Station Name	Position	Elevation (m)	Observation	Time Resolution
RWP	54399	116.29° E; 39.99° N	46.9	WS, WD	1 hour
RWP	54511	116.47° E; 39.81° N	32.5	WS, WD	1 hour
SND	Beijing (BJ)	116.47° E; 39.81° N	31.3	T, P (to calculate PT)	launched twice a day at 0715 and 1915 BJT
MEE	Haidian (HD)	116.32° E; 39.99° N	-	$PM_{2.5}$	1 hour
MEE	Aoti (AT)	116.41° E; 40.00° N	-	$PM_{2.5}$	1 hour
MEE	Guanyuan (GY)	116.36° E; 39.94° N	-	$PM_{2.5}$	1 hour
MEE	Dongsi (DS)	116.43° E; 39.95° N	-	$PM_{2.5}$	1 hour
MEE	Wanshou (WS)	116.37° E; 39.87° N	-	$PM_{2.5}$	1 hour
MEE	Nongzhanguan (NZG)	116.47° E; 39.97° N	-	$PM_{2.5}$	1 hour
MEE	Tiantan (TT)	116.43° E; 39.87° N	-	$PM_{2.5}$	1 hour

Note that RWP (Radar Wind Profiler), MEE (Ministry of Ecology and Environment), and SND (Radiosonde) stand for wind-profiler site, air quality site of the Ministry of Ecology and Environment, and radiosonde site, respectively. (WS: wind speed; WD: wind direction; Temperature: T; Pressure: P; Potential temperature: PT; Aerodynamic diameter smaller than 2.5 µm: $PM_{2.5}$).

2.4. Radiosonde and Other Meteorological Data

The radiosonde soundings routinely measured in Beijing (116.47° E; 39.80° N, marked by the black cross in Figure 1b) were also collected to characterize the temperature inversion in association with aerosol pollution. As stated in our previous studies [29,48], the sounding balloons in China are launched twice per day at around 0800 and 2000 Beijing time (BJT = UTC + 8 h). It follows that the sounding measurements at 0800 BJT were compared with the hourly RWP and $PM_{2.5}$ data at 0800 BJT. As illustrated in Figure 1b, all these meteorological stations and $PM_{2.5}$ monitoring sites are evenly distributed to represent the hourly meteorological conditions in the whole urban area of Beijing well.

2.5. Air Mass Back Trajectory Model

Air masses related to regional or synoptic meteorological conditions could be responsible for the atmospheric transport of aerosol particles in the vertical and horizontal directions [26,49]. As such, we identified the main transport pathways of aerosol pollutants from surrounding regions to Beijing using the Hybrid Single-Particle Lagrangian Integrated Trajectory Model (HYSPLIT) [50]. The HYSPLIT developed by the National Ocean and Atmospheric Administration (NOAA)'s Air Resources Laboratory has been extensively used for the analyses of transboundary transport and dispersion of aerosol [51]. Based on the HYSPLIT and The GDAS (Global Data Assimilation System) reanalysis data, the frequency distribution and cluster mean of 24-h backward trajectories of each day for the period from December 1, 2018, to February 28, 2019, were calculated, respectively. The trajectory endpoint was set in the urban area of Beijing (116.37 °E, 40.09 °N) with a height of 100 m above ground level (AGL).

2.6. Methodology

In order to avoid the effects induced by the seasonal variation of aerosol at specific sites and by the spatial variation among different sites within the same region, hourly $PM_{2.5}$ concentrations for a given site are normalized using the monthly mean of each site and year according to the approach by Wang et al. [52]. The $PM_{2.5}$ dataset is then grouped into three subsets, each of which has the same number of samples. The lower and upper terciles of normalized $PM_{2.5}$ refer to the clean (bottom 1/3) and polluted (top 1/3) conditions, respectively. In this way, comparison analysis can be performed between clean and polluted atmospheric conditions, and good sampling statistics can be maintained [11,53], even though the critical threshold of normalized $PM_{2.5}$, which is used to distinguish between the clean and polluted categories, differs by various time scales. Besides, hourly wind anomalies with 120 m resolution in the vertical are calculated and then used in the subsequent analyses in diurnal association with ground-level $PM_{2.5}$ concentration. This allows us to do a more detailed analysis of the wind variation for various $PM_{2.5}$ pollution levels.

The VWS has been found to play important roles in the dispersion of air pollutants, and thus was calculated here to check its effects on $PM_{2.5}$ variability. Therefore, the bulk shear, which refers to the magnitude of the bulk vector difference (top minus bottom) divided by height [54–56], is calculated as follows:

$$VWS = \frac{\sqrt{(u_{z1} - u_{z2})^2 + (v_{z1} - v_{z2})^2}}{(z1 - z2)} \times 1000 \tag{1}$$

where VWS is the vertical wind shear (units: $m/(s \cdot km)$), u_{z1} and u_{z2} represent the zonal wind at the height of $z1$ and $z2$, respectively; and v_{z1} and v_{z2} represent the meridional wind at the height of $z1$ and $z2$. $z1$ is the top height and $z2$ is the bottom height.

To enhance the visual interpretation, daily 24-h period is divided into eight sub-period at 3-hour intervals, which is defined as follows [57,58]: late night (0000 – 0300 BJT), early morning (0300–0600 BJT), morning (0600–0900 BJT), late morning (0900–1200 BJT), early afternoon (1200–1500 BJT), late afternoon (1500–1800 BJT), evening (1800–2100 BJT), and night (2100–2400 BJT).

Additionally, the bivariate polar plot has been used, combining wind measurements from RWP and $PM_{2.5}$ measurements in Beijing, which is expected to provide insight into the sophisticated relationship between wind and $PM_{2.5}$ [59,60].

3. Results and Discussion

3.1. Thermodynamic and Meteorological Variables Related To $PM_{2.5}$

Figure 2 shows the time series of observed daily $PM_{2.5}$ concentrations with vertically thermodynamic (temperature) and dynamic (wind) variables in Beijing simultaneously observed for the period from December 2018 to February 2019. The heavy $PM_{2.5}$ pollution tended to occur more frequently on the days with low near-surface wind speed, and warmer air at the top of PBL (Figure 2b,c), given the wintertime climatological PBL height of 1–1.5 km in Beijing [61]. This generally does not favor the vertical ventilation and horizontal dispersion of aerosols. For example, during the pollution episode from December 14, to 18 of 2018, the near-surface wind speed in Beijing was significantly lower than those of pre- and post-periods, which was accompanied by strong thermal inversion layer. The southerly winds prevailed in the lowest 1–2 km of PBL (Figure 2a), which tended to transport aerosol particles from Hebei, a significant source region of aerosol emission to the south of Beijing (Figure 1a).

The most severe haze episode occurred during the study period persisted at least three days, starting from 12 January 2019 until 14 January 2019, during which $PM_{2.5}$ exceeded 100 µg m^{-3}. A most extremely high concentration of greater than 200 µg m^{-3} was observed as well. Coincidently, there existed weak southwesterly winds, and strong temperature inversion in the PBL, both of which contributed to this atmospheric pollution.

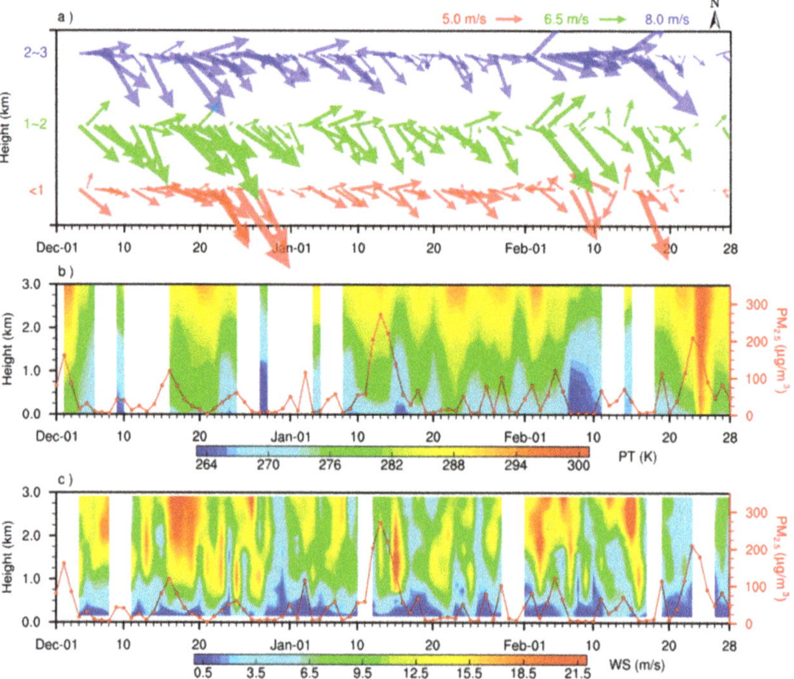

Figure 2. (a) Time series of the horizontal averaged wind vectors as derived from the Radar Wind Profiler (RWP) in Beijing for the altitude ranges from the surface to 1 km (SFC–1 km), 1–2 km, and 2–3 km above ground level (AGL), which are denoted as the red, green, and blue vectors, respectively. The wind arrow is the direction towards which the wind is blowing, and the width of the wind vector is proportional to the wind speed. Time-height cross-sections of (b) potential temperature (PT, color shaded) from radiosonde measurements and (c) horizontal wind speed (color shaded) from RWP, overlaid with observed ground-level $PM_{2.5}$ concentration (red lines). All these measurements were obtained at 0800 BJT during the period December 1, 2018, to February 28, 2019.

3.2. Synoptic-Scale Circulation and Backward Trajectory Statistical Analysis

In this section, we will examine the role of synoptic-scale meteorology underlying the polluted and clean episodes observed during the study period in Beijing. The first step to accomplish this is to determine the climatological wintertime winds at 925 hPa and 850 hPa pressure levels over Beijing and its surroundings. As shown in Figure 3a,c, weak westerly or southwesterly winds dominated both 850 hPa and 925 hPa pressure levels during the high aerosol-loading winter days in Beijing. By comparison, the wind fields at 850 hPa and 925 hPa were characterized by strong northwesterly winds over Beijing, which generally led to frequent intrusion of cold air mass (Figure 3b,d).

This cold advection could bring in cold and clean air from the northern regions without much anthropogenic emission sources (Figure 1a), thus resulting in low $PM_{2.5}$ concentration in Beijing, such as the extremely clean atmospheric episode occurring during December 27 to 30, 2018 (Figure 2). By contrast, less lapse rate of temperature and southwesterly winds featured the synoptic conditions favoring the accumulation of $PM_{2.5}$ (Figure 3a,c), well corroborating the vertical wind measurements in Beijing shown in Figure 2. Our findings were broadly consistent with the relationships between $PM_{2.5}$ concentration, temperature, and wind speed in winter found in other regions of the NCP [62].

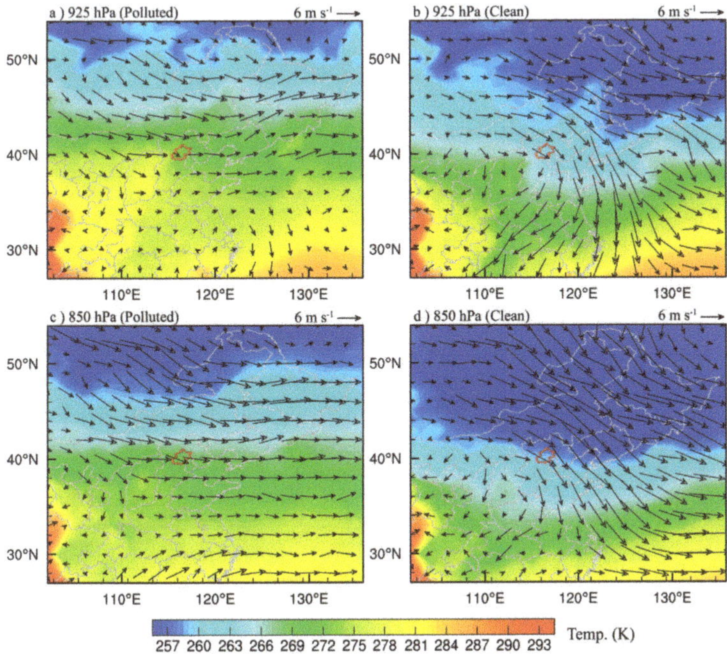

Figure 3. Spatial distribution of the wind field (black arrows, vector), superimposed by temperature (shaded) at 925 hPa (**a**,**b**) and 850 hPa (**c**,**d**) pressure level under polluted (left column) and clean (right column) conditions, respectively. All data are from the National Center for Environmental Prediction (NCEP) global Final (FNL) reanalysis. The areas highlighted with red lines represent the region of interest (Beijing), which is the same as the region highlighted with black lines in Figure 1a.

To further our understanding of the long-range transport to particulate matter pollution in Beijing, the 24-h backward trajectories were calculated and clustered. As illustrated in Figure 4, the prevailing northwesterly winds dominated the contribution in terms of transboundary transport (72% of all 24-h back trajectories). Interestingly, a small fraction of the trajectories (28%) came from the south, which was linked to most of the polluted episodes in Beijing during December 1, 2018, to February 28, 2019.

3.3. Diurnal Variations in Vertical Winds

Figure 5 illustrates the diurnal variations of wind speed and direction in Beijing for the heights ranging from ground-surface up to 3 km AGL under mean, polluted and clean conditions and their corresponding hodographs, and so do the anomalies of wind profile under polluted and clean conditions relative to the average wintertime winds in Beijing.

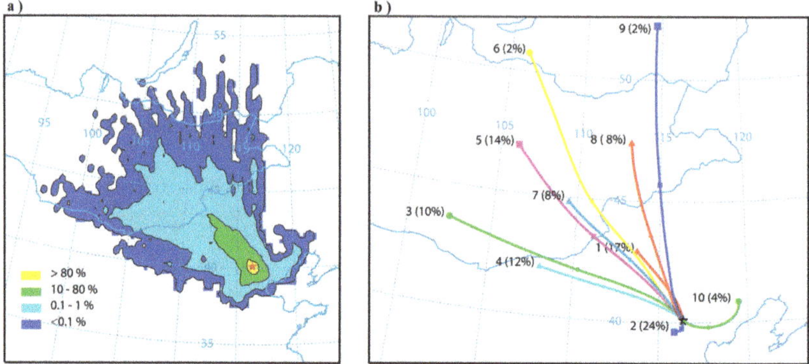

Figure 4. The spatial distribution of the trajectory frequency (**a**), and cluster-mean results of 24-h backward trajectories (**b**) calculated by the Hybrid Single-Particle Lagrangian Integrated Trajectory Model (HYSPLIT) ending at Beijing (116.37 °E; 40.09 °N, pentagram) at 100 m height AGL during winter from December 2018 to February 2019. The dots on the trajectories represent time node of 12h, and the percentage represents the ratio of the number of back trajectories in each cluster to the total number of back trajectories.

Overall, the mean wind speed was found to increase with height (Figure 5a). Meanwhile, the mean wind exhibited a pronounced diurnal cycle in the lower PBL (i.e., from the ground surface to 1 km AGL), and much stronger winds occurred during nighttime near the ground-surface compared with daytime (Figure 5a), which could be due to much reduced turbulence-related friction. This was also likely associated with the radiative (nocturnal) cooling in the night, which tended to reduce the eddy viscosity and momentum transfer from the upper levels, and in turn, lead to decreased wind speed [63]. However, the wind at the top of PBL or low troposphere exhibited double peaks: one is midnight and another in the later morning (1000 BJT). Under clean conditions, wind vector veered with the height between the near ground surface and 1 km AGL, followed by significant backing above 2 km (Figure 5b). Veering winds in the lowest layers of the atmosphere are most likely the result of friction-related processes while the backing winds are indicative of cold advection [64], which is mainly contributed by prevailing northwest winds (Figure 5a).

Under both polluted and clean conditions, the wind profiles showed significant diurnal variation at all heights, and its amplitude and sign differed greatly in the vertical during a daily cycle (Figure 5c,e), indicating that the lower part of PBL was characterized by prevailing southerly and northerly winds, respectively. The hodograph for polluted conditions exhibited smaller vertical shear at most heights, irrespective of the time of a day (Figure 5d). On the contrary, the curvature of the anticyclone rotation at above approximately 2 km AGL was significantly larger under clean conditions, resulting in larger wind shear (Figure 5f), which was most likely due to many more cold waves (Figures 2 and 3d).

Also, there existed significant wind anomalies under polluted and clean conditions (Figure 5g,i). Coincidently, a clockwise turning of the wind with the height was observed near the ground surface under polluted condition (Figure 5h), which was indicative of a warm air advection from the south. In contrast, the backing winds were found near the ground surface under clean conditions (Figure 5j), confirming the notion of northerlies-induced decreases in aerosol concentration. This highlights the urgency of consideration of VWS and wind direction in the aerosol pollution and its formation causes. In particular, under polluted conditions, negative wind anomalies prevailed at almost all times of the day in the PBL over Beijing, especially during the nighttime, indicating aerosol-induced changes in radiation reaching the surface could be linked to the dramatical reduction of wind speed. By comparison, positive southerly wind anomalies emerged during 1000–2000 BJT in the troposphere above 1.75 km, indicating that there were strong elevated $PM_{2.5}$ transport paths, which were located at the heights above the PBL at this time period. Under clean conditions (Figure 5c), positive northerly wind anomalies prevailed at almost all height, with the exception of negative wind anomalies above the PBL at roughly 0600, 1200, and 1900 BJT. This suggested that less aerosol tended to be associated with northerly winds, which further strengthened the role of cold advection from the northwestern parts of Beijing in reducing ground-level $PM_{2.5}$.

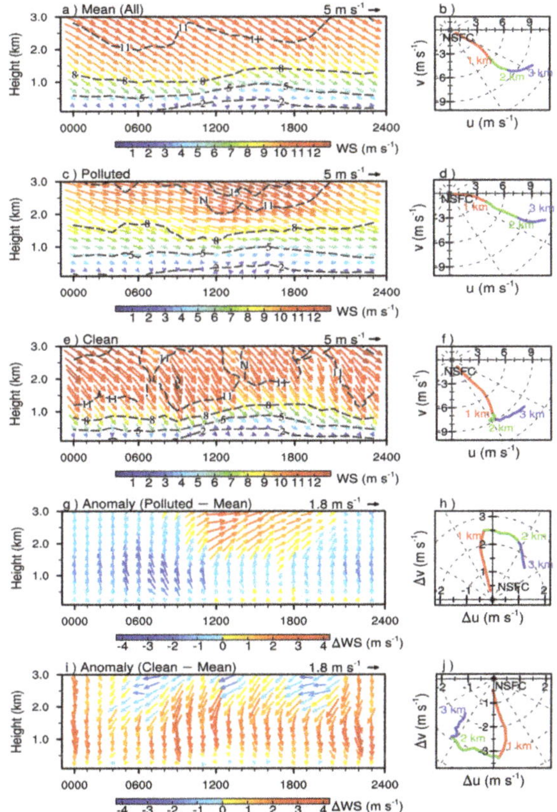

Figure 5. Height-resolved diurnal variation of horizontal wind vector under all-sky mean (**a**), polluted (**c**) and clean (**e**) conditions, and their corresponding anomalies (relative to mean wind) under polluted (**g**) and clean (**i**) conditions in Beijing from December 1, 2018, to February 28, 2019. Also shown are their corresponding hodographs in the panels (**b,d,f,h,j**) on the right-hand sides. Note the vectors in panels (**g**)–(**i**) show the resultant wind anomaly direction, and the vector length and color indicate the magnitude of the resultant wind anomaly relative to the average wintertime winds in Beijing.

3.4. Vertical Wind Shear Under Polluted And Clean Condition

The vertical wind shear is known to be able to strongly influence the vertical mixing process and resultant changes in aerosol pollutants in the PBL [26]. Figures 6 and 7 show the vertical distribution of VWS under polluted and clean conditions, respectively. The leading diagonal (top right to bottom left) denotes the local VWS at each level, whereas the shading in color indicates the magnitude of VWS at least two consecutive vertical levels. Note that VWS distribution does not take into account the shear direction.

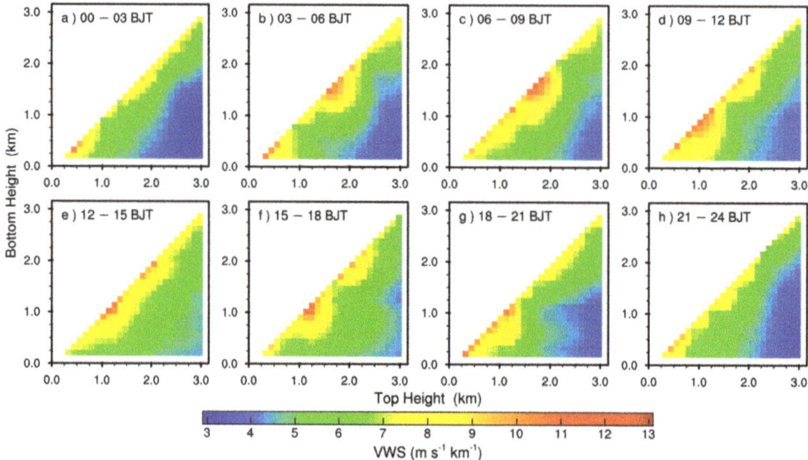

Figure 6. Three-hourly averaged vertical wind shear (VWS) computed between different heights under polluted conditions in Beijing for (**a**) 0000–0300 BJT, (**b**) 0300–0600 BJT, (**c**) 0600–0900 BJT, (**d**) 0900–1200 BJT, (**e**) 1200–1500 BJT, (**f**) 1500–1800 BJT, (**g**) 1800–2100 BJT and (**h**) 2100–2400 BJT, during the period from December 2018 to February 2019. The X and Y axes represent the top and bottom height of the VWS bulk, respectively.

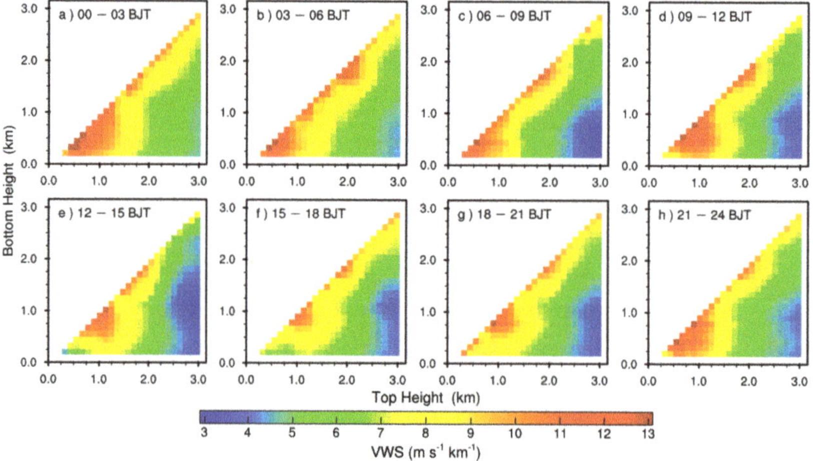

Figure 7. Same as Figure 6 but under clean conditions.

Generally, the pattern of the diurnal cycle did not change much when the atmosphere evolved from clean to polluted conditions, except for the magnitude of VWS. The magnitude of VWS was found to be much smaller under polluted conditions than that under clean conditions, indicative of weaker vertical mixing in the presence of high aerosol concentration in the PBL. This, in turn, led to a stronger accumulation of ground-level $PM_{2.5}$. On average, the local VWS for the polluted condition was ~8 m s^{-1} km^{-1}, as compared to as high as ~10.5 m s^{-1} km^{-1} for the clean condition. However, note that the VWS for polluted conditions between 2.5–3 km and <1.5 km was greater than that under clean conditions during the afternoon (1200–1800 BJT), which could be related with the positive southerly wind vector anomalies above 2 km during 1200–1800 BJT (Figure 5b). It implied that strong wind in the upper level and in the PBL tended to transport $PM_{2.5}$ from southern Hebei province to Beijing, further deteriorating the air quality in Beijing through these strongly vertical mixing exchanges (i.e., larger VWS) in the upper level and in the PBL. This finding was in agreement with the observational evidence from Hong Kong [26]. In addition, the magnitude of VWS (> 5 m s^{-1} km^{-1}) was small during the night to morning (1800–1200 BJT) but large during the afternoon (1200–1800 BJT) under polluted conditions. On the contrary, under clean conditions, the VWS (> 5 m s^{-1} km^{-1}) was relatively larger during the night (2100–0600 BJT) than during the day. These differences in diurnal variations of VWS (> 5 m s^{-1} km^{-1}) under polluted/clean conditions were probably associated with different integrated effects of the local circulation (mountain-valley and urban heat island circulations) [14,32] and synoptic patterns [65,66].

Figure 8 presents the correlation between VWS at different layers and ground-level PM$_{2.5}$ under polluted and clean conditions, respectively. It generally exhibits marked differences (even with opposite signs) in the lowest part of PBL in both conditions. In particular, the correlation coefficients seemed to be positive for the altitudes from the ground surface up to 2.5 km AGL under polluted conditions (Figure 8a), which meant the weak VWS near the ground surface or lower part of PBL observed in Figure 6 favored the accumulation of aerosol. In contrast, the correlation coefficient shifted from positive to negative as the VWS occurred upward, suggesting that the stronger VWS above the PBL was linked to lower ground-level PM$_{2.5}$ concentration. Interestingly, under clean conditions, a ubiquitous negative correlation was found between VWS and ground-level PM$_{2.5}$ in the almost whole lower atmosphere except in the height of 1 km and that above 2.5 km and beyond (Figure 8b). The increase of VWS tended to be accompanied by enhanced vertical mixing of aerosol, leading to reduced ground-level PM$_{2.5}$, which could account for this negative correlation observed for the clean condition. The exception in height above 2.5 km could be associated with air mass intrusion of long-range transported aerosol episodes [67], given the dominant height of long-range transboundary transport being generally above the PBL [68–70].

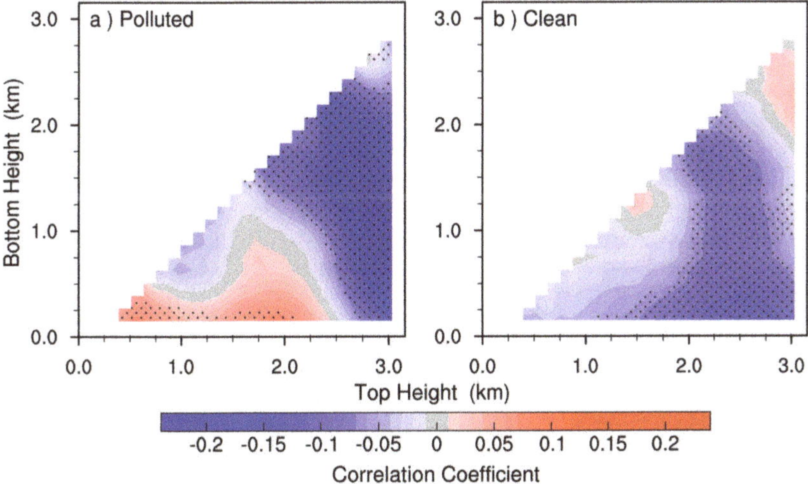

Figure 8. The correlation coefficient distribution between normalized ground-level PM$_{2.5}$ and height-revolved VWS under (**a**) polluted and (**b**) clean conditions in Beijing. Gray dots indicate the Pearson correlation coefficient that is statistically significant at the 90% confidence level. The X and Y axes represent the top and bottom heights of the VWS bulk, respectively.

3.5. The Dependency of Ground-Level PM$_{2.5}$ On Vertically Resolved Winds

The bivariate polar plots in Figure 9 showed that the normalized PM$_{2.5}$ concentration (hereinafter referred to as NPM$_{2.5}$) in Beijing varied by wind direction and speed at different heights, including ground-surface–1km, 1–2 km and 2–3 km. Specifically, at the height of ground surface–1km (Figure 9a), high aerosol concentration episodes (HEP, $NPM_{2.5}$ > 120%) were observed when the cardinal winds occurred along the NNE–NE, and NNW–WNW directions with speeds of greater than 6 m s^{-1} and even up to 10 m s^{-1}. Additionally, the polluted cases were observed when the prevailing cardinal winds were in the WSW–SSE sector with wind speed 4–6 m s^{-1}, consistent with previous findings [71]. At the heights of 1–2 km, extreme HEPs (i.e., $NPM_{2.5}$ >200 %) were observed mainly when winds blew along the SW–SE sector with a speed of <6 m s^{-1} (Figure 9b). By comparison, HPEs occurred when the NE–N winds prevailed at moderate to high wind speeds (4–12 m s^{-1}). As the atmospheric height increased to 2–3km, HEP mainly occurred when the winds were coming from NE–N sector at high wind speed (12–20 m s^{-1}), and also occurred in the NW sector with a wind speed of 14 m s^{-1} (Figure 9c). The surprisingly high $NPM_{2.5}$ tended to occur in Beijing when the winds blew from WSW–S at weaker wind speeds (2–6 m s^{-1}). In contrast, the $NPM_{2.5}$ in the NW sector was found to be significantly reduced. In general, it was found that a smaller range of direction angle was accompanied by smaller wind shear at the lower part of the PBL, whereas stronger southerly wind prevailed with larger wind shear in the upper PBL and even above the PBL under heavy polluted conditions (Figures 5 and 7).

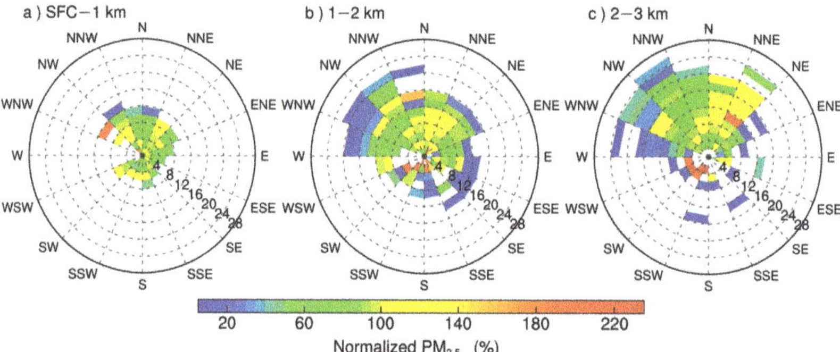

Figure 9. Bivariate polar plot of normalized $PM_{2.5}$ concentration (in percent) for the altitude ranges of (**a**) ground surface to 1 km (SFC–1km), (**b**) 1–2 km and (**c**) 2–3 km AGL during the period from December 1, 2018, to February 28, 2019. The wind directions are denoted by 16 compass direction: N, NNE, NE, ENE, E, ESE, SE, SSE, S, SSW, SW, WSW, W, WNW, NW, NNW. The radial axis represents wind speed in m s^{-1}, which increases radially outward. The concentration of $PM_{2.5}$ is scaled by colors.

The region to the south of Beijing was previously recognized to be a key emission source region for the $PM_{2.5}$ pollution episodes in Beijing, especially southerly or southwesterly wind prevailed [72–74]. This, in part, was attributed to transboundary $PM_{2.5}$ transport [28,75]. The results presented here provided convincing evidence that regional sources largely contribute to $PM_{2.5}$ below 3 km, where existed kind of main $PM_{2.5}$ transport path from south to Beijing, which basically agreed with the findings from model simulation analysis [76].

Another striking feature we observed here was that a few aerosol pollution episodes occurred even as the strong northerly or northwesterly winds dominated from the ground surface up to 3 km AGL (Figure 9). This could be likely linked to the long-range transported aerosol from northwestern or northeasten China, which was verified or corroborated in previous observational and model investigations [77].

4. Concluding Remarks

Based on continuous fine-resolution radar wind profiler (RWP) observations, radiosonde measurements during the winter for the period December of 2018 to February of 2019, the height-resolved wind vectors were analyzed, along with the impact of vertical wind shear on $PM_{2.5}$ pollution in Beijing. The main findings are summarized as follows:

Overall, the diurnal variations in wind profiles were found to differ greatly when classified by different ground-level $PM_{2.5}$ concentrations. Specifically, the southerly wind anomalies dominated throughout the whole PBL or even beyond the PBL under polluted conditions during the course of a day, in sharp contrast to the northerly wind anomalies in the PBL under clean condition. More strikingly, under pollution conditions, the positive anomaly of southerly wind speed mainly occurred at 1.75 km AGL during 1000–2000 BJT. This favored the transboundary transport originated from significant aerosol emission source to the south of Beijing, thereby leading to high ground-level $PM_{2.5}$ concentration in Beijing.

Besides, the ground-level $PM_{2.5}$ pollution exhibited a strong dependence on the vertical variation of the wind direction. The VWS tended to be much weaker in the lower PBL under polluted conditions, compared with under clean conditions, which could be strong ground-level $PM_{2.5}$ accumulation induced by weak vertical mixing in the PBL. Notably, the $PM_{2.5}$ pollution mainly appeared within the PBL as weak southerly winds prevailed when the relatively small VWS was observed as well. Above the PBL, strong northerlies winds also favored the long-range transport of aerosols, which in turn deteriorated the air quality in Beijing as well. This was well corroborated by the results from synoptic-scale circulation and backward trajectory analysis.

In summary, not only wind profiling but also the VWS at various heights could significantly modulate the ground-level $PM_{2.5}$ concentrations. Also, the present work highlighted the role that the height-resolved wind shear plays in better understanding the wintertime aerosol pollution episodes in Beijing. To increase the generalizability of the reported associations between aerosol and VWS; nevertheless, more efforts have to be made to include a much longer time series of observations at larger spatial domains in the future. More importantly, more wind and VWS measurements from RWP are desperately needed to be assimilated into the air quality model, which is expected to have great implications for improving the wintertime air quality forecast in China.

Author Contributions: Conceptualization, J.G.; Methodology, J.G. and Y.Z.; Validation, J.G., Y.Y. and S.H.L.Y.; Formal Analysis, Y.Z., J.G. and Y.Y.; Data Curation, J.G., Y.Z.; Writing–Original Draft Preparation, Y.Z. and J.G.; Review & Editing, J.G., Y.Y., Y.Z. and Y.W.; Visualization, Y.Z.; Supervision, J.G. and Y.W.; Resources, J.G. and S.H.L.Y.; Funding Acquisition, J.G. All authors have read and agreed to the published version of the manuscript.

Funding: This work was supported by the National Natural Science Foundation of China (Grant 41771399), Ministry of Science and Technology of China (Grant 2017YFC1501401), Chinese Academy of Sciences (Grant GXDA20040502), and Chinese Academy of Meteorological Sciences (Grants 2017Z005).

Acknowledgments: We appreciated greatly the National Meteorological Information Center of China Meteorological Administration for providing the radar wind profiler (RWP) data (https://data.cma.cn/en/). We also sincerely appreciated the $PM_{2.5}$ data made publicly accessible by the Ministry of Ecology and Environment of China (http://www.cnemc.cn/en/). Last but not least, the authors would like to thank the editor and three anonymous reviewers for their constructive comments which help improve the quality of our manuscript.

Conflicts of Interest: The authors declare no conflict of interest.

References

1. Qu, Y.; Han, Y.; Wu, Y.; Gao, P.; Wang, T. Study of PBLH and its correlation with particulate matter from one-year observation over Nanjing, southeast China. *Remote Sens.* **2017**, *9*, 668. [CrossRef]
2. Zhao, X.; Zhao, P.; Xu, J.; Meng, W.; Pu, W.; Dong, F.; He, D.; Shi, Q. Analysis of a winter regional haze event and its formation mechanism in the North China Plain. *Atmos. Chem. Phys.* **2013**, *13*, 5685–5696. [CrossRef]
3. Lee, K.H.; Kim, J.E.; Kim, Y.J.; Kim, J.; von Hoyningen-Huene, W. Impact of the smoke aerosol from Russian forest fires on the atmospheric environment over Korea during May 2003. *Atmos. Environ.* **2005**, *39*, 85–99. [CrossRef]
4. Tie, X.; Cao, J. Aerosol pollution in China: Present and future impact on environment. *Particuology* **2009**, *7*, 426–431. [CrossRef]
5. Wang, L.; Wang, H.; Liu, J.; Gao, Z.; Yang, Y.; Zhang, X.; Li, Y.; Huang, M. Impacts of the near-surface urban boundary layer structure on $PM_{2.5}$ concentrations in Beijing during winter. *Sci. Total Environ.* **2019**, *669*, 493–504. [CrossRef] [PubMed]

6. Li, Z.; Lau, W.M.; Ramanathan, V.; Wu, G.; Ding, Y.; Manoj, M.; Liu, J.; Qian, Y.; Li, J.; Zhou, T. Aerosol and monsoon climate interactions over Asia. *Rev. Geophys.* **2016**, *54*, 866–929. [CrossRef]

7. Li, Z.; Wang, Y.; Guo, J.; Zhao, C.; Cribb, M.C.; Dong, X.; Fan, J.; Gong, D.; Huang, J.; Jiang, M.; et al. East Asian Study of Tropospheric Aerosols and their Impact on Regional Clouds, Precipitation, and Climate (EAST-AIRCPC). *J. Geophys. Res. Atmos.* **2020**, *124*, 13026–13054. [CrossRef]

8. Guo, J.; Deng, M.; Lee, S.S.; Wang, F.; Li, Z.; Zhai, P.; Liu, H.; Lv, W.; Yao, W.; Li, X. Delaying precipitation and lightning by air pollution over the Pearl River Delta. Part I: Observational analyses. *J. Geophys. Res. Atmos.* **2016**, *121*, 6472–6488. [CrossRef]

9. Guo, J.; Xia, F.; Zhang, Y.; Liu, H.; Li, J.; Lou, M.; He, J.; Yan, Y.; Wang, F.; Min, M. Impact of diurnal variability and meteorological factors on the $PM_{2.5}$-AOD relationship: Implications for $PM_{2.5}$ remote sensing. *Environ. Pollut.* **2017**, *221*, 94–104. [CrossRef]

10. Guo, J.; Su, T.; Li, Z.; Miao, Y.; Li, J.; Liu, H.; Xu, H.; Cribb, M.; Zhai, P. Declining frequency of summertime local-scale precipitation over eastern China from 1970 to 2010 and its potential link to aerosols. *Geophys. Res. Lett.* **2017**, *44*, 5700–5708. [CrossRef]

11. Guo, J.; Liu, H.; Li, Z.; Rosenfeld, D.; Jiang, M.; Xu, W.; Jiang, J.H.; He, J.; Chen, D.; Min, M. Aerosol-induced changes in the vertical structure of precipitation: A perspective of TRMM precipitation radar. *Atmos. Chem. Phys.* **2018**, *18*, 13329–13343. [CrossRef]

12. Rosenfeld, D.; Sherwood, S.; Wood, R.; Donner, L. Climate effects of aerosol-cloud interactions. *Science* **2014**, *343*, 379–380. [CrossRef] [PubMed]

13. Wang, Q.; Li, Z.; Guo, J.; Zhao, C.; Cribb, M. The climate impact of aerosols on the lightning flash rate: Is it detectable from long-term measurements? *Atmos. Chem. Phys.* **2018**, *18*, 12797–12816. [CrossRef]

14. Yang, Y.; Zheng, Z.; Yim, S.Y.; Roth, M.; Ren, G.; Gao, Z.; Wang, T.; Li, Q.; Shi, C.; Ning, G. $PM_{2.5}$ Pollution Modulates Wintertime Urban-Heat-Island Intensity in the Beijing-Tianjin-Hebei Megalopolis, China. *Geophys. Res. Lett.* **2020**, *47*. [CrossRef]

15. Yang, Y.-J.; Wang, H.; Chen, F.; Zheng, X.; Fu, Y.; Zhou, S. TRMM-Based Optical and Microphysical Features of Precipitating Clouds in Summer Over the Yangtze–Huaihe River Valley, China. *Pure Appl. Geophys.* **2019**, *176*, 357–370. [CrossRef]

16. Gu, Y.; Wong, T.W.; Law, C.; Dong, G.H.; Ho, K.F.; Yang, Y.; Yim, S.H.L. Impacts of sectoral emissions in China and the implications: Air quality, public health, crop production, and economic costs. *Environ. Res. Lett.* **2018**, *13*, 084008. [CrossRef]

17. Yin, P.; Guo, J.; Wang, L.; Fan, W.; Lu, F.; Guo, M.; Moreno, S.; Wang, Y.; Wang, H.; Zhou, M.; et al. Higher risk of cardiovascular disease associated with smaller size-fractioned particulate matter. *Environ. Sci. Technol. Lett.* **2020**. [CrossRef]

18. Liu, C.; Chen, R.; Sera, F.; Vicedo-Cabrera, A.M.; Guo, Y.; Tong, S.; Coelho, M.S.Z.S.; Saldiva, P.H.N.; Lavigne, E.; Matus, P.; et al. Ambient Particulate Air Pollution and Daily Mortality in 652 Cities. *N. Engl. J. Med.* **2019**, *381*, 705–715. [CrossRef]

19. Cohen, A.J.; Brauer, M.; Burnett, R.; Anderson, H.R.; Frostad, J.; Estep, K.; Balakrishnan, K.; Brunekreef, B.; Dandona, L.; Dandona, R.; et al. Estimates and 25-year trends of the global burden of disease attributable to ambient air pollution: An analysis of data from the Global Burden of Diseases Study 2015. *Lancet* **2017**, *389*, 1907–1918. [CrossRef]

20. Li, Z.; Guo, J.; Ding, A.; Liao, H.; Liu, J.; Sun, Y.; Wang, T.; Xue, H.; Zhang, H.; Zhu, B. Aerosol and boundary-layer interactions and impact on air quality. *Natl. Sci. Rev.* **2017**, *4*, 810–833. [CrossRef]

21. Zhang, Q.; Zheng, Y.; Tong, D.; Shao, M.; Wang, S.; Zhang, Y.; Xu, X.; Wang, J.; He, H.; Liu, W. Drivers of improved $PM_{2.5}$ air quality in China from 2013 to 2017. *Proc. Natl. Acad. Sci. USA* **2019**, *116*, 24463–24469. [CrossRef] [PubMed]

22. Chen, H.; Wang, H. Haze days in North China and the associated atmospheric circulations based on daily visibility data from 1960 to 2012. *J. Geophys. Res. Atmos.* **2015**, *120*, 5895–5909. [CrossRef]

23. Chen, S.; Guo, J.; Song, L.; Li, J.; Liu, L.; Cohen, J.B. Inter-annual variation of the spring haze pollution over the North China Plain: Roles of atmospheric circulation and sea surface temperature. *Int. J. Climatol.* **2019**, *39*, 783–798. [CrossRef]

24. Miao, Y.; Guo, J.; Liu, S.; Liu, H.; Zhang, G.; Yan, Y.; He, J. Relay transport of aerosols to Beijing-Tianjin-Hebei region by multi-scale atmospheric circulations. *Atmos. Environ.* **2017**, *165*, 35–45. [CrossRef]

25. Yang, Y.; Zheng, X.; Gao, Z.; Wang, H.; Wang, T.; Li, Y.; Lau, G.N.; Yim, S.H. Long-term trends of persistent synoptic circulation events in planetary boundary layer and their relationships with haze pollution in winter half year over eastern China. *J. Geophys. Res. Atmos.* **2018**, *123*, 10,991–11,007. [CrossRef]

26. Yang, Y.; Yim, S.H.; Haywood, J.; Osborne, M.; Chan, J.C.; Zeng, Z.; Cheng, J.C. Characteristics of Heavy Particulate Matter Pollution Events Over Hong Kong and Their Relationships With Vertical Wind Profiles Using High-Time-Resolution Doppler Lidar Measurements. *J. Geophys. Res. Atmos.* **2019**, *124*, 9609–9623. [CrossRef]

27. Tai, A.P.; Mickley, L.J.; Jacob, D.J. Correlations between fine particulate matter ($PM_{2.5}$) and meteorological variables in the United States: Implications for the sensitivity of $PM_{2.5}$ to climate change. *Atmos. Environ.* **2010**, *44*, 3976–3984. [CrossRef]

28. Li, J.; Du, H.; Wang, Z.; Sun, Y.; Yang, W.; Li, J.; Tang, X.; Fu, P. Rapid formation of a severe regional winter haze episode over a mega-city cluster on the North China Plain. *Environ. Pollut.* **2017**, *223*, 605–615. [CrossRef] [PubMed]

29. Guo, J.; Li, Y.; Cohen, J.B.; Li, J.; Chen, D.; Xu, H.; Liu, L.; Yin, J.; Hu, K.; Zhai, P. Shift in the temporal trend in boundary layer height trend in China using long-term (1979–2016) radiosonde data. *Geophys. Res. Lett.* **2019**, *46*, 6080–6089. [CrossRef]

30. Lou, M.; Guo, J.; Wang, L.; Xu, H.; Chen, D.; Miao, Y.; Lv, Y.; Li, Y.; Guo, X.; Ma, S. On the relationship between aerosol and boundary layer height in summer in China under different thermodynamic conditions. *Earth Space Sci.* **2019**, *6*, 887–901. [CrossRef]

31. Zheng, Z.; Li, Y.; Wang, H.; Ding, H.; Li, Y.; Gao, Z.; Yang, Y. Re-evaluating the variation in trend of haze days in the urban areas of Beijing during a recent 36-year period. *Atmos. Sci. Lett.* **2019**, *20*, e878. [CrossRef]

32. Zheng, Z.; Ren, G.; Wang, H.; Dou, J.; Gao, Z.; Duan, C.; Li, Y.; Ngarukiyimana, J.P.; Zhao, C.; Cao, C. Relationship between fine-particle pollution and the urban heat island in Beijing, China: Observational evidence. *Bound. Layer Meteor.* **2018**, *169*, 93–113. [CrossRef]

33. Yang, Q.; Yuan, Q.; Li, T.; Shen, H.; Zhang, L. The relationships between $PM_{2.5}$ and meteorological factors in China: Seasonal and regional variations. *Inter. J. Env. Res. Pub. Heal.* **2017**, *14*, 1510. [CrossRef] [PubMed]

34. He, L.J.; Lin, A.W.; Chen, X.X.; Zhou, H.; Zhou, Z.G.; He, P.P. Assessment of MERRA-2 Surface $PM_{2.5}$ over the Yangtze River Basin: Ground-based Verification, Spatiotemporal Distribution and Meteorological Dependence. *Remote Sens.* **2019**, *11*, 460. [CrossRef]

35. Miao, Y.; Liu, S.; Guo, J.; Yan, Y.; Huang, S.; Zhang, G.; Zhang, Y.; Lou, M. Impacts of meteorological conditions on wintertime PM 2.5 pollution in Taiyuan, North China. *Environ. Sci. Pollut. R.* **2018**, *25*, 21855–21866. [CrossRef]

36. Quan, J.; Tie, X.; Zhang, Q.; Liu, Q.; Li, X.; Gao, Y.; Zhao, D. Characteristics of heavy aerosol pollution during the 2012–2013 winter in Beijing, China. *Atmos. Environ.* **2014**, *88*, 83–89. [CrossRef]

37. Tie, X.; Zhang, Q.; He, H.; Cao, J.; Han, S.; Gao, Y.; Li, X.; Jia, X.C. A budget analysis of the formation of haze in Beijing. *Atmos. Environ.* **2015**, *100*, 25–36. [CrossRef]

38. Zhu, X.; Tang, G.; Guo, J.; Hu, B.; Song, T.; Wang, L.; Xin, J.; Gao, W.; Münkel, C.; Schäfer, K. Mixing layer height on the North China Plain and meteorological evidence of serious air pollution in southern Hebei. *Atmos. Chem. Phys.* **2018**, *18*, 4897–4910. [CrossRef]

39. Chen, Y.; An, J.L.; Lin, J.; Sun, Y.L.; Wang, X.Q.; Wang, Z.F.; Duan, J. Observation of nocturnal low-level wind shear and particulate matter in urban Beijing using a Doppler wind lidar. *Atmos. Ocean. Sci. Lett.* **2017**, *10*, 411–417. [CrossRef]

40. Liu, X.; Li, J.; Qu, Y.; Han, T.; Hou, L.; Gu, J.; Chen, C.; Yang, Y.; Liu, X.; Yang, T. Formation and evolution mechanism of regional haze: A case study in the megacity Beijing, China. *Atmos. Chem. Phys.* **2013**, *13*, 4501–4514. [CrossRef]

41. Liu, B.; Ma, Y.; Guo, J.; Gong, W.; Zhang, Y.; Mao, F.; Li, J.; Guo, X.; Shi, Y. Boundary layer heights as derived from ground-based Radar wind profiler in Beijing. *IEEE Trans. Geosci. Remote Sens.* **2019**, *57*, 8095–8104. [CrossRef]

42. Cai, W.; Li, K.; Liao, H.; Wang, H.; Wu, L. Weather conditions conducive to Beijing severe haze more frequent under climate change. *Nat. Clim. Chang.* **2017**, *7*, 257. [CrossRef]

43. Pei, L.; Yan, Z.; Sun, Z.; Miao, S.; Yao, Y. Increasing persistent haze in Beijing: Potential impacts of weakening East Asian winter monsoons associated with northwestern Pacific sea surface temperature trends. *Atmos. Chem. Phys.* **2018**, *18*, 3173–3183. [CrossRef]

44. Guo, J.; Deng, M.; Fan, J.; Li, Z.; Chen, Q.; Zhai, P.; Dai, Z.; Li, X. Precipitation and air pollution at mountain and plain stations in northern China: Insights gained from observations and modeling. *J. Geophys. Res. Atmos.* **2014**, *119*, 4793–4807. [CrossRef]
45. Miao, Y.; Guo, J.; Liu, S.; Wei, W.; Zhang, G.; Lin, Y.; Zhai, P. The climatology of low-level jet in Beijing and Guangzhou, China. *J. Geophys. Res. Atmos.* **2018**, *123*, 2816–2830. [CrossRef]
46. Wei, W.; Zhang, H.; Ye, X. Comparison of low-level jets along the north coast of China in summer. *J. Geophys. Res. Atmos.* **2014**, *119*, 9692–9706. [CrossRef]
47. *Operating Manual, TEOM Series 1400a Ambient Particulate Monitor*; Thermo Fisher Scientific Inc.: Franklin, MA, USA, 2007.
48. Guo, J.; Miao, Y.; Zhang, Y.; Liu, H.; Li, Z.; Zhang, W.; He, J.; Lou, M.; Yan, Y.; Bian, L. The climatology of planetary boundary layer height in China derived from radiosonde and reanalysis data. *Atmos. Chem. Phys.* **2016**, *16*, 13309. [CrossRef]
49. Guo, J.; Zhang, X.; Cao, C.; Che, H.; Liu, H.; Gupta, P.; Zhang, H.; Xu, M.; Li, X. Monitoring haze episodes over Yellow Sea by combining multi-sensor measurements. *Int. J. Remot. Sens.* **2010**, *31*, 4743–4755. [CrossRef]
50. Draxler, R.; Hess, G. An overview of the HYSPLIT_4 modelling system for trajectories. *Aust. Meteorol. Mag.* **1998**, *47*, 295–308.
51. Stein, A.F.; Draxler, R.R.; Rolph, G.D.; Stunder, B.J.; Cohen, M.D.; Ngan, F. NOAA's HYSPLIT atmospheric transport and dispersion modeling system. *B. Am. Meteorol. Soc.* **2015**, *96*, 2059–2077. [CrossRef]
52. Wang, X.; Dickinson, R.E.; Su, L.; Zhou, C.; Wang, K. PM$_{2.5}$ pollution in China and how it has been exacerbated by terrain and meteorological conditions. *Bull. Am. Meteoro. Soc.* **2018**, *99*, 105–119. [CrossRef]
53. Koren, I.; Altaratz, O.; Remer, L.A.; Feingold, G.; Martins, J.V.; Heiblum, R.H. Aerosol-induced intensification of rain from the tropics to the mid-latitudes. *Nat. Geosci.* **2012**, *5*, 118. [CrossRef]
54. Li, X.; Guo, X.; Fu, D. TRMM-retrieved cloud structure and evolution of MCSs over the northern South China Sea and impacts of CAPE and vertical wind shear. *Adv. Atmos. Sci.* **2013**, *30*, 77–88. [CrossRef]
55. Markowski, P.; Richardson, Y. On the classification of vertical wind shear as directional shear versus speed shear. *Weather Forecast.* **2006**, *21*, 242–247. [CrossRef]
56. Thompson, R.L.; Mead, C.M.; Edwards, R. Effective storm-relative helicity and bulk shear in supercell thunderstorm environments. *Weather Forecast.* **2007**, *22*, 102–115. [CrossRef]
57. Guo, J.; Zhai, P.; Wu, L.; Cribb, M.; Li, Z.; Ma, Z.; Wang, F.; Chu, D.; Wang, P.; Zhang, J. Diurnal variation and the influential factors of precipitation from surface and satellite measurements in Tibet. *Int. J. Climatol.* **2014**, *34*, 2940–2956. [CrossRef]
58. Singh, P.; Nakamura, K. Diurnal variation in summer precipitation over the central Tibetan Plateau. *J. Geophys. Res. Atmos.* **2009**, *114*. [CrossRef]
59. Carslaw, D.C.; Beevers, S.D.; Ropkins, K.; Bell, M.C. Detecting and quantifying aircraft and other on-airport contributions to ambient nitrogen oxides in the vicinity of a large international airport. *Atmos. Environ.* **2006**, *40*, 5424–5434. [CrossRef]
60. Westmoreland, E.J.; Carslaw, N.; Carslaw, D.C.; Gillah, A.; Bates, E. Analysis of air quality within a street canyon using statistical and dispersion modelling techniques. *Atmos. Environ.* **2007**, *41*, 9195–9205. [CrossRef]
61. Zhang, W.; Guo, J.; Miao, Y.; Liu, H.; Li, Z.; Zhai, P. Planetary boundary layer height from CALIOP compared to radiosonde over China. *Atmos. Chem. Phys.* **2016**, *16*, 9951–9963. [CrossRef]
62. Zhang, H.; Wang, Y.; Hu, J.; Ying, Q.; Hu, X.-M. Relationships between meteorological parameters and criteria air pollutants in three megacities in China. *Environ. Res.* **2015**, *140*, 242–254. [CrossRef] [PubMed]
63. Yu, R.; Li, J.; Chen, H. Diurnal variation of surface wind over central eastern China. *Climate Dyn.* **2009**, *33*, 1089. [CrossRef]
64. Doswell III, C.A. A review for forecasters on the application of hodographs to forecasting severe thunderstorms. *Natl. Wea. Dig.* **1991**, *16*, 2–16.
65. Miao, Y.; Guo, J.; Liu, S.; Liu, H.; Li, Z.; Zhang, W.; Zhai, P. Classification of summertime synoptic patterns in Beijing and their associations with boundary layer structure affecting aerosol pollution. *Atmos. Chem. Phys.* **2017**, *17*, 3097–3110. [CrossRef]
66. Liu, L.; Guo, J.; Miao, Y.; Li, J.; Chen, D.; He, J.; Cui, C. Elucidating the relationship between aerosol concentration and summertime boundary layer structure in central China. *Environ. Pollut.* **2018**, *241*, 646–653. [CrossRef]

67. Raveh-Rubin, S. Dry intrusions: Lagrangian climatology and dynamical impact on the planetary boundary layer. *J. Climate* **2017**, *30*, 6661–6682. [CrossRef]

68. Schaub, D.; Weiss, A.; Kaiser, J.; Petritoli, A.; Richter, A.; Buchmann, B.; Burrows, J. A transboundary transport episode of nitrogen dioxide as observed from GOME and its impact in the Alpine region. *Atmos. Chem. Phys.* **2005**, *5*, 23–37. [CrossRef]

69. Huang, J.; Guo, J.; Wang, F.; Liu, Z.; Jeong, M.J.; Yu, H.; Zhang, Z. CALIPSO inferred most probable heights of global dust and smoke layers. *J. Geophys. Res. Atmos.* **2015**, *120*, 5085–5100. [CrossRef]

70. Gu, Y.; Yim, S. The air quality and health impacts of domestic trans-boundary pollution in various regions of China. *Environ. Int.* **2016**, *97*, 117–124. [CrossRef] [PubMed]

71. Chen, Y.; Schleicher, N.; Fricker, M.; Cen, K.; Liu, X.-l.; Kaminski, U.; Yu, Y.; Wu, X.-f.; Norra, S. Long-term variation of black carbon and $PM_{2.5}$ in Beijing, China with respect to meteorological conditions and governmental measures. *Environ. Pollut.* **2016**, *212*, 269–278. [CrossRef] [PubMed]

72. Lv, B.; Liu, Y.; Yu, P.; Zhang, B.; Bai, Y. Characterizations of $PM_{2.5}$ pollution pathways and sources analysis in four large cities in China. *Aerosol Air Qual. Res.* **2015**, *15*, 1836–1843. [CrossRef]

73. Wang, L.; Liu, Z.; Sun, Y.; Ji, D.; Wang, Y. Long-range transport and regional sources of $PM_{2.5}$ in Beijing based on long-term observations from 2005 to 2010. *Atmos. Res.* **2015**, *157*, 37–48. [CrossRef]

74. Zhang, Y.; Chen, J.; Yang, H.; Li, R.; Yu, Q. Seasonal variation and potential source regions of $PM_{2.5}$-bound PAHs in the megacity Beijing, China: Impact of regional transport. *Environ. Pollut.* **2017**, *231*, 329–338. [CrossRef] [PubMed]

75. Yang, Y.; Russell, L.M.; Lou, S.; Liao, H.; Guo, J.; Liu, Y.; Singh, B.; Ghan, S.J. Dust-wind interactions can intensify aerosol pollution over eastern China. *Nat. Commun.* **2017**, *8*, 15333. [CrossRef]

76. Chen, H.; Li, J.; Ge, B.; Yang, W.; Wang, Z.; Huang, S.; Wang, Y.; Yan, P.; Li, J.; Zhu, L. Modeling study of source contributions and emergency control effects during a severe haze episode over the Beijing-Tianjin-Hebei area. *Sci. China Chem.* **2015**, *58*, 1403–1415. [CrossRef]

77. Zhang, R.; Jing, J.; Tao, J.; Hsu, S.-C.; Wang, G.; Cao, J.; Lee, C.S.L.; Zhu, L.; Chen, Z.; Zhao, Y. Chemical characterization and source apportionment of $PM_{2.5}$ in Beijing: Seasonal perspective. *Atmos. Chem. Phys.* **2013**, *13*, 7053–7074. [CrossRef]

Article

Role and Mechanisms of Black Carbon Affecting Water Vapor Transport to Tibet

Min Luo [1], Yuzhi Liu [1,*], Qingzhe Zhu [1], Yuhan Tang [1] and Khan Alam [2]

[1] Key Laboratory for Semi-Arid Climate Change of the Ministry of Education, College of Atmospheric Sciences, Lanzhou University, Lanzhou 730000, China; lzu_luom@lzu.edu.cn (M.L.); zhuqzh18@lzu.edu.cn (Q.Z.); tangyh14@lzu.edu.cn (Y.T.)

[2] Department of Physics, University of Peshawar, Peshawar 25000, Pakistan; khanalam@uop.edu.pk

* Correspondence: liuyzh@lzu.edu.cn; Tel.: +86-139-1991-5375

Received: 10 December 2019; Accepted: 7 January 2020; Published: 9 January 2020

Abstract: Although some studies reported the impact of black carbon (BC) on the climate over the Tibetan Plateau (TP), the contribution and mechanisms of BC affecting the water vapor transport to Tibet are not fully understood yet. Here, utilizing the satellite observations and reanalysis data, the effects of BC on the climate over the TP and water vapor transport to the Tibet were investigated by the Community Earth System Model (CESM 2.1.0). Due to the addition of BC, a positive net heat forcing (average is 0.39 W/m^2) is exerted at the surface, which induces a pronounced warming effect over the TP and consequently intensifies the East Asian Summer monsoon (EASM). However, significant cooling effects in northern India, Pakistan, Afghanistan and Iran are induced due to the BC and related feedbacks, which reduces significantly the meridional land–sea thermal contrast and finally weakens the South Asian summer monsoon (SASM). Consequently, the water vapor transport to the south border will be decreased due to addition of BC. Moreover, through affecting the atmospheric circulation, the BC could induce an increase in the imported water vapor from the west and east borders of the TP, and an increase outflowing away from the north border of the TP. Overall, due to the BC, the annual mean net importing water vapor over TP is around 271 Gt, which could enhance the precipitation over the TP. The results show that the mean increase in the precipitation over TP is about 0.56 mm/day.

Keywords: black carbon; Tibetan plateau; water vapor transport; South Asian summer monsoon; East Asian summer monsoon

---

1. Introduction

The black carbon (BC) aerosol emitted from the combustion of some biomass and fossil fuel has a strong "Greenhouse Effect" by absorbing solar radiation and longwave radiation [1,2] in the atmosphere with a few days' lifetime [3,4]. Moreover, the BC can also affect the earth–atmosphere energy balance through indirect and semi-direct effects [5], having a profound influence on the hydrological cycle and climate [2,6,7]. However, the uncertainties in estimating the magnitude of the hydrological cycle and regional climate responses to the BC are still pronounced [8].

During recent years, the aerosol emissions, including the BC over Asia are obviously increasing, and these particles can be transported to the Tibetan Plateau (TP) by atmospheric circulations [9–11]. The BC could be deposited into snow and exert a pronounced "snow darkening" effect [12], and further affect the radiation budget [13,14]. The annual mean snow albedo direct radiative forcing of BC may reach 2.9 W/m^2 [15], which will reduce the snow albedo and accelerate the melting of glaciers [16–19]. Moreover, the BC over the TP can affect the properties of cloud [20–22], precipitation [22–24] and the monsoon circulations [25].

Furthermore, the BC-in-snow effect can induce an increase in surface temperature over the TP and cause the earlier onset of the South Asian summer monsoon (SASM) [26]. Besides, the "Elevated Heat Pump" (EHP) effect suggests that the BC can induce an updraft motion with a warm anticyclone circulation in the upper atmosphere over the TP in late spring or early summer [27,28], the EHP effect could reduce the SASM in summertime by its dynamic and thermal forcing [28,29]. Additionally, the decreased meridional temperature gradient from the Indian Ocean to Northern India caused by absorbing aerosols could also reduce the SASM significantly [6,30,31]. On the other hand, the BC-in-snow effect could enhance the East Asian summer monsoon (EASM) by increasing the land–sea thermal contrast [26]. The changes in EASM are closely related to the aerosols disturbing the thermal contrast between land and ocean [32]. Li et al. [33] observed that the greenhouse gases and aerosols could increase the thermal contrast between the East China and the adjacent sea, and hence the EASM. The warming effect of BC over East Asia is the main factor, which could induce the enhanced EASM [34].

Generally, the SASM and EASM are the main dynamic factors in terms of the water vapor transport. The water vapor can be carried from the ocean to the land by the circulations of SASM and EASM [35], in which the northward transport is mainly caused by the lower southerly wind [36]. The TP, which is named the "Asian water tower", is feeding several major rivers in Asia and providing fresh water for more than one third of the populations of the world [37], and has been receiving much attention [37–42]. Generally, the water vapor may be gathered in the western and southern TP, and advected to the rest of the TP [40]. The water vapor could be transported to the TP by upslope transport and up-and-over patterns [37,38]. In addition, the perturbed cyclone and anticyclone over Lake Baikal are closely related to the water situations of TP, and the warming in the northwestern Atlantic Ocean is the key factor contributing to the wetting TP [42]. However, under the global dryland expansion and warming [43,44], the potential role and mechanism of the BC affecting the water vapor transport and the burden over TP is poorly studied.

Although previous researchers have revealed that the BC has a pronounced climate effect on the SASM and EASM, which are closely related to the water vapor transport from the ocean to the TP, there are few studies focused on the effects of BC on the water vapor transport from the surrounding to the TP. In this study, the role and mechanisms of BC affecting the water vapor budget over the TP are investigated by utilizing the CESM, which is fully coupled including atmosphere, ocean, sea ice, land and land–ice components.

2. Data and Methods

2.1. The Uncertainties and Applicability of Each Data

To evaluate the model performance on the climate and water vapor transport over the TP, the product of Multi-angle Imaging Spectro Radiometer (MISR) and several reanalysis data were used to compare with simulations. The resolutions of observations and reanalysis data sets depend on the assimilation system and the accuracy of the equipment. Before the analysis, we have interpolated the results of simulations whose resolution is 0.9° (latitude) × 1.25° (longitude) to the resolutions of observations and reanalysis data sets.

2.1.1. Multi-Angle Imaging Spectro Radiometer (MISR)

The MISR measures the aerosol optical depth (AOD) at a spatial resolution from 275 to 1100 m globally. Because the atmospheric path contribution from the surface-leaving radiance can be removed by taking advantage of differences in multi-angular signatures, the MISR aerosol retrieval algorithm is less sensitive to surface type especially over the bright surfaces [45]. Therefore, compared with the ground-based remote sensing, the AOD products from MISR have good accuracy over the TP. However, MISR has a limited swath coverage that is much too narrow, which has a lower frequency of observations at the given ground-based site in each orbital cycle. In this study, the MISR-3 product derived from multiple orbits monthly with a resolution of 0.5° × 0.5° [46] was used to evaluate the simulated AOD by a model.

2.1.2. Cloud and Earth's Radiant Energy System (CERES)

The CERES is used to investigate the cloud/radiation feedback, the data sets are measured by the broadband scanning radiometers [47]. The all sky surface net radiation fluxes (longwave and shortwave radiation, W/m^2) were obtained from the CERES 4.0 product, whose spatial and temporal resolutions are 1.25° × 1.25° and monthly, respectively. The CERES data sets have a strong correlation with meteorological station data [48]. For the CERES products, the uncertainties in surface net radiation are attributable to the environmental parameters, including surface water vapor pressure, surface temperature, the Normalized Difference Vegetation Index (NDVI) and surface albedo. Studies show that the errors of each radiation component of the CERES product are within 20 W/m^2 at the monthly scale [49].

2.1.3. ERA-Interim

The ERA-interim data sets are obtained from the European Center for Medium-Range Weather Forecasts (ECMWF) covering the period from 1979 to the present. It has a good performance on describing the actual atmosphere over Asia [50]. In this study, the monthly mean skin temperatures (K) with a spatial resolution of 0.75° × 0.75° are used. Compared with the observations, the root mean square error (RMSE) of temperature is about 3.2 °C, and the correlation coefficient is 0.709. Generally, the ERA-interim is closer to the ground observations than many other reanalysis data over Asia [51].

2.1.4. Global Precipitation Climatology Project (GPCP)

The GPCP is a merged precipitation reanalysis data which incorporates information from the low-orbit-satellite microwave, the geosynchronous-orbit-satellite infrared, and the rain observations. The GPCP can figure out the temporal and spatial features of precipitation (mm/month) quite well, having a good performance over Asia [52]. Here, the monthly mean accumulated precipitation data at the surface from GPCP-2 with a spatial resolution of 2.5° × 2.5° were used. Some studies reported that the GPCP data may overestimate the precipitation, especially when the precipitation rate increases. This observed uncertainty may be due to the shortcomings of GPCP for the retrieval of summer precipitation over land, such as mistaking higher clouds as precipitation clouds [53,54].

2.1.5. Modern-Era Retrospective Analysis for Research and Applications (MERRA)

The MERRA-2 is an assimilation which includes different ground-based and space-based remote sensing information. The monthly mean surface mass concentration of BC (kg/m^3) is obtained from the version 2 of MERRA (MERRA-2), which has a spatial resolution of 2.5° × 2.5°. The MERRA data can describe the distribution of the BC mass concentration over Asia very well [55]. The aerosol AODs from the MERRA-2 data have a high correlation but low bias relative to the ground-based observations (e.g. sun photometer) [56]. Here, we use the surface concentration of BC from the MERRA-2 data.

2.1.6. National Centers for Environment Prediction (NCEP)

The NCEP reanalysis data covers the information of satellite, and it is produced by a forecast model together with a data assimilation system. Because the NCEP data sets have a good performance on describing the winds and free atmosphere of the temperate region in the Northern Hemisphere [57], the horizontal wind (u and v component, m/s) and specific humidity (kg/kg) in NCEP data sets are used to evaluate the simulated water vapor in the atmosphere. The spatial resolution of the monthly NCEP reanalysis data used in this study is 2.5° × 2.5° [58]. For the NCEP data, due to the topographical height and systematic deviations of the assimilation model, it was a false trend on the longer time scale in the middle and lower troposphere over the TP [59].

2.1.7. Emission Data of Aerosols and Greenhouse Gases

In this study, the emission data of aerosols and greenhouse gases for the year 2000 are used to be the background reference. The anthropogenic aerosol emissions from industrial production, agriculture activities and human activities are derived from the emission data of the Intergovernmental Panel on Climate Change (IPCC) Fifth Assessment Report (AR5) [60]. The emissions of BC together with sulfur dioxide are updated from Smith et al. [61]. Besides, the aerosol size distribution is classified as three lognormal modes: Aitken (ranging from 0.015 mm to 0.053 mm), accumulation (ranging from 0.058 mm to 0.27 mm), and coarse modes (ranging from 0.80 mm to 3.65 mm) [62]. The concentrations of greenhouse gases for the year 2000 are derived from the specific concentration [63].

2.2. Methods of Calculating Asian Summer Monsoon and Water Vapor Transport Changes

2.2.1. Estimation of Summer Monsoon

To estimate the strength of the summer monsoon, the dynamical normalized seasonality index (DNS) was used in this paper [64], and can be calculated as using the following Equation (1):

$$DNS = \frac{\left\| \vec{V1} - \vec{V}(m,n) \right\|}{\left\| \vec{V} \right\|} - 2 \tag{1}$$

where the $\vec{V1}$ is the climatological wind vector in January, the $\vec{V}$ is the mean wind vector of January and July, while the $V(m,n)$ is the wind vector in the year 'n' and month 'm'.

2.2.2. Changes of Water Vapor

The water vapor flux in the whole atmospheric layer (Q) is calculated from the land surface to 100 hPa and the water vapor budget (N) [42] are calculated as follows:

$$Q = -\frac{1}{g} \int_{p_s}^{p} q\, \vec{V} dp \tag{2}$$

$$N = \oint Q dl \tag{3}$$

where q denotes the specific humidity (kg/kg), $\vec{V}$ denotes the horizontal wind vector (u and v component, m/s), g denotes the constant of gravity acceleration (9.8 m/s^2). Here, the atmospheric pressure at the top of atmosphere (TOA) is set as 100 hPa, and l denotes the border of the TP. In this study, the border of TP is considered as the east, south, west and north borders separately.

2.3. Description of Model Setting

In this study, CESM (version 2.1.0) is used to study the effects of BC on the climate over the TP. The atmospheric, land and ocean models are the Community Atmosphere Model (CAM) 5.0, Community Land Model (CLM) 4.5 and three-dimensional active Parallel Ocean Program (POP) 2.0, respectively, in CESM.

The default aerosol configuration coupled to the Modal Aerosol Model (MAM) is adopted to investigate the climate effects of BC [65]. The CAM 5.0 of CESM includes BC, dust, precursors of sulfate (SO$_2$ and SO$_4$), sea salt, particulate organic matter, and secondary organic aerosol in the MAM3 aerosol module [65]. The dust and sea salt modes are merged into a coarse mode in MAM3 for the online calculating of the emissions. The dry deposition process of aerosols is calculated by a parameterization which includes the information of land use and land cover [66], while the wet deposition process is calculated by the wet removal routine [67]. Meanwhile, the revised cloud macrophysics processes are considered in CESM-CAM5, in terms of the cloud fractions, and the interconversion rates between water vapor and cloud condensation.

The parameterization of cloud microphysics is utilized to calculate the droplet and ice number concentration [68]. Besides, the CAM 5.0 forms the main atmospheric components in CESM and has the capacity of simulating the aerosol–cloud interaction, which includes the cloud droplet activation by aerosols, precipitation process affected by particles and the radiative interactions of cloud particles [69–71].

In this study, two experiments are carried out to study the climatic effects of BC. The control experiment, named the 'ALL' experiment, considers the emissions of BC, particulate organic matter (POM), dust, SO_2, SO_4 and second organic aerosol gases (SOAG). The contrast experiment, named as the 'ALL-BC' experiment, is run with the same emissions as the ALL experiment, except for the BC emissions. The simulations started from January 2000 to December 2005 with a spin-up time in the year of 2000. The finite volume (FV) dynamical core with a resolution of $0.9° \times 1.25°$ and a hybrid sigma-p vertical coordinate of 30 layers from the land surface to 3.64 hPa are utilized in CESM. The details of the model setup and experiment design are presented in Table 1.

Table 1. Physical and chemical schemes used in the Community Earth System Model (CESM) simulation.

Model Settings	Configurations
Horizontal resolutions	$0.9° \times 1.25°$
Vertical resolutions	30 layers from the surface to 3.64 hPa
Physical schemes	Used schemes
Radiation schemes (Longwave, Shortwave)	Rapid radiative transfer model RRTMG
Cloud Microphysics scheme	MG scheme
Cloud Macrophysics scheme	Park scheme
Chemical schemes	Used schemes
Aerosol scheme	A 3-mode modal aerosol scheme MAM3
Aerosol data sets	IPCC AR5 emissions of 2000
Dry deposition	A resistance-based parameterization
Wet deposition	The wet removal routine

2.4. The Study Area

Considering the typical topography of the TP and distribution of BC, we focused on the region with latitude and longitude coordinate ranges of 60° E–140° E and 5° N–45° N (see Figure 1). It covers most parts of China, the whole Indo-China Peninsula, India and Pakistan. The areas marked by the blue and the red rectangle in Figure 1 are adopted to calculate the changes in the intensity of the SASM and the EASM, respectively. Besides, the climatic effects caused by BC over the TP are analyzed for the summertime (June–July–August). Before analyzing water vapor changes due to BC, the assessment for the model ability of CESM is performed first.

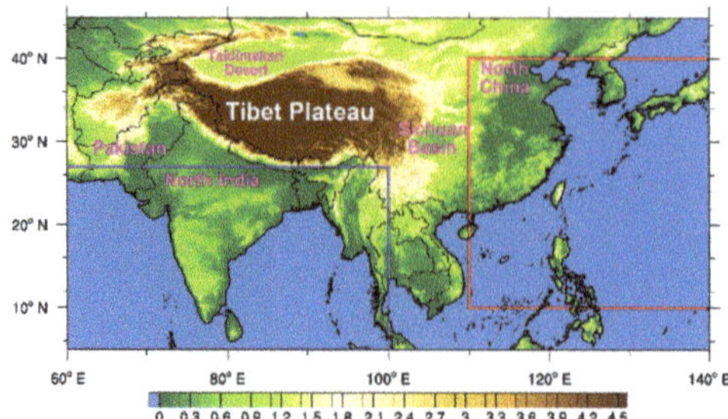

Figure 1. The topography (km) of study area, red and blue rectangles correspond to the East Asian Summer monsoon (EASM) and the South Asian summer monsoon (SASM) regions, respectively.

3. Results

3.1. Model Assessment

Figure 2 shows the comparisons of the simulated net surface radiation budget, surface temperature (ST) and precipitation rate with observations. It shows that the "northern high and southern low" pattern of all sky surface net radiation (integrated by shortwave and longwave radiation) is simulated well (Figure 2a,d). The values of net surface radiation over South Asia, together with East Asia, the Taklimakan Desert and Pakistan range from 150–300W/m^2, 300–400 W/m^2 and 400–450 W/m^2, respectively. While the net surface radiation is underestimated over TP. Because of the bias in surface net radiation over TP, the simulated ST is also underestimated slightly (Figure 2b,e). Generally, the distributions of simulated ST are consistent with the ERA-interim. Overall, the heavy precipitation values are found over the Western India, Bay of Bengal and Indo-China Peninsula in model simulations. The areas of simulated heavy precipitation are consistent with the results from GPCP. In Figure 2c,f, over the south slopes of TP, the model overestimates the precipitation compared with the GPCP results.

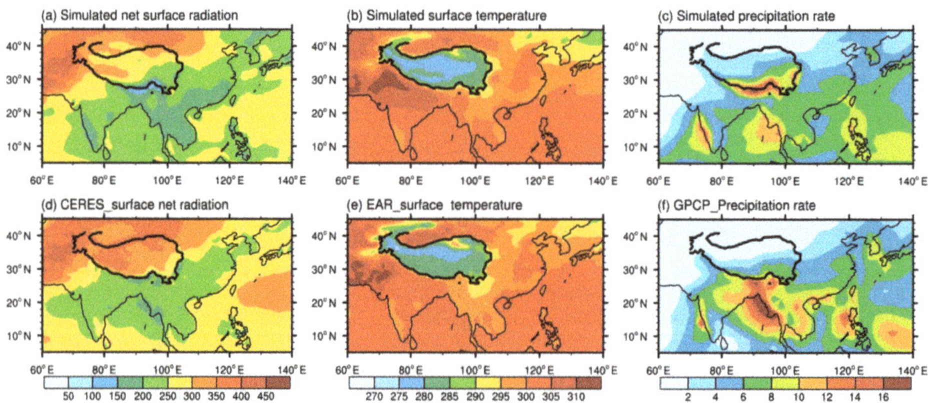

Figure 2. Evaluations of simulated summer all sky net surface radiation (W/m^2), surface temperature (ST) (K) and precipitation rate (mm/day) by the Community Earth System Model (CESM) model for the period from 2001–2005. Where, (**a**–**c**) are model simulations, while the (**d**–**f**) are derived from the data of CERES, EAR-interim and the Global Precipitation Climatology Project (GPCP), respectively. The down direction in (**a**) and (**d**) is defined as the positive value, otherwise the upward direction is defined as the negative value. The area of the Tibetan Plateau (TP) is enclosed by a black line.

Furthermore, the comparisons of simulated water vapor transports and that from the NCEP are presented in Figure 3. In general, both the simulated and observed water vapor burdens over the

TP ranges from 5 to 10 kg/m^2, with almost the same water vapor gradients over the borders of the TP. Over the TP, the westerly winds control the water vapor transportation. As shown in Figure 3, the water vapors are mainly imported from the south and west borders and exported from the east borders. It shows that the southwest water vapor transport channel from the Arabian Sea to the TP is important, which is consistent with the result reported by [72].

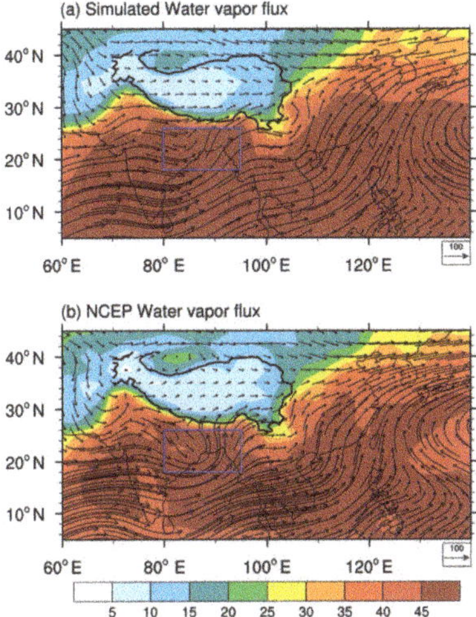

Figure 3. Evaluation of simulated atmospheric water vapor flux (kg/(m·s)) and column water vapor burden (kg/m^2) from surface to 100 hPa by CESM model in the summer for period from 2001–2005. (**a**) Denotes the model simulations. (**b**) Derived from the NCEP. The area of TP is enclosed by a black line. The blue rectangle indicates the region of southwest vortex.

Besides, as given in Figure 3, comparing the water vapor fluxes from the model simulation and NCEP reanalysis over the Bay of Bengal, a weaker, southwest vortex (blue rectangle in Figure 3) is found in the simulation (Figure 3a). In our model simulation, the weaker southwest vortex induced more water vapor northward transport to the south border of the TP, resulting in more precipitation over the south slopes of the TP (Figure 2c).

Before analyzing the BC's impact on the Tibet climate further, an assessment of aerosol optical depth (AOD) and BC concentration distributions is needed. Figure 4 shows the comparisons of AOD from model simulations and MISR satellite observations, and BC distributions from simulations and MERRA-2 reanalysis data. Figure 4a,c, show a similar distribution feature of AOD over the East China, Northern India, Arabian Sea and Taklimakan Desert, both in model simulation and MISR data. Overall, the simulated values of AOD are higher than those from MISR, especially over the Taklimakan Desert. In contrary, the simulated AOD over other regions is lower than the reanalysis data, and that is similar to the features of surface concentration. It is found that the simulated distributions of the surface BC concentration are also consistent with the results from MERRA-2. Both the model and the reanalysis data indicated three centers with a high concentration of BC over Northern India, the Sichuan Basin and Northern China. The model simulated BC concentration over Northern India, the Sichuan Basin and Northern China ranges from 0.5–2.0 μg/m^3, 2.5–3.0 μg/m^3 and 1–1.5 μg/m^3 (Figure 4b,d). Due to the lower concentration simulated by the model, the effects of BC on the water vapor transport and climate may be underestimated over the TP. Generally, the coupled model has a good simulated ability on the precipitation, ST, water vapor flux, BC concentration and AOD distributions over the TP. Based on the model simulation, the effects of BC on water vapor transport and its mechanism are further investigated.

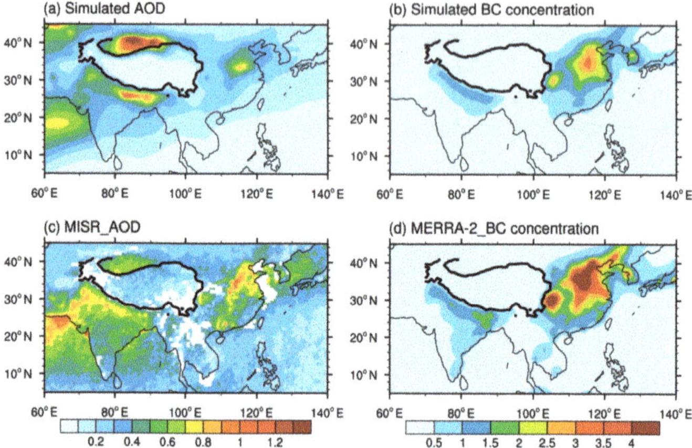

Figure 4. Distributions of aerosol optical depth (AOD) and black carbon (BC) mass concentration ($\mu g/m^3$) in the summer for a period from 2001 to 2005, (**a**,**b**) simulated by the CESM model. (**c**,**d**) are derived from the data of Multi-angle Imaging Spectro Radiometer (MISR) satellite observations and MERRA-2 reanalysis data sets. The area of the TP is enclosed by a black line.

3.2. The Radiative Forcing

In this study, both the direct and indirect effects of BC are considered in the CESM model. Figure 5 shows the clear sky direct radiative forcing (DRF), latent heat (LH) at the surface and the surface sensible heat (SH) fluxes. The results revealed that the shortwave and longwave DRFs at TOA in clear sky are positive with the values about 2.05 and 0.72 W/m^2, respectively (Figure 5a,b). The positive shortwave DRF at TOA indicated that the earth–atmosphere system absorbs more solar radiation, while the positive longwave DRFs are due to less outgoing longwave radiation caused by the absorption of BC. In this study, the results of the TOA DRFs are in agreement with the previous researches, in which some studies show the clear sky DRFs at TOA ranges 0.3–2.4 W/m^2 over east Asia [73,74] and 2–6 W/m^2 over East China [75,76]. In contrast, the surface shortwave DRFs of clear sky is negative with a mean value of -0.86 W/m^2 (Figure 5g), while the longwave DRF is positive with an average of about 0.29 W/m^2 (Figure 5h). Obviously, the shortwave radiative forcing plays a dominant role. Because of the absorption of BC, the reduction of solar radiation at the surface is obvious, resulting in a cooling effect. However, the surface longwave radiative forcing is positive, the reason of which is that the cooled surface emits less longwave radiation, while the warmed atmosphere emits more longwave radiation downward to the surface. In addition, the BC-in-snow effect could reduce the surface longwave albedo, which contributes to the surface positive longwave radiative forcing also.

It is found that both the shortwave and longwave DRFs in the atmosphere are positive with the mean values of 2.9 W/m^2 and 0.44 W/m^2, respectively (Figure 5d,e). The positive DRFs in the atmosphere presents a pronounced warming effect of BC upon the atmosphere. In addition, due to the BC, the LH at the surface is about 0.82 W/m^2 (Figure 5c), and the SH is about 0.13 W/m^2 (Figure 5f). The positive heat flux at the surface could significantly offset the surface cooling effect. The net surface heat change is about 0.39 W/m^2 (Figure 5i), dominated by the disturbed latent heat flux (0.82 W/m^2), and followed by the longwave radiative forcing (0.29 W/m^2).

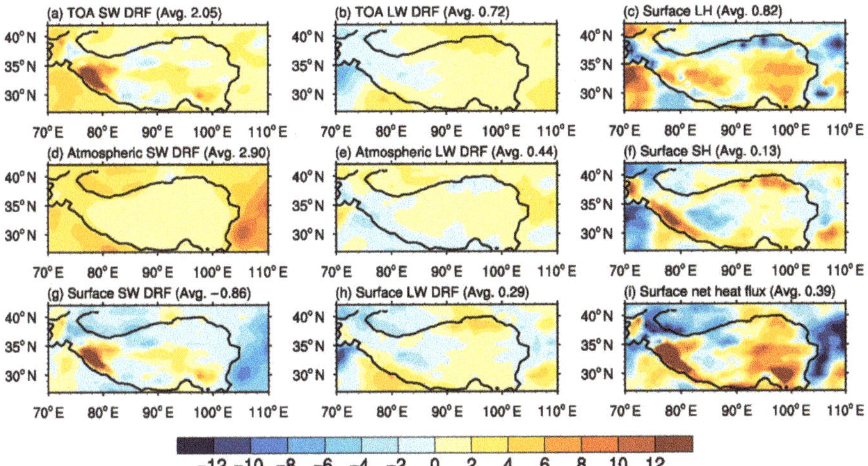

Figure 5. Shortwave and longwave direct radiative forcing (DRF) (W/m^2) at the top of atmosphere (TOA) (**a,b**), in the atmosphere (**d,e**) and on the surface (**g,h**), latent heat flux (**c**), sensible heat flux (**f**) and the net heat flux (**i**) during the summer for period from 2001–2005. The area of the TP is enclosed by a black line. the down direction is defined as the positive value, otherwise the upward direction is defined as the negative value.

Simultaneously, the effects of BC on the cloud radiative forcing (CRF) is another important aspect for the radiation transfer process [7]. Here, we mainly analyze the changes in shortwave and longwave CRFs caused by BC. It is found that the mean change in shortwave CRF is about 1.75 W/m^2 (Figure 6a), while the longwave one is about −1.62 W/m^2 (see Figure 6b), thus the comprehensive change in CRF is warming the earth–atmosphere system. Furthermore, the shortwave and longwave CRFs showed an opposite variation. The change in shortwave CRF is positive over the eastern TP but negative over the western TP. In contrary, the change in longwave CRF is positive over the western TP and negative over the eastern TP. Overall, the net positive changes in CRF denotes a warming effect. The changes in CRF are associated with the cloud microphysics. Due to the indirect effects, BC can affect the cloud microphysics and hence the cloud albedo. In the semi-direct effects, the BC could evaporate the cloud by absorbing more solar radiation. As a result, the increased cloud fraction over the western TP could strengthen the cloud longwave radiative forcing and reduce the shortwave radiative forcing. The decreased cloud is mainly distributed over the southeast TP, that may increase the shortwave radiative forcing while reducing the longwave radiative forcing.

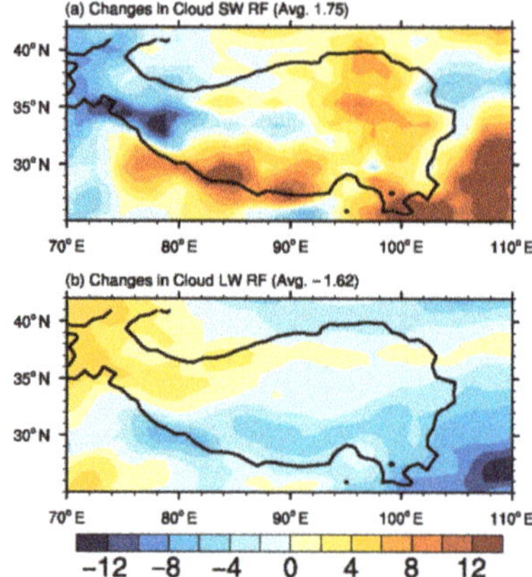

Figure 6. Changes in cloud radiative forcing (W/m^2) during summer for the period (2001–2005) for (**a**) Shortwave and (**b**) Longwave. The area of the TP is enclosed by a black line.

3.3. Changes in Surface Temperature

The anomalies in the earth–atmosphere heat equilibrium caused by BC are reflected by the perturbed ST. Based on the former studies, the changes in sea surface temperature (SST) gradients are important in terms of the monsoon dynamics [6,30], especially over the areas of tropical seas [33]. As some previous studies reported [6,30,31], the anomaly meridional SST gradients over the region (5°–27°N, 60°–100°E) tightly relates to the SASM. In this study, the region (5° N–27° N, 60° E–100° E), enclosed by the blue rectangle in Figure 7, is used to calculate the SASM changes. For EASM, the latitudinal land land–sea thermal contrast is more important compared with the meridional thermal contrast [77]. The area enclosed by the red rectangle in Figure 7 is taken as the critical area to calculate the change of EASM.

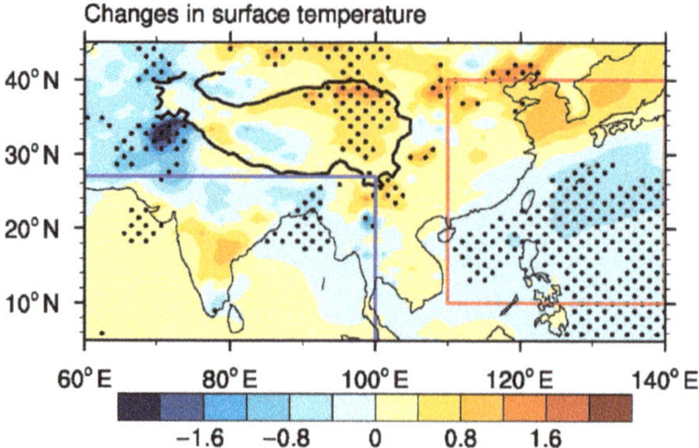

Figure 7. Changes in surface temperature (K) in the summer for period from 2001 to 2005. The area of the TP is enclosed by a black line. The dots denote the changes in surface temperatures that are significant above the 90% confidence level. The blue and red rectangles denote the areas to calculate the changes in the SASM and EASM indices, respectively.

Figure 7 shows that, because of the addition of BC, the surface temperature increases by 0.8–1.6 K over most parts of the TP. The warmer TP is closely related to the surface positive heat flux and the reduced surface albedo (Figure 8).

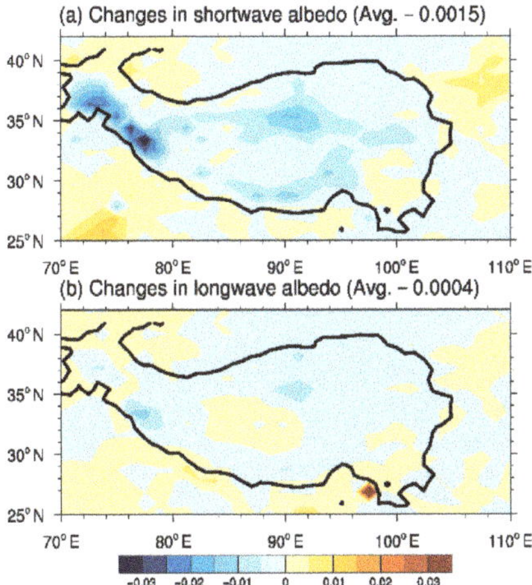

Figure 8. Changes in (**a**) surface shortwave albedo and (**b**) longwave albedo.

The reduced surface albedo means more solar radiation will be absorbed by the land surface. Besides, the increased SST is mainly distributed in the equatorial Arabian Sea and the Southern Bay of Bengal because of the positive heat flux at the surface. In contrast, a pronounced decrease is found over Northern India, Pakistan, Afghanistan and Iran because of the dimming effect of BC. Consequently, the surface temperature shows a decreased meridional gradient. On the other hand, the surface temperature increases over East China, but decreases over the Northwest Pacific Ocean, which could enhance the land-sea thermal contrast.

Generally, the anomalous meridional temperature gradient and the enhanced land–sea thermal contrast could affect the SASM and EASM circulations and further the water vapor transport significantly. More details would be discussed in the following sections.

3.4. Changes in Atmospheric Circulation

Figure 9 describes the altitude–latitude cross-section of zonal mean (60° E–120° E) air temperature and wind field changes. As shown in Figure 9a, over South Asia, the upper atmosphere which is marked by the red rectangle has a warming affect. Meanwhile, an anticyclonic circulation is stimulated due to the addition of BC over the area from 15° N to 25° N, which is consistent with the "Elevated Heat Pump" (EHP) effect of BC reported by Lau and Kim [28]. Due to the EHP effect induced by BC, the anomalous circulation appears opposite to the Hadley circulation, resulting in a northerly wind at the lower atmosphere, which may further reduce the water vapor transport from the Indian Ocean to the TP.

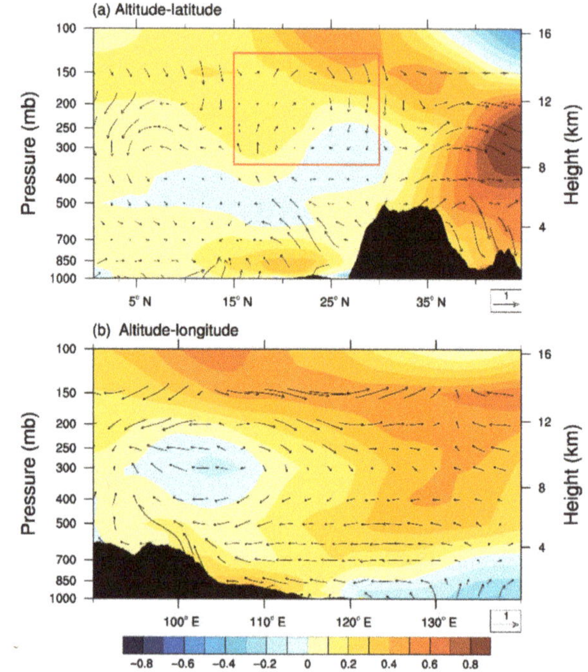

Figure 9. (**a**) Altitude–latitude cross-section of the changes in zonal mean air temperature (colors, K) and wind field (arrows) averaged along meridian region of 60° E–120° E in the latitude zone of 27° N–42° N in the summer for period from 2001 to 2005. (**b**) Same as (**a**) but for altitude–longitude cross-section of meridional means along latitude region of 27° N–42° N in the longitude zone of 60° E–120° E. Red rectangle indicates the area of anticyclonic circulation induced by BC.

Figure 9b describes the altitude–longitude cross-section of changes in meridional mean (27° N–42° N) air temperature and wind field over East Asia. Generally, East Asia is affected by the subtropical summer monsoon, which attributes the intensity mainly to the latitudinal thermal contrast and secondarily to the meridional thermal contrast [77]. Thus, we analyzed the latitudinal land–sea thermal contrast from 100° E to 140° E. Besides, the pattern of "western warm eastern cold" induces easterlies from the Northwest Pacific Ocean to the east border of the TP, leading to an enhanced EASM.

To estimate the changes in SASM and EASM due to the BC, the DNS indices are calculated according to Li et al. [64], as shown in Figure 10. Figure 10a shows that the SASM indices under the ALL experiment are smaller than that under the ALL-BC experiment, indicating a weak SASM. Conversely, the EASM index has the opposite variations, in which an enhanced EASM is caused by including BC in the model (Figure 10b).

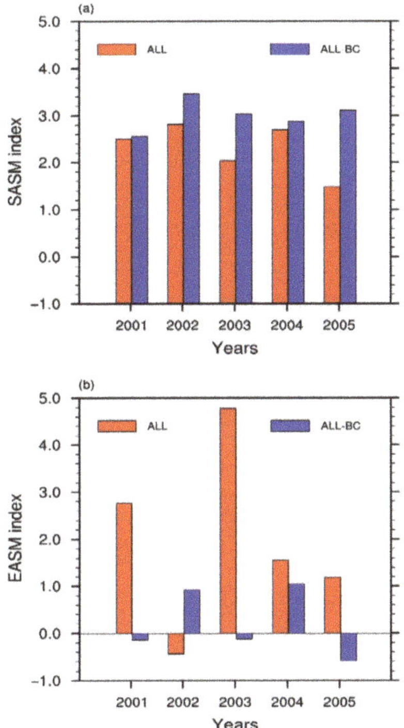

Figure 10. Changes in the (**a**) SASM and (**b**) EASM index from 2001–2005. The red and blue bars denote the simulations under the 'ALL' and 'ALL-BC' experiments, respectively.

3.5. Effects of BC on the Water Vapor Budget

It should be noted that the EHP effect can reduce the SASM in summertime. The anticyclone in the upper atmosphere, which is caused by BC, can induce a northerly wind in the lower atmosphere from the south slope of the Tibetan Plateau to Northern India (see Figure 9a). The northerly wind is contrary to the southerly wind over the region of 0° N–15° N, where can represent the SASM. Thus, the EHP effect can reduce the SASM in summertime. In addition, the strong cooling effect over North India, Pakistan, Afghanistan and Iran, and the warming effect over South India, together with the surrounding ocean area, could reduce the meridional temperature gradient and hence weaken the SASM. Meanwhile, the warmer TP caused by BC can increase the land–sea thermal contrast and the EASM. Overall, the BC could induce a weak SASM but an enhanced EASM and further affect the water vapor transport from the ocean to the TP.

Water vapor plays a significant role in adjusting the air temperature and precipitation. As reported by previous researchers [78], the water vapor from the west and southwest borders contributed mostly to the precipitation over the TP region. Besides, the net importing water vapor to the TP is mainly attributed to the east and west borders with the values of 1991.47 kg/(m·s) and 160.13 kg/(m·s), respectively. Likewise, the exporting water vapor away from the TP is attributed mainly to the north and south borders with the values of 267.77 kg/(m·s) and 381.97 kg/(m·s), respectively [37]. Generally, the four borders of the TP have different characteristics on the water vapor transport. The importing water vapor from the west border serves as the primary contribution to the water vapor over the TP, and that from the southern border is the second contribution [37,79]. Besides, the eastern border has the main export channels, which are closely related to the atmospheric circulations [42].

Figure 11a shows that a pronounced anomalous cyclone over the Northwest Pacific Ocean and a weak cyclone system over Pakistan and Afghanistan are stimulated because of the addition of BC in the simulation. The TP, located at the north side of the cyclone over the Northwest Pacific Ocean, was affected by the perturbed east airflows extending from the Northwest Pacific Ocean to the east border of the TP. Over the western TP, the anomalous southeast flow exerted an impact on the TP because of the weak cyclone system over Pakistan and Afghanistan.

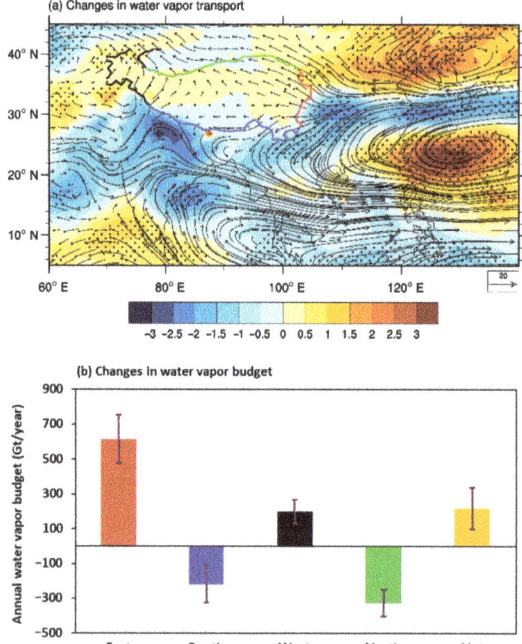

Figure 11. Effects of BC on (**a**) water vapor flux (kg/m/s) and (**b**) water vapor budget (Gt/year). The red, green, black and blue lines denote the west, east, north and south borders of the TP, respectively. The dots in (**a**) denote the changes in water vapor that are significant above the 90% confidence level. The purple short lines in the bar in (**b**) denote the error ranges from a negative to a positive standard deviation.

Over the south and north borders of the TP, it was controlled by the anomalous north and southeast flows, respectively. Consequently, more water vapor was exported from the TP. The mean flux of water vapor imported from the west and east borders are 6.3×10^6 kg/s and 19.5×10^6 kg/s, repsectively. The water vapors exported from the south and north border are about 6.9×10^6 kg/s and 10×10^6 kg/s, respectively. Thus, the horizontal net water vapor budget of TP is about 8.6×10^6 kg/s. The net annual mean water vapor budget is about 271Gt/year (see Figure 11b). This net budget is dominated by the east border of TP than others.

4. Discussion

As mentioned in Section 3.3, the pronounced increase in ST over the TP is caused by the positive surface net heat flux with the value of 0.39 W/m^2. The warmer TP is dominated by the surface latent heat flux, followed by the longwave radiative forcing and sensible heat flux.

In Section 3.5, we noted that the net annual mean positive importing water vapor caused by BC over the TP is about 271 Gt/year, in which the positive feature is consistent with the previous study [72]. The anomalous water vapor could partially modify the precipitation pattern. Figure 12 shows the anomalous precipitation rate over the TP. As given in Figure 12, precipitation was mainly increased over the northern and western TP, with a maximum value of 2 mm/day. Simultaneously, the precipitation was decreased over the southern TP with a maximum decreasing about 3 mm/day. The average value of increased precipitation due to BC over the TP is 0.56 mm/day. The increased precipitation over TP is attributed to the anomalous water vapor transport and atmospheric circulations caused by BC [2,3].

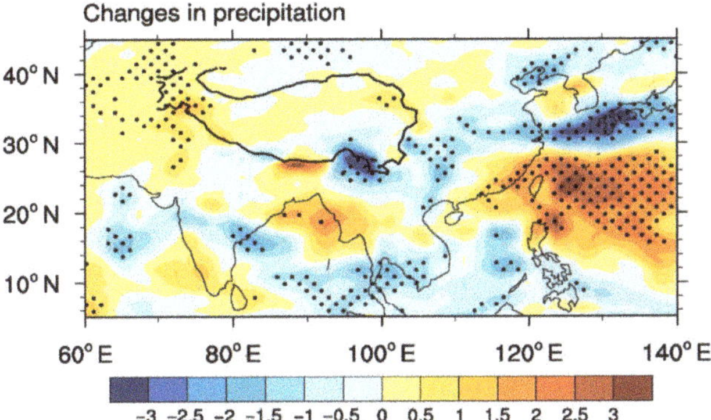

Figure 12. Changes in precipitation rate (mm/day) over the TP. The dots denote the changes in precipitation that are significant above the 90% confidence level.

As shown in Figure 11a, the cyclone induced by the surface cooling effect of BC over Pakistan and Afghanistan is beneficial to increase the precipitation over the Western TP. Besides, the perturbed easterly wind and the orographic lifting are in favor of increasing the precipitation over the Northern TP.

As illustrated above, the BC aerosol can induce a warmer and wetter plateau. The results are consistent with current studies [42,73,80]. Our results suggest that the heating of LH, SH and longwave radiative forcing could offset the surface cooling effect. Besides, the BC-in-snow effect could contribute to the warmer TP also [16–19].

In this study, the net surface radiative forcing has included the BC-in-snow effect already. Based on the previous studies, the "snow darkening" effect of BC could also contribute to the warmer TP [81,82], and leads to a reduced snow cover by about 10%–20%, accompanied with a decreased surface albedo. This feedback could heat the TP significantly [82]. In this study, the changes in surface albedo are shown in Figure 8. The results indicate that the mean shortwave and longwave albedo are decreased by 0.0015 and 0.0004, respectively, because of the addition of black carbon. The magnitude of the reduced shortwave albedo is about four times greater than that of the surface longwave albedo. Besides, the reduced shortwave albedo could reach to −0.03 at the south slope of the TP. It is closely related to the high concentration of BC. Generally, the reduced albedo means that the BC-in-snow could absorb more downward solar radiation and longwave radiation to heat the TP.

Moreover, the BC-in-snow effect could further significantly affect the SASM and EASM by the thermal and dynamical forcing. Based on the current study, the warmer TP could increase the land–sea thermal contrast to enhance the EASM in summertime [26], which is consistent with the result of this study. Besides, the BC-in-snow can strengthen the upward motion over the TP to advance the SASM in pre-monsoon [26], which is similar to the EHP effect of BC [27].

The premise of the EHP is that BC can stack up against the south slope of the TP in the springtime and then induce an anomalous warming anticyclone in the upper atmosphere [27]. Because of the latent heat warming effect over the TP, the meridional temperature gradient and the SASM could be enhanced in springtime. It should be noted that Lau and Kim [27] proposed the EHP by the results from the GCM model, neglecting the indirect effects by the off-line method. On the contrary, our results include the direct and indirect effects and the feedback of the ocean. The results show that EHP could induce a warm anticyclone in the upper atmosphere over BC-contaminated regions from

Northern India to TP. The northerly wind of the anticyclone in the lower atmosphere and the reduced meridional ST gradient could reduce the SASM in summertime. The results are consistent with the previous studies [28,29].

In the following, a mechanism analysis is performed (Figure 13). Because of the absorption of BC, the solar radiation reaching the surface is reduced obviously, leading to a surface cooling effect. Furthermore, the ST increased over the TP because of the addition of BC. Therefore, a warmer TP can be induced by the BC.

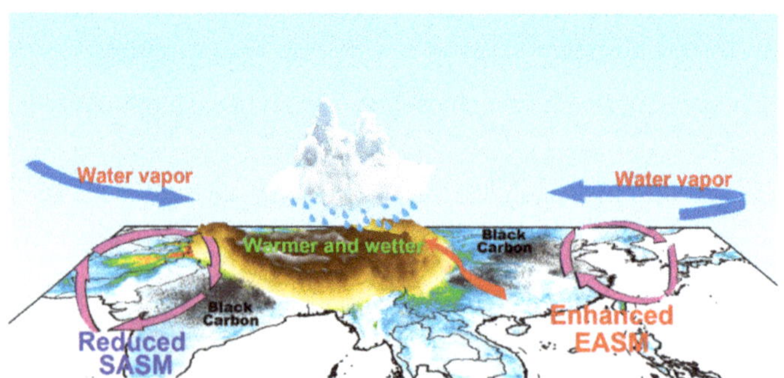

Figure 13. Mechanism of BC affecting water vapor transport over the TP.

Though the thermal effects of BC on the lower atmosphere over the TP are warming, spatial discrepancies over the other regions exist. On the one hand, over the regions surrounding Pakistan, Afghanistan and Northern India, the addition of BC can decrease the ST. However, it can increase the SST of Indian Ocean. Thus, a weak thermal contrast between the land and ocean is induced, leading to a weak SASM, then to less water vapor transporting from the south border of the TP. On the other hand, the land–sea thermal contrast is intensified over East Asia, inducing an intensified EASM. The changes in SASM and EASM dramatically impact the water vapor to the TP. Furthermore, a cyclone is stimulated in the upper atmosphere due to the decreased ST over Pakistan, Afghanistan and the Northwest Pacific Ocean. Thus, the western TP is controlled by the southwest winds, leading to more water vapor being transported to the TP. Besides, the eastern TP is controlled by the east winds which is on the north side of the cyclone, resulting in more water vapor being transported to the TP also. Consequently, due to the BC, though the water vapor imported from south side is weak, and the exported water vapor from north side of TP is enhanced, more water vapor is transported from the east and west to the TP. Therefore, because of BC addition, the TP will be wetter.

The uncertainties in the results apply mostly to the aerosol–clouds interactions (ACI, indirect effects) because of the lack of an accurate resolution to represent ACI. Current understanding and classification of this ACI regime is based on the response of the droplet number concentration (N_c) to the aerosol number concentration (N_a) and vertical velocity (w) [83], and that are represented by empirical parameterizations. The ACI could affect the performance of the climate model strongly [83]. The other uncertainties mainly derived from the short records of the horizontal and vertical BC distributions. An accurate distribution of BC data sets may help to interpret the transport and buildup of the BC well. Recently, using the observations which include aerosol physical and chemical properties and mixing state to constrain the simulations is an effective approach to limit the uncertainties.

5. Conclusions

The BC could significantly affect the climate over the TP especially in the summer. Here, combining satellite observations and reanalysis data, we studied the climate effect of BC over the TP by using a fully coupled model. In this study, we have estimated the results of BC affecting the water vapor transport, and we have revealed the concerning mechanism.

The simulation indicates that the BC can induce the positive radiative effects (including direct and indirect effects) and thus exert a pronounced warming over the TP. Based on a detailed analysis, the mean shortwave and longwave radiative forcing at the TOA over TP are 2.05 W/m^2 and 0.72 W/m^2, respectively.

Besides, considering the LH and SH, the surface net heat forcing presents positive with a value of 0.39 W/m^2. Furthermore, the BC can induce an anomalous temperature and then change the atmospheric circulations. As mentioned above, because of the addition of BC, there is a pronounced decreased temperature over Pakistan and Afghanistan, but an increased surface temperature over Southern India and its surrounding ocean. Such temperature patterns can induce a weak thermal contrast between the land and sea, leading to a weak SASM. Thus, less water vapor could be transported from the Indian Ocean to the TP. Besides, over East Asia, the "western warm and eastern cold" pattern enhanced the land–sea thermal contrast over East Asia and the surrounding ocean, inducing an intensified EASM significantly. Consequently, more water vapor is transported from the east of TP. Overall, due to the BC, the net water vapor is positive over the TP, implying a net import of water vapor from the surroundings to the TP. Furthermore, the increased water vapor is closely related to the anomalies in precipitation over the TP.

Author Contributions: Conceptualization, Y.L.; methodology, Y.L. and M.L.; investigation, Y.L and M.L.; writing—original draft preparation, M.L.; writing—review and editing, Y.L., M.L., K.A. and Y.T.; visualization, M.L., Q.Z. and Y.T.; funding acquisition, Y.L. All authors have read and agreed to the published version of the manuscript.

Funding: This research was mainly supported by the Strategic Priority Research Program of Chinese Academy of Sciences (Grant No. XDA2006010301) and the National Natural Science Foundation of China (91737101 and 91744311). This work was also jointly supported by the Foundation of Key Laboratory for Semi-Arid Climate Change of the Ministry of Education in Lanzhou University.

Acknowledgments: We acknowledge the MISR (http://eosweb.larc.nasa.gov/PRODOCS/misr/table_misr. html), CERES (https://ceres.larc.nasa.gov/), EAR-interim (https://www.ecmwf.int/en/forecasts/datasets/reanalysis-datasets/era-interim), MERRA-2 (https://gmao.gsfc.nasa.gov/reanalysis/MERRA-2/) and NCEP (https://www. esrl.noaa.gov/psd/data/gridded/data.ncep.reanalysis.html) science teams for providing excellent and accessible products which favor this study. We also thank the Intergovernmental Panel on Climate Change (IPCC) Fifth Assessment Report (AR5) for providing us the data sets of anthropogenic aerosol emissions (ftp: //ftp.cgd.ucar.edu/cesm/inputdata/atm/cam/chem/trop_mozart_aero/emis/).

Conflicts of Interest: The authors declare no conflict of interest.

References

1. Roberts, D.L.; Jones, A. Climate sensitivity to black carbon aerosol from fossil fuel combustion. *J. Geophys. Res. Atmos.* **2004**, *109*, D16202. [CrossRef]
2. Wu, J.; Fu, C.; Xu, Y.; Tang, J.P.; Wang, W.; Wang, Z. Simulation of direct effects of black carbon aerosol on temperature and hydrological cycle in Asia by a Regional Climate Model. *Meteorol. Atmos. Phys.* **2008**, *100*, 179–193. [CrossRef]
3. Booth, B.; Bellouin, N. Black carbon and atmospheric feedbacks. *Nature* **2015**, *519*, 167–168. [CrossRef]
4. Ji, D.; Li, L.; Pang, B.; Xue, P.; Wang, Y.; Wang, L.; Wu, Y.; Zhang, H. Characterization of black carbon in an urban-rural fringe area of Beijing. *Environ. Pollut.* **2017**, *223*, 524–534. [CrossRef] [PubMed]
5. Stjern, C.W.; Samset, B.H.; Myhre, G.; Forster, P.M.; Hodnebrog, Ø.; Andrews, T.; Boucher, O.; Faluvegi, G.; Iversen, T.; Kasoar, M.; et al. Rapid Adjustments Cause Weak Surface Temperature Response to Increased Black Carbon Concentrations. *J. Geophys. Res. Atmos.* **2017**, *122*, 462–481. [CrossRef]
6. Ramanathan, V.; Chung, C.; Kim, D.; Bettge, T.; Buja, L.; Kiehl, J.; Washington, W.; Fu, Q.; Sikka, D.; Wild, M. Atmospheric brown clouds: Impacts on South Asian climate and hydrological cycle. *Proc. Natl. Acad. Sci. USA* **2005**, *102*, 5326–5333. [CrossRef]
7. Liu, Y.; Jia, R.; Dai, T.; Xie, Y.; Shi, G. A review of aerosol optical properties and radiative effects. *J. Meteor. Res.* **2014**, *28*, 1003–1028. [CrossRef]
8. Chen, B.; Chen, J.; Bai, Z.; Cui, X.; Andersson, A.; Gustafsson, Ö. Light absorption enhancement of black carbon from urban haze in Northern China winter. *Environ. Pollut.* **2017**, *221*, 418–426. [CrossRef]

9.   Liu, Y.; Sato, Y.; Jia, R.; Xie, Y.; Huang, J.; Nakajima, T. Modeling study on the transport of summer dust and anthropogenic aerosols over the Tibetan Plateau. *Atmos. Chem. Phys.* **2015**, *15*, 12581–12594. [CrossRef]

10. Liu, Y.; Zhu, Q.; Wang, R.; Xiao, K.; Cha, P. Distribution, source and transport of the aerosols over Central Asia. *Atmos. Environ.* **2019**, *210*, 120–131. [CrossRef]

11. Jia, R.; Luo, M.; Liu, Y.; Zhu, Q.; Hua, S.; Wu, C.; Shao, T. Anthropogenic Aerosol Pollution over the Eastern Slope of the Tibetan Plateau. *Adv. Atmos. Sci.* **2019**, *36*, 847–862. [CrossRef]

12. Ming, J.; Xiao, D.; Du, C.; Yang, G. An overview of black carbon deposition in High Asia glaciers and its impacts on radiation balance. *Adv. Water Res.* **2013**, *55*, 80–87. [CrossRef]

13. Liu, Y.; Huang, J.; Shi, G.; Takamura, T.; Khatri, P.; Bi, J.; Shi, J.; Wang, T.; Wang, X.; Zhang, B. Aerosol optical properties determined from sky-radiometer over Loess Plateau of Northwest China. *Atmos. Chem. Phys.* **2011**, *11*, 23883–23910. [CrossRef]

14. Jia, R.; Liu, Y.; Hua, S.; Zhu, Q.; Shao, T. Estimation of the aerosol radiative effect over the Tibetan Plateau based on the latest CALIPSO product. *J. Meteorol. Res.* **2018**, *32*, 707–722. [CrossRef]

15. He, C.; Li, Q.; Liou, K.; Takano, Y.; Gu, Y.; Qi, L.; Mao, Y.; Leung, L. Black carbon radiative forcing over the Tibetan Plateau. *Geophys. Res. Lett.* **2014**, *41*, 7806–7813. [CrossRef]

16. Flanner, M.; Zender, C.; Randerson, J.; Rasch, P. Present-day climate forcing and response from black carbon in snow. *J. Geophys. Res. Atmos.* **2007**, *112*, D11202. [CrossRef]

17. Xu, B.; Cao, J.; Hansen, J.; Yao, T.; Joswia, D.; Wang, N.; Wu, G.; Wang, M.; Zhao, H.; Yang, W.; et al. Black soot and the survival of Tibetan glaciers. *Proc. Natl. Acad. Sci. USA* **2009**, *106*, 22114–22118. [CrossRef]

18. Jacobi, H.; Lim, S.; Ménégoz, M.; Ginot, P.; Laj, P.; Bonasoni, P.; Stocchi, P.; Marinoni, A.; Arnaud, Y. Black carbon in snow in the upper Himalayan Khumbu Valley, Nepal: Observations and modeling of the impact on snow albedo, melting, and radiative forcing. *Cryosphere* **2015**, *9*, 1685–1699. [CrossRef]

19. He, C.; Flanner, M.; Chen, F.; Barlage, M.; Liou, K.; Kang, S.; Ming, J.; Qian, Y. Black carbon-induced snow albedo reduction over the Tibetan Plateau: Uncertainties from snow grain shape and aerosol–snow mixing state based on an updated SNICAR model. *Atmos. Chem. Phys.* **2018**, *18*, 11507–11527. [CrossRef]

20. Hua, S.; Liu, Y.; Jia, R.; Chang, S.; Wu, C.; Zhu, Q.; Shao, T.; Wang, B. Role of Clouds in Accelerating Cold-Season Warming During 2000-2015 over the Tibetan Plateau. *Int. J. Climatol.* **2018**, *38*, 4950–4966. [CrossRef]

21. Liu, Y.; Hua, S.; Jia, R.; Huang, J. Effect of aerosols on the ice cloud properties over the Tibetan Plateau. *J. Geophys. Res. Atmos.* **2019**, *124*, 9594–9608. [CrossRef]

22. Liu, Y.; Zhu, Q.; Huang, J.; Hua, S.; Jia, R. Impact of dust-polluted convective clouds over the Tibetan Plateau on downstream precipitation. *Atmos. Environ.* **2019**, *209*, 67–77. [CrossRef]

23. Liu, Y.; Li, Y.; Huang, J.; Wang, S.; Zhu, Q. Attribution of Tibetan Plateau to the northern drought. *Natl. Sci. Rev.* **2019**, nwz191. [CrossRef]

24. Liu, Y.; Luo, R.; Zhu, Q.; Hua, S.; Wang, B. Cloud Ability to Produce Precipitation over Arid and Semiarid Regions of Central and East Asia. *Int. J. Climatol.* **2019**, 1–14. [CrossRef]

25. Liu, Y.; Wu, C.; Jia, R.; Huang, J. An overview of the influence of atmospheric circulation on the climate in arid and semi-arid region of Central and East Asia. *Sci. China Earth Sci.* **2018**, *61*, 31–42. [CrossRef]

26. Qian, Y.; Flanner, M.; Leung, L.; Wang, W. Sensitivity studies on the impacts of Tibetan Plateau snowpack pollution on the Asian hydrological cycle and monsoon climate. *Atmos. Chem. Phys.* **2011**, *11*, 1929–1948. [CrossRef]

27. Lau, K.; Kim, M.; Kim, K. Asian summer monsoon anomalies induced by aerosol direct forcing: The role of the Tibetan Plateau. *Clim. Dyn.* **2006**, *26*, 855–864. [CrossRef]

28. Lau, W.; Kim, K. Fingerprinting the impacts of aerosols on long-term trends of the Indian summer monsoon regional rainfal. *Geophys. Res. Lett.* **2010**, *37*, 127–137. [CrossRef]

29. D'Errico, M.; Cagnazzo, C.; Fogli, P.; Lau, W.; von Hardenberg, J.; Fierli, F.; Cherchi, A. Indian monsoon and the elevated-heat-pump mechanism in a coupled aerosol-climate model. *J. Geophys. Res. Atmos.* **2015**, *120*, 8712–8723. [CrossRef]

30. Chung, C.; Ramanathan, V. Weakening of North Indian SST gradients and the monsoon rainfall in India and the Sahel. *J. Clim.* **2006**, *19*, 2036–2045. [CrossRef]

31. Meehl, G.A.; Arblaster, J.M.; Collins, W.D. Effects of black carbon aerosols on the Indian monsoon. *J. Clim.* **2008**, *21*, 2869–2882. [CrossRef]

32. Wu, G.; Li, Z.; Fu, C.; Zhang, X.; Zhang, R.; Zhang, R.; Zhou, T.; Li, J.; Zhou, D.; Wu, L.; et al. Advances in studying interactions between aerosols and monsoon in China. *Sci. Chin. Earth Sci.* **2016**, *59*, 1–16. [CrossRef]

33. Li, H.; Dai, A.; Zhou, T.; Lu, J. Responses of EASM to historical SST and atmospheric forcing during 1950–2000. *Clim. Dyn.* **2010**, *34*, 501–514. [CrossRef]

34. Zhuang, B.; Chen, H.; Li, S.; Wang, T.; Liu, J.; Zhang, L.; Liu, H.; Xie, M.; Chen, P.; Li, M.; et al. The direct effects of black carbon aerosols from different source sectors in East Asia in summer. *Clim. Dyn.* **2019**, *53*, 5293–5310. [CrossRef]

35. Renhe, Z. Relations of Water Vapor Transport from Indian Monsoon with That over East Asia and the Summer Rainfall in China. *Adv. Atmos. Sci.* **2001**, *18*, 1005–1017. [CrossRef]

36. Zhou, T.; Yu, R. Atmospheric water vapor transport associated with typical anomalous summer rainfall patterns in China. *J. Geophys. Res. Atmos.* **2005**, *110*, D8. [CrossRef]

37. Ma, Y.; Lu, M.; Chen, H.; Pan, M.; Hong, Y. Atmospheric moisture transport versus precipitation across the Tibetan Plateau: A mini-review and current challenges. *Atmos. Res.* **2018**, *209*, 50–58. [CrossRef]

38. Dong, W.; Lin, Y.; Wright, J.; Xie, Y.; Xu, F.; Xu, W.; Wang, Y. Indian Monsoon Low-Pressure Systems Feed Up-and-Over Moisture Transport to the Southwestern Tibetan Plateau. *J. Geophys. Res. Atmos.* **2017**, *122*, 12–140. [CrossRef]

39. Li, C.; Zuo, Q.; Xu, X.; Gao, S. Water vapor transport around the Tibetan Plateau and its effect on summer rainfall over the Yangtze River valley. *J. Meteorol. Res.* **2016**, *30*, 472–482. [CrossRef]

40. Wu, H.; Li, X.; Zhang, J.; Li, J.; Liu, J.; Tian, L.; Fu, C. Stable isotopes of atmospheric water vapour and precipitation in the northeast Qinghai-Tibetan Plateau. *Hydrol. Process.* **2019**, *33*, 2997–3009. [CrossRef]

41. Zhao, Y.; Zhou, T. Asian water tower evinced in total column water vapor: A comparison among multiple satellite and reanalysis data sets. *Clim. Dyn.* **2019**, 1–15. [CrossRef]

42. Zhou, C.; Zhao, P.; Chen, J. The Interdecadal Change of Summer Water Vapor over the Tibetan Plateau and Associated Mechanisms. *J. Clim.* **2019**, *32*, 4103–4119. [CrossRef]

43. Huang, J.; Yu, H.; Guan, X.; Wang, G.; Guo, R. Accelerated dryland expansion under climate change. *Nat. Clim. Change* **2016**, *6*, 166–171. [CrossRef]

44. Huang, J.; Li, Y.; Fu, C.; Chen, F.; Fu, Q.; Dai, A.; Shinoda, M.; Ma, Z.; Guo, W.; Li, Z.; et al. Dryland climate change Recent progress and challenges. *Rev.Geophys.* **2017**, *55*, 719–778. [CrossRef]

45. Martonchik, J.V.; Diner, D.J.; Crean, K.A.; BullRegional, M.A. Regional aerosol retrieval results from MISR. *IEEE Trans. Geosci. Remote Sens.* **2002**, *40*, 1520–1531. [CrossRef]

46. Diner, D.; Beckert, J.; Reilly, T.; Bruegge, C.; Conel, J.; Kahn, R.; Martonchik, J.; Ackerman, T.; Davies, R.; Gerstl, S.; et al. Multi-angle Imaging SpectroRadiometer (MISR) instrument description and experiment overview. *IEEE Trans. Geosci. Remote Sens.* **1998**, *36*, 1072–1087. [CrossRef]

47. Wielicki, B.; Barkstrom, B.; Harrison, E.; Lee, R.; Smith, G.; Cooper, J. Clouds and the Earth's Radiant Energy System (CERES): An Earth Observing System Experiment. *Bull. Am. Meteorol. Soc.* **1996**, *77*, 853–868. [CrossRef]

48. Almorox, J.; Ovando, G.; Sayago, S.; Bocco, M. Assessment of surface solar irradiance retrieved by CERES. *Int. J. Remote Sens.* **2017**, *38*, 3669–3683. [CrossRef]

49. Kato, S.; Loeb, N.G.; Rutan, D.A.; Rose, F.G.; Sun-Mack, S.; Miller, W.F.; Chen, Y. Uncertainty estimate of surface irradiances computed with MODIS-, CALIPSO-, and CloudSat-derived cloud and aerosol properties. *Surv. Geophys.* **2012**, *33*, 395–412. [CrossRef]

50. Dee, D.P.; Uppala, S.M.; Simmons, A.J.; Berrisford, P.; Poli, P.; Kobayashi, S.; Andrae, U.; Balmaseda, M.A.; Balsamo, G.; Bauer, D.P.; et al. The ERA-Interim reanalysis: Configuration and performance of the data assimilation system. *Q. J. R. Meteorol. Soc.* **2011**, *137*, 553–597. [CrossRef]

51. Wang, S.; Zhang, M.; Sun, M.; Wang, B.; Huang, X.; Wang, Q.; Feng, F. Comparison of surface air temperature derived from NCEP/DOE R2, ERA-Interim, and observations in the arid northwestern China: A consideration of altitude errors. *Theor. Appl. Climatol.* **2015**, *119*, 99–111. [CrossRef]

52. You, Q.L.; Min, J.Z.; Zhang, W.; Pepin, N.; Kang, S.H. Comparison of multiple datasets with gridded precipitation observations over the Tibetan Plateau. *Clim. Dyn.* **2014**, *45*, 791–806. [CrossRef]

53. Gao, Y.; Liu, M. Evaluation of high-resolution satellite precipitation products using rain gauge observations over the Tibetan Plateau. *Hydrol. Earth Syst. Sci.* **2013**, *17*, 837–849. [CrossRef]

54. Adler, R.F.; Huffman, G.J.; Chang, A.; Ferraro, R.; Xie, P.; Janowiak, J.; Rudolf, B.; Schneider, U.; Curtis, S.; Bolvin, D.; et al. The Version-2 Global Precipitation Climatology Project (GPCP) Monthly Precipitation Analysis (1979–Present). *J. Hydrol.* **2003**, *4*, 1147–1167. [CrossRef]

55. Randles, C.; Coauthors. The MERRA-2 aerosol reanalysis, 1980 onward. Part I: System description and data assimilation evaluation. *J. Clim.* **2017**, *30*, 6823–6850. [CrossRef]

56. Gelaro, R.; McCarty, W.; Suarez, M.; Todling, R.; Molod, A.; Takacs, L.; Randles, C.; Darmenov, A.; Bosilovich, M.; Reichle, R.; et al. The Modern-Era Retrospective Analysis for Research and Applications, Version 2 (MERRA-2). *J. Clim.* **2017**, *30*, 5419–5454. [CrossRef]

57. Bao, Q.; Liu, Y.; Shi, J.C.; Wu, G.X. Comparisons of Soil Moisture Datasets over the Tibetan Plateau and Application to the Simulation of Asia Summer Monsoon Onset. *Adv. Atmos. Sci.* **2010**, *27*, 303–314. [CrossRef]

58. Kanamitsu, M.; Ebisuzaki, W.; Woollen, J.; Yang, S.; Hnilo, J.; Fiorino, M.; Potter, G. NCEP-DOE AMIP-II reanalysis (R-2). *Bull. Am. Meteorol. Soc.* **2002**, *83*, 1631–1643. [CrossRef]

59. Kalnay, E.; Kanamitsu, M.; Kistler, R.; Collins, W.; Deaven, D.; Gandin, L.; Iredell, M.; Saha, S.; White, G.; Woollen, J.; et al. The NCEP/NCAR 40-Year Reanalysis Project. *Bull. Am. Meteor. Soc.* **1996**, *77*, 461–468. [CrossRef]

60. Lamarque, J.; Bond, T.; Eyring, V.; Granier, C.; Heil, A.; Klimont, Z.; Lee, D.; Liousse, C.; Mieville, A.; Owen, B.; et al. Historical (1850–2000) gridded anthropogenic and biomass burning emissions of reactive gases and aerosols: Methodology and application. *Atmos. Chem. Phys.* **2010**, *10*, 7017–7039. [CrossRef]

61. Smith, S.J.; Pitcher, H.; Wigley, T.M.L. Global and regional anthropogenic sulfur dioxide emissions. *Glob. Planet. Change* **2001**, *29*, 99–119. [CrossRef]

62. Yang, Y.; Russell, L.M.; Lou, S.; Lamjiri, M.A.; Liu, Y.; Singh, B.; Ghan, S.J.; Richland, W.A. Changes in Sea Salt Emissions Enhance ENSO Variability. *J. Clim.* **2016**, *29*, 8575–8588. [CrossRef]

63. Kiehl, J.; Hack, J.; Bonan, G.; Boville, B.; Williamson, D.; Rasch, P. The National Center for Atmospheric Research Community Climate Model: CCM3. *J. Clim.* **1998**, *11*, 1131–1149. [CrossRef]

64. Li, J.P.; Feng, J.; Li, Y. A possible cause of decreasing summer rainfall in northeast Australia. *Int. J. Climatol.* **2011**, *32*, 995–1005. [CrossRef]

65. Liu, X.; Easter, R.; Ghan, S.; Zaveri, R.; Rasch, P.; Shi, X.; Lamarque, J.; Gettelman, A.; Morrison, H.; Vitt, F.; et al. Toward a minimal representation of aerosols in climate models: Description and evaluation in the Community Atmosphere Model CAM5. *Geosci. Model Dev.* **2012**, *5*, 709–739. [CrossRef]

66. Zhang, L.M.; Gong, S.L.; Padro, J.; Barrie, L. A size-segregated particle dry deposition scheme for an atmospheric aerosol module. *Atmos. Environ.* **2001**, *35*, 549–560. [CrossRef]

67. Rasch, P.J.; Barth, M.C.; Kiehl, J.T.; Schwartz, S.E.; Benkovitz, C.M. A description of the global sulfur cycle and its controlling processes in the National Center for Atmospheric Research Community Climate Model, Version 3. *J. Geophys. Res. Atmos.* **2000**, *105*, 1367–1385. [CrossRef]

68. Morrison, H.; Curry, J.A.; Khvorostyanov, V.I. A new double-moment microphysics parameterization for application in cloud and climate models. part I: Description. *J. Atmos. Sci.* **2005**, *62*, 1665–1677. [CrossRef]

69. Morrison, H.; Gettelman, A. A new two-moment bulk stratiform cloud microphysics scheme in the NCAR Community Atmosphere Model (CAM3), Part I: Description and numerical tests. *J. Clim.* **2008**, *21*, 3642–3659. [CrossRef]

70. Bretherton, C.S.; Park, S. A new moist turbulence parameterization in the community atmosphere model. *J. Clim.* **2009**, *22*, 3422–3448. [CrossRef]

71. Park, S.; Bretherton, C. The university of washington shallow convection and moist turbulence schemes and their impact on climate simulations with the community atmosphere model. *J. Clim.* **2009**, *22*, 3449–3469. [CrossRef]

72. Zhang, W.; Zhou, T.; Zhang, L. Wetting and greening Tibetan Plateau in early summer in recent decades. *J. Geophys. Res. Atmos.* **2017**, *122*, 11–5808. [CrossRef]

73. Yang, Y.; Wang, H.; Smith, S.; Ma, P.; Rasch, P. Source attribution of black carbon and its direct radiative forcing in China. *Atmos. Chem. Phys.* **2017**, *17*, 4319–4336. [CrossRef]

74. Gnanadesikan, A.; Scott, A.; Pradal, M.; Seviour, W.; Waugh, D. Regional Responses to Black Carbon Aerosols: The Importance of Air-Sea Interaction. *J. Geophys. Res. Atmos.* **2017**, *122*, 12–982. [CrossRef]

75. Zhuang, B.; Liu, Q.; Wang, T.; Yin, C.; Li, S.; Xie, M.; Jiang, F.; Mao, H. Investigation on semi-direct and indirect climate effects of fossil fuel black carbon aerosol over China. *Theor. Appl. Climatol.* **2013**, *114*, 651–672. [CrossRef]

76. Li, K.; Liao, H.; Mao, Y.; Ridley, D.A. Source sector and region contributions to concentration and direct radiative forcing of black carbon in China. *Atmos. Environ.* **2016**, *124*, 351–366. [CrossRef]
77. He, J.; Qi, L.; Wei, J.; Chi, Y. Reinvestigations on the East Asian Subtropical Monsoon and Tropical Monsoon. *Chin. J. Atmos. Sci.* **2007**, *31*, 1257–1265.
78. Zhang, C.; Tang, Q.; Chen, D. Recent changes in the moisture source of precipitation over the Tibetan Plateau. *J. Clim.* **2017**, *30*, 1807–1819. [CrossRef]
79. Lin, C.; Chen, D.; Yang, K.; Ou, T. Impact of model resolution on simulating the water vapor transport through the central Himalayas: Implication for models' wet bias over the Tibetan Plateau. *Clim. Dyn.* **2018**, *51*, 3195–3207. [CrossRef]
80. Kuang, X.; Jiao, J. Review on climate change on the Tibetan Plateau during the last half century. *J. Geophys. Res. Atmos.* **2016**, *121*, 3979–4007. [CrossRef]
81. Yang, S.; Xu, B.; Cao, J.; Zender, C.; Wang, M. Climate effect of black carbon aerosol in a Tibetan Plateau glacier. *Atmos. Environ.* **2015**, *111*, 71–78. [CrossRef]
82. Yang, J.; Kang, S.; Ji, Z.; Chen, D. Modeling the Origin of Anthropogenic Black Carbon and Its Climatic Effect Over the Tibetan Plateau and Surrounding Regions. *J. Geophys. Res. Atmos.* **2018**, *123*, 671–692. [CrossRef]
83. Reutter, P.; Su, H.; Trentmann, M.; Simmel, D.; Rose, S.; Gunthe, H.; Wernli, M.; Andreae, M.O.; Pöschl, U. Aerosol-and updraft-limited regimes of cloud droplet formation: Influence of particle number, size and hygroscopicity on the activation of cloud condensation nuclei (CCN). *Atmos. Chem. Phys.* **2009**, *9*, 7067–7080. [CrossRef]

Article

Overview of the New Version 3 NASA Micro-Pulse Lidar Network (MPLNET) Automatic Precipitation Detection Algorithm

Simone Lolli [1,2], Gemine Vivone [3,*], Jasper R. Lewis [4], Michaël Sicard [5,6], Ellsworth J. Welton [7], James R. Campbell [8], Adolfo Comerón [5], Leo Pio D'Adderio [9], Ali Tokay [4], Aldo Giunta [1] and Gelsomina Pappalardo [1]

[1]   CNR-IMAA, Consiglio Nazionale delle Ricerche, Contrada S. Loja snc, 85050 Tito PZ, Italy; simone.lolli@imaa.cnr.it (S.L.); aldo.giunta@imaa.cnr.it (A.G.); gelsomina.pappalardo@imaa.cnr.it (G.P.)
[2]   Science Systems and Applications-NASA GSFC, Inc., Lanham, MD 20706, USA
[3]   Deptartment of Informatic Engineering, Electric Engineering and Applied Mathematics, University of Salerno, 84084 Fisciano, Italy
[4]   JCET-UMBC, Baltimore, MD 21228, USA; jasper.r.lewis@nasa.gov (J.R.L.); ali.tokay-1@nasa.gov (A.T.)
[5]   CommSensLab, Deptartment of Signal Theory and Communications, Universitat Politècnica de Catalunya, 08034 Barcelona, Spain; msicard@tsc.upc.edu (M.S.); comeron@tsc.upc.edu (A.C.)
[6]   Ciències i Tecnologies de l'Espai, Centre de Recerca de l'Aeronàutica i de l'Espai/Institut d'Estudis Espacials de Catalunya (CTE-CRAE/IEEC), Universitat Politècnica de Catalunya, 08034 Barcelona, Spain
[7]   NASA GSFC, Code 612, Greenbelt, MD 20771, USA; ellsworth.j.welton@nasa.gov
[8]   Naval Research Laboratory, Monterey, CA 93940, USA; james.campbell@nrlmry.navy.mil
[9]   CNR-ISAC, Consiglio Nazionale delle Ricerche, Roma, Via del Fosso del Cavaliere, 100 Roma RM, Italy; leopio.dadderio@artov.isac.cnr.it
*   Correspondence: gvivone@unisa.it

Received: 31 October 2019; Accepted: 21 December 2019; Published: 24 December 2019

Abstract: Precipitation modifies atmospheric column thermodynamics through the process of evaporation and serves as a proxy for latent heat modulation. For this reason, a correct precipitation parameterization (especially for low-intensity precipitation) within global scale models is crucial. In addition to improving our modeling of the hydrological cycle, this will reduce the associated uncertainty of global climate models in correctly forecasting future scenarios, and will enable the application of mitigation strategies. In this manuscript we present a proof of concept algorithm to automatically detect precipitation from lidar measurements obtained from the National Aeronautics and Space Administration Micropulse lidar network (MPLNET). The algorithm, once tested and validated against other remote sensing instruments, will be operationally implemented into the network to deliver a near real time (latency <1.5 h) rain masking variable that will be publicly available on MPLNET website as part of the new Version 3 data products. The methodology, based on an image processing technique, detects only light precipitation events (defined by intensity and duration) such as light rain, drizzle, and virga. During heavy rain events, the lidar signal is completely extinguished after a few meters in the precipitation or it is unusable because of water accumulated on the receiver optics. Results from the algorithm, in addition to filling a gap in light rain, drizzle, and virga detection by radars, are of particular interest for the scientific community as they help to fully characterize the aerosol cycle, from emission to deposition, as precipitation is a crucial meteorological phenomenon accelerating atmospheric aerosol removal through the scavenging effect. Algorithm results will also help the understanding of long term aerosol–cloud interactions, exploiting the multi-year database from several MPLNET permanent observational sites across the globe. The algorithm is also applicable to other lidar and/or ceilometer network infrastructures in the framework of the Global Aerosol Watch (GAW) aerosol lidar observation network (GALION).

Keywords: lidar; aerosol; aerosol–cloud interactions; MPLNET; image processing; precipitation; network; infrastructure; virga

---

1. Introduction

Human life is strongly dependent on the water cycle [1]. In particular, precipitation is a key-player in pairing the Earth–atmosphere water and energy cycle, through modulating atmospheric column latent heat and affecting cloudiness and cloud lifetime. For this reason, long-term precipitation datasets are needed to analyze spatial and temporal trends and variability, especially at the global scale [2]. In the last two decades, thanks to the internet, a ground-based network of instruments has started to develop and measure important climate-related variables [3], including columnar and atmospheric profiles of aerosol optical and micro-physical properties through passive and active optical sensors (i.e., sunphotometers and lidars). Nevertheless, elastic [4] multi-wavelength and Doppler lidar observations containing raining events are usually unjustifiably disregarded in standard monitor activities, even if light rain events are clearly detectable on lidar data [5].

Some recent studies show that a correct precipitation parameterization [6] will drastically improve global climate models to forecast future scenarios that can help define the best mitigation and adaptation strategies. Further, precipitation studies are crucial to assessing aerosol indirect and semi-direct effects, since aerosols influence both cloud formation and precipitation that in turn removes aerosols from the atmosphere by scavenging effect. Isolated case studies using lidar data (together with ancillary instrumentation) to quantitatively assess the atmospheric profile of precipitation micro-physical and optical characteristics are shown in [5,7–10]. Nevertheless, these efforts, due to their intrinsic complexity, are not suitable to be operationally implemented in a network of instruments.

Several studies showing the development of aerosol [11,12] and cloud [13] masking algorithms exist, but, to our knowledge, none demonstrate automatically detecting light rain events using lidar observations. In this paper we present a proof-of-concept rain masking algorithm and report results of an intercomparison with a disdrometer to prove the efficacy of the algorithm in detecting light rain, drizzle, and virga events from lidar observations. The algorithm, once extensively tested and validated also against other remote sensing instruments (i.e., high-frequency radars) will be implemented in the National Administration and Space Agency (NASA) Micropulse lidar network (MPLNET https://mplnet.gsfc.nasa.gov/; see Section 3) and will provide, when available, a new complimentary rain mask variable that can be used either as the starting point to further investigating scientifically interesting-precipitation cases (i.e., to assess their optical and micro-physical properties, [5,7–9]) or simply to better characterize precipitation patterns and its variability at different spatial scales.

The developed algorithm, based on image processing techniques, applies morphological filters on composite plots of the Volume Depolarization Ratio (VDR) variable, as defined in [14] and this algorithm will permit MPLNET to fill the gap left by the joint NASA and Japan Aerospace Exploration Agency (JAXA) missions, as the Tropical Rainfall Measuring Mission (TRMM) followed by the Global Precipitation Measurement (GPM) [15], in detecting low intensity precipitation [5], especially at mid and higher latitudes [16]. Our paper is outlined as follows: in Section 3 we describe in detail the NASA MPLNET network and its products used as input by the rain masking algorithm. Section 4 shows the algorithm flowchart and all the different phases from input to output are carefully described. Section 5 reports the algorithm intercomparison and validation through co-located ground-based observations by disdrometer, while in Section 6 discussion and future perspectives are reported.

2. Materials and Methods

The lidar full dataset is publicly available at NASA MPLNET website (https://mplnet.gsfc.nasa.gov), while the 18 events of disdrometer data are available upon request.

3. The MicroPulse Lidar Network

The NASA MPLNET network [17], active since 1999, is a federated network of commercially–available Micropulse lidar (MPL) systems [18], produced by LEICA Geosystems, Lanham, MD, USA (formerly SigmaSpace). The instruments, active optical devices that have developed since CO_2 laser invention [19], are globally deployed to support the NASA Earth Observing System (EOS) program [20]. The MPLNET lidar network continuously monitors the atmosphere every 60 s, from the surface up to 30 km with a software-adjustable vertically-resolved spatial resolution (depending on the station, 0.030–0.075 km), under any meteorological condition and to the limit of laser signal attenuation. Both temporary and permanent observational sites are globally deployed, and are located at polar, mid-latitudes, tropical, and equatorial regions to retrieve the aerosol [21–23], cloud optical [24,25], and geometrical properties together with their radiative effects. The single-wavelength MPL lidar system, is co-located, when possible, together with the NASA Aerosol Robotic Network (AERONET; [26]) sunphotometer to reduce error in retrievals [27]. MPLNET products are freely and publicly available at the MPLNET website, which follows the modified EOS convention as: Level 1, Level 1.5, and Level 2, are all available in near real time (NRT). The only difference between L1 and L15 products are that data failing to meet the L15 Quality Assurance (QA) criteria are screened and replaced with Not a Number (NaN) in the files. The primary difference between L2 and L1/15 files is that L2 may have additional post-calibrations applied as well as corrections to instrument temperatures.

Since 2017, MPLNET has fully integrated polarized MPL systems into the network, which provides information about particle shape. Each instrument relies on the collection of two-channel measurements (i.e., the signal measure $P_{co}(z)$ and $P_{cr}(z)$). A detailed description of the depolarization channel can be found in [14]. Even a half degree tilting of the lidar instrument with respect to the vertical direction, needed to avoid cirrus cloud specular reflection, is sufficient to substantially increase the VDR of the precipitation (even for spherical raindrops), as shown in [28]. This rain enhanced contrast in lidar VDR composite images facilitates its detection. Figure 1 shows a front descent on 27 March 2018 with multiple rainfall episodes showing higher VDR values (green bins). The start of the precipitation event is highlighted by red arrows.

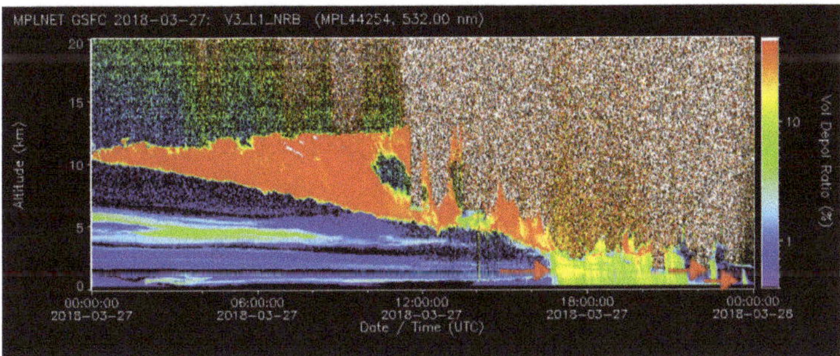

Figure 1. 27 March 2017 Micropulse lidar network (MPLNET) Version 3 Volume Depolarization Ratio (VDR) variable (L15 MPLNET Normalized Relative Backscattering (NRB) product). Red arrows highlight the starting point of the precipitation events.

The proposed algorithm, uses, as the input composite image, the new Version 3 (V3) VDR variable, paired with the cloud mask [13] variable found in the L15 Normalized Relative Backscatter (NRB) and Cloud (CLD) data products [27,29], respectively. The cloud mask is an array of integer numbers where cloudy bins are labeled as 2, non-cloudy bins as 1, while bins with an indistinguishable signal-to-noise ratio are labeled as 4. For image-based detection techniques, the L15 NRB VDR variable is preferred to the L15 NRB variable (i.e., the backscattered energy by the atmosphere) as in the VDR variable composite images the rainfall bins show a higher volume depolarization ratio. This translates into higher contrast, as shown in Figure 2b.

It is clearly visible that, with respect to the strong red depolarizing structures (VDR>35%) (e.g., clouds containing ice), the signal assumes a well-defined rectangular shape that can be identified as rainfall. In contrast, during non-rain episodes, the signal does not assume a particular shape and the VDR shows lower values.

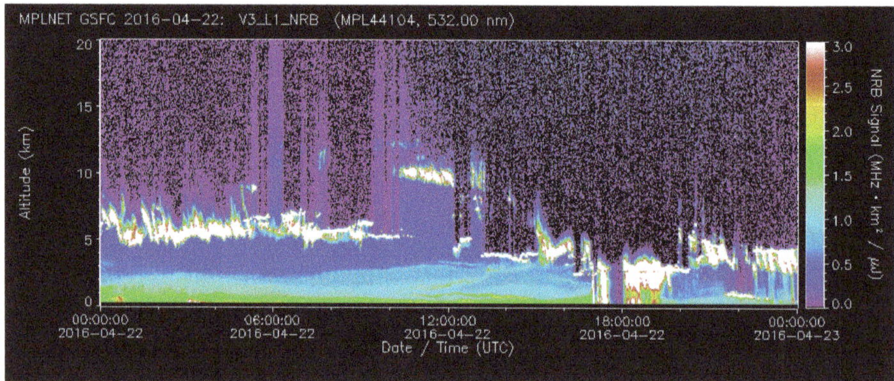

(**a**) NRB variable composite image on 22 April 2016 observed at National Administration and Space Agency (NASA) Goddard Space Flight Center permanent MPLNET observational site.

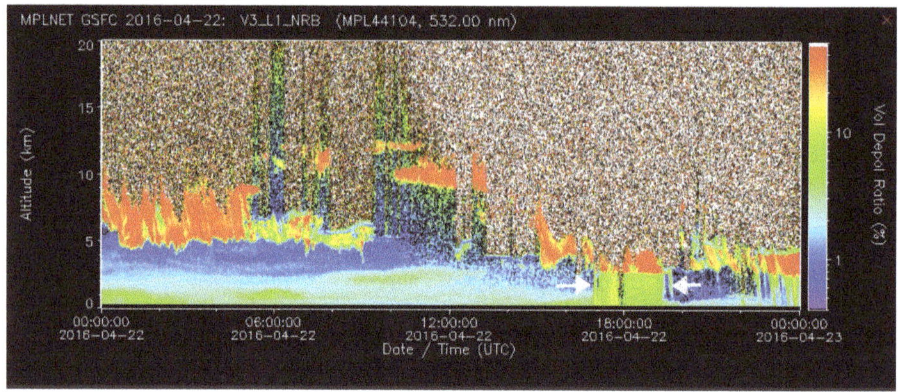

(**b**) VDR variable composite image on 22 April 2016 observed at NASA Goddard Space Flight Center permanent MPLNET observational site. The rainfall event has a different color with respect to the background aerosol (greener tone), while clouds are represented in red.

Figure 2. Precipitation event detected on 22 April 2016. With respect to (**a**) NRB, the precipitation on (**b**) VDR has more defined and sharp contours. The precipitation under the cloud (in red) has a different green tone. For this reason, the detection is easier on (**b**).

4. Version 3 Image-Based Rain Detection Algorithm

4.1. Processing Chain for Rain Detection

The proposed algorithm (flowchart shown in Figure 3) is based on image processing techniques.

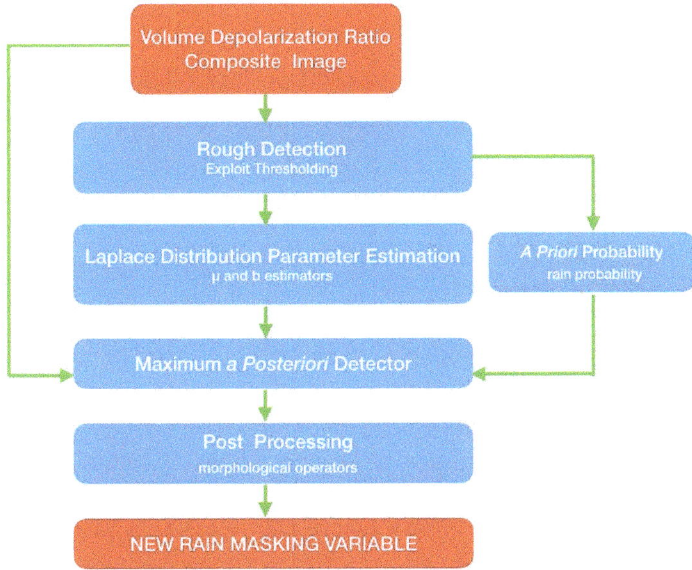

Figure 3. Flowchart of V3 MPLNET rain detection algorithm. Red boxes indicate data input/output, while blue boxes represent the processing steps. A detailed description of all the steps can be found in the text.

The algorithm processes VDR composite images from the MPLNET L15 NRB product. Then, the VDR composite plot is paired with the cloud masking variable found in L15 CLD product, as precipitation detection is uniquely carried out under the cloud base, given by the cloud mask [13] (natural rain does not exist under clear sky conditions). Thus, the proposed rain detection algorithm will analyze only the VDR bins under the clouds. Afterwards, these bins are preliminarily labeled as "rain" if, and only if, they are above a certain threshold value (0.07, see Section 4.2.1). Non-rainy bins are those either not topped by a cloud or with a VDR value below the threshold (0 < VDR < 0.07). This preliminary rain mask is used to estimate the parameters of the Laplace distributions under the hypotheses *rain* and *non-rain* and the rain a priori probability in order to use a maximum a posteriori (MAP) detector for estimating an accurate rain mask, see Section 4.2. Finally, a post-processing phase based on morphological operators is applied to reduce the image noise due to the signal extinction above the clouds and to remove any non-rectangular shaped detection, thus producing the final rain mask.

The algorithm is currently set-up to detect rain events for cloud bases at least 400 m above the surface and for rainfall episodes that last a minimum of 7 min. As a final step, all the detected rainfall events without any physical meaning are filtered out (i.e., precipitation not originating from the cloud base).

4.2. Maximum a Posterior Detector

To detect rainfall events, we use the previously described volume depolarization ratio composite image, which can be represented in a vectorial form as $\mathbf{VDR} = [VDR_1, \ldots, VDR_i, \ldots, VDR_n]$ with $VDR_i \in \mathcal{R}^+$, where n is the total number of bins. We also denote $\mathbf{c} = [c_1, \ldots, c_i, \ldots, c_n]$, as the vector of the labels in the set $\mathcal{C} = \{non\text{-}rain, rain\}$, where *rain* means that a generic VDR_i is classified as "rain", otherwise it belongs to the class "non-rain". The detection problem is formalized into the Bayesian framework. Hence, the minimum Bayesian risk is achieved by the maximum a posterior probability (MAP) rule [30], i.e., we have

$$\widehat{\mathbf{c}} = \arg\max_{\mathbf{c}} p(\mathbf{c} \mid \mathbf{VDR}) = \arg\max_{\mathbf{c}} \left[p(\mathbf{VDR} \mid \mathbf{c}) \cdot p(\mathbf{c}) \right], \tag{1}$$

where $p(\mathbf{VDR} \mid \mathbf{c})$ is the likelihood and $p(\mathbf{c})$ represents thea priori probability. The likelihood $p(\mathbf{VDR} \mid \mathbf{c})$ in Equation (1) can be simplified by the additional assumption of conditional independence among data. This yields its factorized form as follows

$$\widehat{\mathbf{c}} = \arg\max_{c_1,\dots,c_n} \left[\prod_{i=1}^{n} p(VDR_i \mid c_i) \cdot p(c_i) \right]. \tag{2}$$

Thus, given $\widehat{\mathbf{c}} = [\widehat{c}_1, \dots, \widehat{c}_i, \dots, \widehat{c}_n]$, the problem of its maximization can be simply solved by maximizing each term, i.e., for each pixel $i \in [1, \dots, n]$, we have:

$$\widehat{c}_i = \arg\max_{c_i} \left[p(VDR_i \mid c_i) \cdot p(c_i) \right]. \tag{3}$$

This leads to a binary hypothesis test that can be written as

$$\Lambda(VDR_i) = \frac{p(VDR_i \mid c_i = rain)}{p(VDR_i \mid c_i = non-rain)} \underset{non-rain}{\overset{rain}{\gtrless}} \frac{p(c_i = non-rain)}{p(c_i = rain)}, \tag{4}$$

where the likelihood ratio $\Lambda(VDR_i)$ is compared with the threshold $\frac{p(c_i=non-rain)}{p(c_i=rain)}$. If $\Lambda(VDR_i) > \frac{p(c_i=non-rain)}{p(c_i=rain)}$, VDR_i is classified as *rain*, otherwise VDR_i is associated to the class *non-rain*.

Under the assumption of Laplace distributed data for both the hypotheses (which have been experimentally validated, see Figure 4), the solution for each VDR_i is

$$\frac{|VDR_i - \mu^{nr}|}{b^{nr}} - \frac{|VDR_i - \mu^{r}|}{b^{r}} \underset{non-rain}{\overset{rain}{\gtrless}} \gamma, \tag{5}$$

where

$$\gamma = \log\left(\frac{b^{nr} \cdot p(c_i = non-rain)}{b^{r} \cdot p(c_i = rain)} \right), \tag{6}$$

log is the natural logarithm, μ^{r} and b^{r} are the location and scale parameters, respectively, which characterize the Laplace distribution under the *rain* hypothesis and μ^{nr} and b^{nr} are the related parameters for the *non-rain* hypothesis.

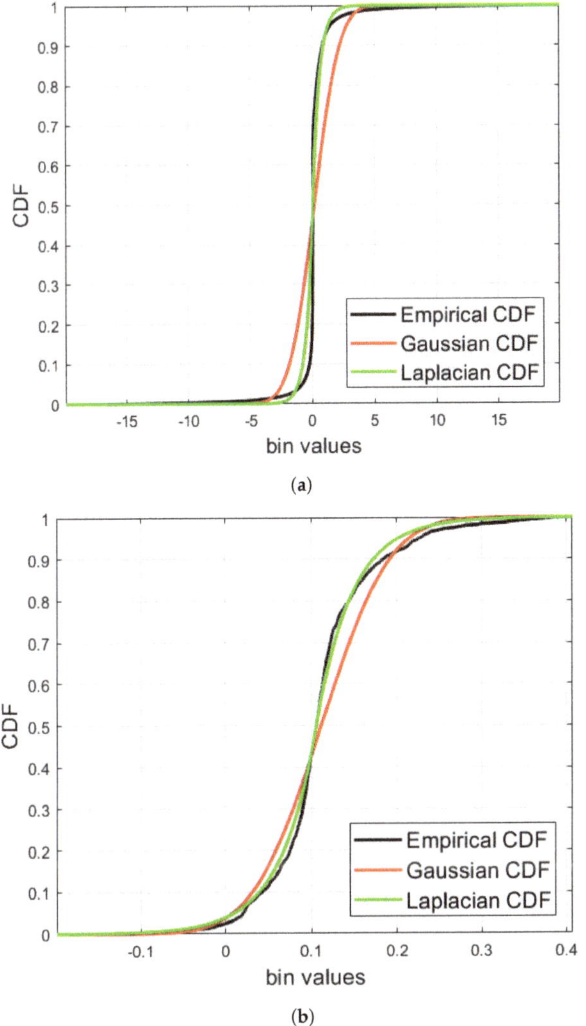

Figure 4. Cumulative distribution functions (CDFs) for (**a**) *non-rain* VDR data and (**b**) *rain* VDR data. The empirical CDFs are the estimated CDFs from the data, i.e., considering *non-rain* data in (**a**) and *rain* data in (**b**).

4.2.1. Parameter Estimation

In order to estimate all the parameters in Equation (5), we exploit the lidar image under analysis and we find a first rough detection map, i.e., roughly identifying the rain bins. Thus, we apply a thresholding of the image's bins to solve the rain detection problem. The used threshold has been experimentally set to 0.07. After this preliminary step, we use the data labeled either as *rain* or as *non-rain* to estimate the parameters for both classes. Thus, given that n^r independent and identically distributed samples belonging to a class *rain*, obtained selecting in **VDR** the *rain* bins, i.e., the ones detected as *rain* by the above-mentioned thresholding approach, we have that $\mathbf{VDR}^r = [VDR^r_1, \ldots, VDR^r_i, \ldots, VDR^r_N]$. In order to estimate the values of the a priori probability for both the classes, we count all these bins, which belong to the class *rain* in the rough detection map.

Hence, if count$(\cdot)$ is the counter operator, $p(c_i = rain) = \frac{\text{count}(\mathbf{VDR}^r)}{n}$. $p(c_i = non - rain)$ is simply equal to $1 - p(c_i = rain)$.

The other parameters to be estimated are the ones related to the Laplace distribution. Focusing on the problem of the parameter estimations for the Laplace distribution under the class *rain* (all the considerations are equivalent even in the case of the estimations of the same parameters under the hypothesis of *non-rain*), the maximum likelihood estimators (MLEs) for μ^r and b^r are:

- The MLE, $\hat{\mu}^r$, of μ^r is $\hat{\mu}^r = \text{med}(\mathbf{VDR}^r)$, where med$(\cdot)$ is the sample median operator.
- The MLE, $\hat{b}^r$, of b^r is the mean absolute deviation from the median, i.e.,

$$\hat{b}^r = \frac{1}{n^r} \sum_{i=1}^{n^r} |VDR_i^r - \hat{\mu}^r|. \tag{7}$$

4.3. Post-Processing Based on Morphological Operators

In this section we highlight and describe the post-processing step to raise the MAP detector performance presented in Section 4.2. Morphological operators basis are first described referring to their application to a generic image, **I**, see Section 4.3.1. Their use for the problem at hand is instead described in Section 4.3.2.

4.3.1. Basics of Morphological Operators

An image $\mathbf{I} : E \subseteq \mathbb{Z}^2 \rightarrow V \subseteq \mathbb{Z}$ is analyzed by the morphological operators through the so-called structuring element (SE) B [31], which can be defined through its spatial support $N_B(\mathbf{x})$, i.e., the neighborhood with respect to the position $\mathbf{x} \in E$ in which B is centered, and by its values. It is possible to characterize *flat SEs* by unitary values and the only free parameters for defining B in this case are the origin and N_B.

Erosion $\varepsilon_B[\mathbf{I}]$ and *Dilation* $\delta_B[\mathbf{I}]$ are the two basic operators defined for each point $\mathbf{x} \in \mathbf{I}$, as:

$$\varepsilon_B[\mathbf{I}](\mathbf{x}) = \bigwedge_{\mathbf{y} \in N_B(\mathbf{x})} \mathbf{I}(\mathbf{y}) \; ; \delta_B[\mathbf{I}](\mathbf{x}) = \bigvee_{\mathbf{y} \in N_B(\mathbf{x})} \mathbf{I}(\mathbf{y}), \tag{8}$$

where $\bigwedge_S$ and $\bigvee_S$ are the infimum and supremum values within the set S, respectively.

The erosion (respectively, dilation) application has a filtering effect that suppresses bright (respectively, dark) regions smaller than B and the enlargement of dark (respectively, bright) ones. For bright and dark regions we mean that the local contrast in a certain region has intensity values respectively greater or lower with respect to the surrounding ones. Erosion and dilation operators can be recast into minimum and maximum operators on B, respectively, if **I** is a binary image. We also introduce, for convenience, the *opening* and *closing* that correspond to the two possible sequential compositions of erosion and dilation:

$$\gamma_B[\mathbf{I}] = \delta_{\check{B}}[\varepsilon_B[\mathbf{I}]] \, , \phi_B[\mathbf{I}] = \varepsilon_{\check{B}}[\delta_B[\mathbf{I}]] \, , \tag{9}$$

with $\check{B}$ denoting the SE obtained by reflecting B with respect to its origin. A closing removes dark regions smaller than B while an opening suppresses bright ones. For further details about morphological operators, the interesting readers can refer to the related literature [31].

4.3.2. Use of Morphological Operators for Rain Detection Post-Processing

Two kinds of post-processing based on morphological operators are applied to the rain detection map coming from the MAP detector described in Section 4.2. In particular, given the detection map (*rain*/*non-rain*) in matrix form, i.e., $\widehat{\mathbf{C}}$, we first apply a low-pass morphological filter to remove salt and pepper noise into $\widehat{\mathbf{C}}$ to achieve a new low-pass version image denoted as $\widehat{\mathbf{C}}^{LP}$. The low-pass morphological filter can be formulated by sequentially applying a closing operator and an opening operator with the same structuring element B_1, i.e.,

$$\widehat{\mathbf{C}}^{LP} = \phi_{B_1}\left[\gamma_{B_1}\left[\widehat{\mathbf{C}}\right]\right],\tag{10}$$

where B_1 is a disk structuring element with a radius experimentally set to 4.

Last, but not least, the post-processing operation is based on an opening operator with a rectangular structuring element B_2, i.e.,

$$\widehat{\mathbf{C}}^{PP} = \gamma_{B_2}\left[\widehat{\mathbf{C}}^{LP}\right],\tag{11}$$

where $\widehat{\mathbf{C}}^{PP}$ is the final map after the morphological post-processing. This last processing is performed to delete detected objects that do not follow a rectangular shape with some constraints about the minimal size. In particular, the minimal sizes of the sides of the rectangle are tunable parameters that are selected to 7 (i.e., 7 min) that is the minimal temporal resolution to consider that is raining (i.e., we are not able to detect rains that last less than 7 min), and about 2.67 for the other side. This value takes into account of the minimal vertical dimension of the rains and coming from the fact that the data has a spatial resolution of 75 m, thus, we have a minimum vertical dimension of the rains equal to 200 m. Therefore, only these kinds of detected objects have the right spatial features to be considered rain. These parameters have been tuned for this specific problem and type of data. Obviously, considering lidar data with different features will lead to a new tuning of parameters in order to achieve high performance.

5. Results

5.1. Intercomparison with Ground-Based Disdrometer Measurements

The algorithm results are compared against rain intensity measurements obtained from a co-located disdrometer, an in-situ measurement device designed to measure the drop size distribution (DSD) [10], represented as the number of drops per unit of volume and per unit of raindrop diameter. Disdrometers can be based on different measurement principles (high-speed cameras, Doppler effect, laser-optical, impact, etc.). The second generation Parsivel (Parsivel^2) laser-optical disdrometer manufactured by OTT [32] is used in this work. Parsivel systems were originally developed by PM Tech Inc., Germany. The instrument has a laser diode (emitting wavelength of 780 nm) generating a horizontal flat beam with a measurement area of 54 cm^2.

The disdrometer principle of operation is based on laser technology. When a hydro-meteor passes through a volume uniformly illuminated by a laser beam , it produces a temporal attenuation proportional to its size with a duration depending on its fall speed. A relationship between the laser beam occlusion by the falling particle is applied to estimate the particle size. Parsivel instruments can measure particle diameters up to about 25 mm classifying them in 32 size classes of different widths. The instrument also estimates the hydro-meteor fall velocity by measuring the time necessary for the particle to pass through the laser beam, and thus stores particles in 32 × 32 matrices. The disdrometers high temporal resolution (60 s for this work) permits study in great detail of physical precipitation variability.

5.2. Rain Detection Algorithm Working under Simpler and Complex Meteorological Conditions

In this section we show how the rain masking algorithm works in different meteorological conditions, i.e., for light and stronger precipitation intensities. The comparison results are carried out at the NASA Goddard Space Flight Center (GSFC) MPLNET permanent observational site (Lat: 38.9 N, Lon: 76.3 W, Alt: 50 m a.s.l.), where measurement data from a co-located Parsivel^2 disdrometer are also available. The rain masking algorithm can also detect the so-called virga streaks (Figure 5), a kind of precipitation that, due to the strong evaporation, will never reach the ground [5]. As can be easily understood, virga episodes cannot be detected by the disdrometers.

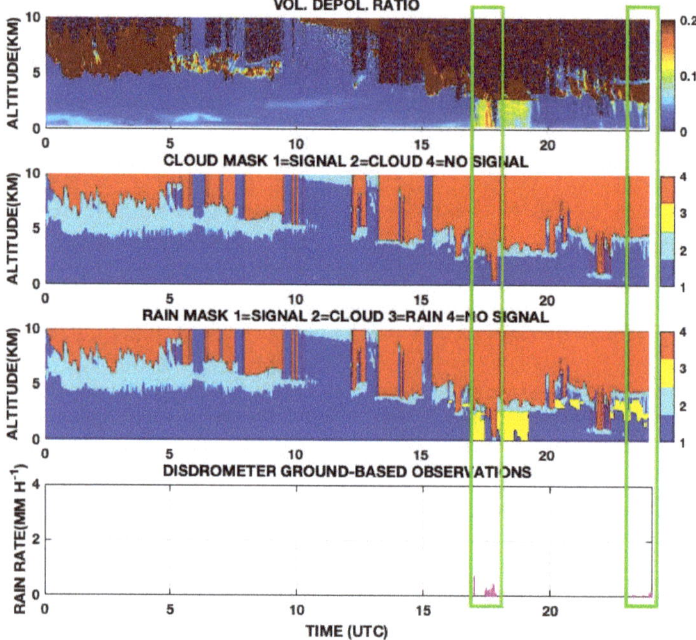

Figure 5. Rain mask variable from rain detection algorithm on 22 April 2016 from MPLNET observation at NASA GSFC (Lat: 38.99 N, Lon: 76.38 W, Alt: 50 m a.s.l.). The different steps to obtain the rain mask are shown in the flowchart in Figure 3: **Upper**: Volume depolarization ratio variable obtained from MPLNET NRB L15 signal product. **Middle 1**: Cloud mask variable from MPLNET L15 CLD product. The algorithm retrieves precipitation only on blue regions topped by clouds (cyan bins). **Middle 2**: New rain mask product (rain plotted in yellow). **Lower**: Co-located precipitation intensity measured by disdrometer. The green rectangular shapes help in visualizing the detected rain events.

5.2.1. A Simple Case: 22 April 2016

As described in Section 4.2, the rain detection algorithm first step consists in pairing VDR composite images with the L15 cloud masking variable. Then, a first guess of rain probability is produced only for the VDR signal above a certain threshold and below a cloud base, i.e., on deep blue regions topped by cyan cloud regions of upper middle plots of Figure 5. The detection in this case does not show any critical aspects. The cloud base is never below 400 m and the precipitation intensity is very low, i.e., 0.25 mm h⁻¹ on daily average. Those intensities cannot completely attenuate the lidar signal. Two virga streaks are detected by the algorithm in the second half of the day. The retrieved disdrometer rain rate (Figure 5; bottom), shows very low values, with a maximum of 0.76 mm h⁻¹ at 1701 UTC. Globally, there is a partial agreement between the disdrometer and the rain masking algorithm: After 1815 UTC rain intensity drops so much that the disdrometer is unable to detect the precipitation, while after 2200 UTC, the rain masking algorithm fails to detect the rainfall up to the ground. Precipitation events from lidar data are then necessary to fill a gap in detecting very low precipitation intensity events (<0.05 mm h⁻¹) that are crucial to study the aerosol effects and interactions on clouds and rainfall [33].

5.2.2. Intermittent Rain: 12 April 2016

The meteorological situation is more complex with respect to the previous case. In the upper composite plot (the volume depolarization ratio) of Figure 6 we can notice the transit of a front over the observational site with a progressive descent of the cloud base from 0000 UTC to about 0530 UTC. The precipitation is intermittent and at least four events are detected by the algorithm. The rainfall intensity is still low, as shown by in-situ disdrometer observations (Figure 6): the average intensity during the day is 0.9 mm h^{-1}, with a peak of 12 mm h^{-1} around 1300 UTC that lasts only less than 5 min. The precipitation event at 0600 UTC in the rain mask is not detected by the disdrometer as the precipitation intensity is below its sensitivity (0.05 mm h^{-1}) .

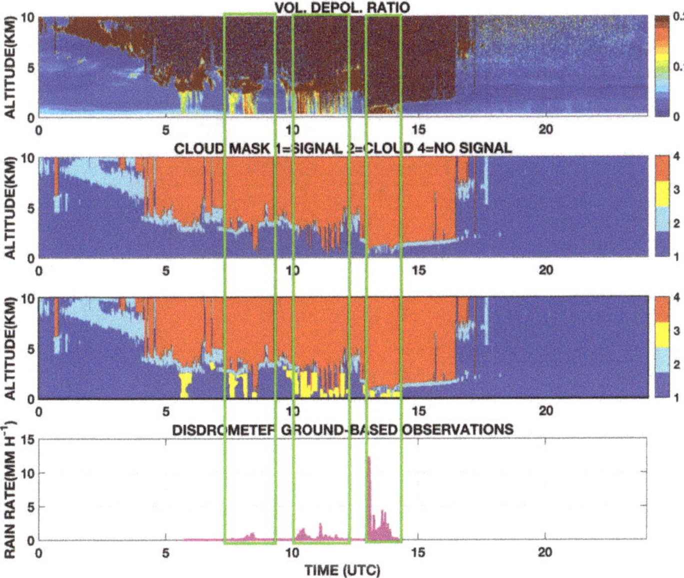

Figure 6. Similar to Figure 5, but for 12 April 2016. This case presents different short precipitation originating from the transit of a front.

5.2.3. Lighter Rain Followed by Stronger Rain: Case of 10 November 2015

During this day, the precipitation is more intense with respect to the the previous two cases, with an average intensity recorded, from 0000 UTC to 0700 UTC, of about 3.7 mm h^{-1}. Higher peaks of rain rates of about 8.8 mm h^{-1}, lasting more than 30 min, completely soak the telescope and receiver optics making it impossible to perform any further detection by the algorithm, as shown in Figure 7. The disdrometer shows that precipitation, even with a lower intensity, lasted almost the whole day. But after 0600 UTC the cloud base drops below 400 m making the detection by the algorithm impossible. This rainfall event highlights the limits of the lidar technology in detecting precipitation under very low signal to noise ratio [34]. From Figure 7 it is possible to fix a detection limit threshold depending on rain intensity and duration, i.e., rain rate > 8 mm h^{-1} lasting > 30 min.

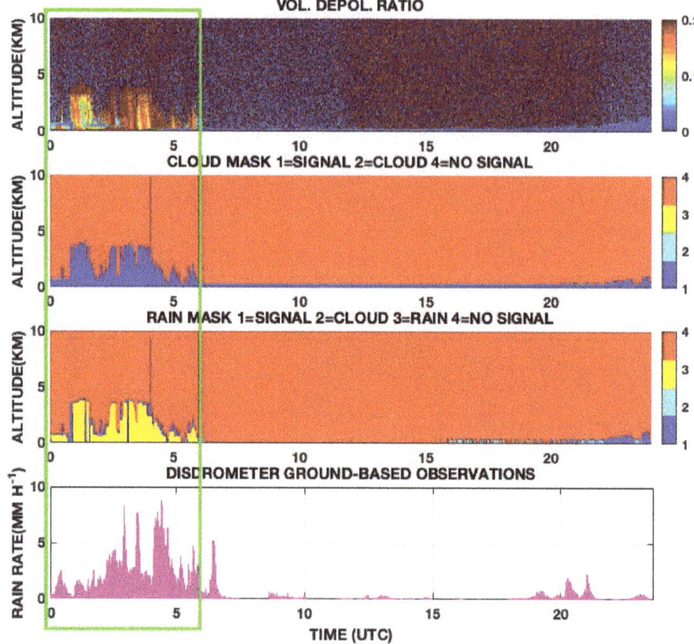

Figure 7. Same as Figure 6, but for 10 November 2015. After 0600 UTC, the signal is completely extinguished making any detection impossible.

5.3. Overall Intercomparison

The rain detection algorithm has been intercompared with ground-based disdrometer observations over 7 days between November 2015 and April 2016. The intercomparison, as stated before, was carried out at NASA Goddard Space Flight Center. Table 1 summarizes all the precipitation events, together with their lengths used, for the intercomparison.

As shown in Table 1, the algorithm performances are tested vs. the disdrometer observations. The detected precipitation events by the rain masking algorithm and the ground based disdrometers have different sensitivities, i.e., the lidar instrument can detect rainfall episodes not reaching the ground while the disdrometer can not detect intensities lower than 0.05 mm h^{-1}. On the contrary, higher intensity precipitation (at least 30 min with a rain rate > 8 mm h^{-1}, see Figure 7) can not be detected by the rain algorithm as the lidar signal is completely extinguished. It is important to stress that the rain masking algorithm has a 100% (14/14) success rate in detecting disdrometer observed precipitation. A detection is considered successful if the disdrometer observation and the algorithm detection share at least one minute in common, independently of the total precipitation duration, which can be different for the reasons previously explained.

Analysis from Table 1, shows that the algorithm is more sensitive at detecting lower intensity precipitation by 22% (4/18). If we examine the bias with respect to precipitation start and stop (see Table 1, fifth column), we found that the detection algorithm precedes the disdrometer observations by about 9 min, while the opposite is true in detection of the precipitation end, where the disdrometer lasts 5.5 min longer. This can be partially attributed to the water soaking the lidar telescope window with a consequent fully attenuated signal.

Table 1. Precipitation events observed by Parsivel[2] at NASA GSFC used to validate the rain detection algorithm. The three values reported in the fifth column are the differences in minutes between disdrometer and algorithm in detecting precipitation. The first value represents the difference in minutes with respect to the precipitation start, the second one the same but for the precipitation end, while the third represents the difference in minutes on overall precipitation duration.

Event nr.	Day	Disdrometer	Lidar	Diff. (min)
1	09 November 2015	2034–2051	2026–2042	$-8/-9/-1$
2	09 November 2015	2140–0000	2214–0000	$+34/0/-34$
3	10 November 2015	0000–0600	0000–0600	$0/0/0$
4	30 November 2015	1300–1351	1328–1347	$-28/-4/-32$
5	30 November 2015	1431–1445	1431–1459	$0/+14/+14$
6	17 December 2015	1216–1259	1212–1432	$-4/+93/+97$
7	12 April 2016	0547	0534–0605	$-13/18/+30$
8	12 April 2016	0735–0843	0730–0743	
8	12 April 2016	0735–0843	0803–0815	$-5/+3/-19$
8	12 April 2016	0735–0843	0822–0846	
9	12 April 2016	1005–1223	1000–1214	$-5/-9/-4$
10	12 April 2016	1255–1413	1246–1415	$-9/+2/+11$
11	22 April 2016	1727–1753	1659–1731	$-28/-22/+6$
12	22 April 2016	———	1809–1842	———
13	22 April 2016	———	1858–1915	———
14	22 April 2016	2315–0000	2230–0000	$-45/0/+45$
15	23 April 2016	0000–0110	0000–0101	$0/-9/-9$
16	23 April 2016	———	1423–1458	———
17	23 April 2016	1247–1320	1300–1320	$-13/0/-13$
18	23 April 2016	———	1728–1937	———

Overall, the disdrometer measured 1084 min of precipitation vs. 1095 min of detected precipitation by the algorithm. We also calculated the root mean square error (RMSE) with respect to the precipitation global duration as defined in [4], and we found that the average absolute difference in precipitation total duration detected by the algorithm and measured by the disdrometer is of about 10.23 min. Considering the different characteristics of the ground-based instrument and the detection algorithm, the agreement is within reason.

6. Discussion and Conclusions

Automated networks of instruments started to develop in the last two decades aiming to continuously monitor crucial atmospheric physical, thermodynamic, geometrical, and optical variables. Among them, the NASA MicroPulse Lidar NETwork (MPLNET), active since 1999, has globally deployed more than 21 worldwide observational sites in the tropics, mid-latitudes, and polar regions in both hemispheres to automatically retrieve 24/24 the geometrical and optical properties of aerosol and cloud atmospheric profiles under any meteorological conditions. Despite that lidar has proven to be very effective in detecting especially light precipitation and drizzle, lidar data containing precipitation episodes are currently unjustifiably disregarded. As a proof of concept, in this study we developed a rain masking algorithm, based on the volume depolarization ratio variable, which is proven to be effective in detecting light rain, drizzle, and virga episodes. Once rigorously validated and operationally implemented into the NASA MPLNET lidar network, the rain masking algorithm will consistently help in understanding how light precipitation contributes to cloud formation and will fill a gap left by TRMM and GPM missions in detecting low intensity rainfall episodes. This is crucial to improving global climate model forecasts and for aerosol–clouds and in turn, precipitation interactions. Finally, as future development, the algorithm will be also tested on simpler elastic lidar instruments without the depolarization channel, i.e., the ceilometers, to assess the rain detection feasibility. In more detail, precipitation is a fundamental meteorological phenomenon that is the principal responsible for atmospheric aerosol removal. Analyzing a large database of lidar measurements will help in fully characterizing the aerosol cycle, from emission to deposition, and validate global model observations that show a strong negative correlation between Aerosol Optical Depth (AOD) and precipitation due to wet scavenging [35]. A synergy between both passive and active satellite NASA missions, i.e., the Moderate Resolution Imager Spectrometer (MODIS; [36] and the Cloud-Aerosol Lidar and Infrared Pathfinder Satellite Observation (CALIPSO; [37]), and the ground based lidar networks, such as the European Aerosol LIdar NETwork (EARLINET; [38]) part of the Aerosols, Clouds and Trace gases Research Infrastructure (ACTRIS http://www.actris.eu) and in North America as MPLNET, and more in general in the frame of Global Aerosol Watch (GAW) aerosol lidar observation network (GALION; [39]) will strongly contribute to quantitatively assess how the above cloud aerosol load influences clouds and then rainfalls. The synergy will then assess the "all-sky" aerosol contribution to clouds and precipitation.

The image-based technique methodology used in developing the proposed algorithm, will be tested in a future work over different instruments, i.e., ceilometers, where precipitation still looks like a higher-contrasted feature in the range corrected backscattered energy.

Author Contributions: Conceptualization, S.L., G.V., J.R.L., E.J.W.; methodology, S.L., G.V., J.R.L., E.J.W., M.S., and A.C.; software, G.V.; validation, S.L., G.V., J.R.L., E.J.W., M.S., A.C., and L.P.D.; formal analysis, S.L. ; investigation, S.L., J.R.L., E.J.W., M.S. and A.C.; resources, E.J.W., G.P. and A.T.; data curation, A.T., L.P.D. E.J.W. and A.G.; writing–original draft preparation, S.L.; writing–review and editing, S.L., J.R.C., A.C., M.S., L.P.D. G.P. and A.G.; visualization, G.P.; supervision, G.P.; project administration, A.C., E.J.W., G.P.; funding acquisition, S.L. and E.J.W. All authors have read and agreed to the published version of the manuscript.

Funding: This research was funded by the Italian Research Council (CNR) Short Term Mobility Program. The NASA Micro-Pulse Lidar Network is supported by the NASA Earth Observing System (S. Platnick) and Radiation Sciences Program (H. Maring).

Conflicts of Interest: The authors declare no conflict of interest.

References

1. Bosilovich, M.G.; Schubert, S.D.; Walker, G.K. Global Changes of the Water Cycle Intensity. *J. Clim.* **2005**, *18*, 1591–1608. [CrossRef]
2. Koster, R.D.; Suarez, M.J.; Heiser, M. Variance and Predictability of Precipitation at Seasonal-to-Interannual Timescales. *J. Hydrometeorol.* **2000**, *1*, 26–46. [CrossRef]
3. Lolli, S.; Di Girolamo, P. Principal component analysis approach to evaluate instrument performances in developing a cost-effective reliable instrument network for atmospheric measurements. *J. Atmos. Ocean. Technol.* **2015**, *32*, 1642–1649. [CrossRef]
4. Lolli, S.; Delaval, A.; Loth, C.; Garnier, A.; Flamant, P. 0.355-micrometer direct detection wind lidar under testing during a field campaign in consideration of ESA's ADM-Aeolus mission. *Atmos. Meas. Tech.* **2013**, *6*, 3349–3358. [CrossRef]
5. Lolli, S.; D'Adderio, L.; Campbell, J.; Sicard, M.; Welton, E.; Binci, A.; Rea, A.; Tokay, A.; Comerón, A.; Baldasano, R.B.J.M.; et al. Vertically Resolved Precipitation Intensity Retrieved through a Synergy between the Ground-Based NASA MPLNET Lidar Network Measurements, Surface Disdrometer Datasets and an Analytical Model Solution. *Remote Sens.* **2018**, *10*, 1102. [CrossRef]
6. Campbell, J.R.; Ge, C.; Wang, J.; Welton, E.J.; Bucholtz, A.; Hyer, E.J.; Reid, E.A.; Chew, B.N.; Liew, S.C.; Salinas, S.V.; et al. Applying advanced ground-based remote sensing in the Southeast Asian Maritime Continent to characterize regional proficiencies in smoke transport modeling. *J. Appl. Meteorol. Climatol.* **2016**, *55*, 3–22. [CrossRef]
7. Westbrook, C.; Hogan, R.; O'Connor, E.; Illingworth, A. Estimating drizzle drop size and precipitation rate using two-colour lidar measurements. *Atmos. Meas. Tech.* **2010**, *3*, 671–681. [CrossRef]
8. Lolli, S.; Welton, E.J.; Campbell, J.R. Evaluating light rain drop size estimates from multiwavelength micropulse lidar network profiling. *J. Atmos. Ocean. Technol.* **2013**, *30*, 2798–2807. [CrossRef]
9. Lolli, S.; Di Girolamo, P.; Demoz, B.; Li, X.; Welton, E. Rain Evaporation Rate Estimates from Dual-Wavelength Lidar Measurements and Intercomparison against a Model Analytical Solution. *J. Atmos. Ocean. Technol.* **2017**, *34*, 829–839. [CrossRef]

10. D'Adderio, L.; Porcù, F.; Tokay, A. Evolution of drop size distribution in natural rain. *Atmos. Res.* **2018**, *200*, 70–76. [CrossRef]

11. Omar, A.H.; Winker, D.M.; Vaughan, M.A.; Hu, Y.; Trepte, C.R.; Ferrare, R.A.; Lee, K.P.; Hostetler, C.A.; Kittaka, C.; Rogers, R.R.; et al. The CALIPSO automated aerosol classification and lidar ratio selection algorithm. *J. Atmos. Ocean. Technol.* **2009**, *26*, 1994–2014. [CrossRef]

12. Papagiannopoulos, N.; Mona, L.; Amodeo, A.; D'Amico, G.; Gumà Claramunt, P.; Pappalardo, G.; Alados-Arboledas, L.; Guerrero-Rascado, J.L.; Amiridis, V.; Kokkalis, P.; et al. An automatic observation-based aerosol typing method for EARLINET. *Atmos. Chem. Phys.* **2018**, *18*, 15879–15901. [CrossRef]

13. Lewis, J.R.; Campbell, J.R.; Welton, E.J.; Stewart, S.A.; Haftings, P.C. Overview of MPLNET Version 3 Cloud Detection. *J. Atmos. Ocean. Technol.* **2016**, *33*, 2113–2134. [CrossRef]

14. Flynn, C.J.; Mendoza, A.; Zheng, Y.; Mathur, S. Novel polarization-sensitive micropulse lidar measurement technique. *Opt. Express* **2007**, *15*, 2785–2790. [CrossRef]

15. Hou, A.Y.; Kakar, R.K.; Neeck, S.; Azarbarzin, A.A.; Kummerow, C.D.; Kojima, M.; Oki, R.; Nakamura, K.; Iguchi, T. The Global Precipitation Measurement Mission. *Bull. Am. Meteorol. Soc.* **2014**, *95*, 701–722. [CrossRef]

16. Kidd, C.; Joe, P. Importance, identification and measurement of light precipitation at mid-to high-latitudes. In Proceedings of the Joint EUMETSAT Meteorological Satellite Conference and 15th Satellite Meteorology and Oceanography Conference, Amsterdam, The Netherlands, 24–28 September 2007; pp. 24–28.

17. Welton, E.J.; Campbell, J.R.; Spinhirne, J.D.; Stanley Scott, V., III. Global monitoring of clouds and aerosols using a network of micropulse lidar systems. *SPIE Conf. Proc.* **2001**, *4153*, 151–158.

18. Spinhirne, J.D.; Rall, J.A.; Scott, V.S. Compact Eye Safe Lidar Systems. *Rev. Laser Eng.* **1995**, *23*, 112–118. [CrossRef]

19. Ciofini, M.; Lapucci, A.; Lolli, S. Diffractive optical components for high power laser beam sampling. *J. Opt. Pure Appl. Opt.* **2003**, *5*, 186. [CrossRef]

20. Wielicki, B.A.; Cess, R.D.; King, M.D.; Randall, D.A.; Harrison, E.F. Mission to planet Earth: Role of clouds and radiation in climate. *Bull. Am. Meteorol. Soc.* **1995**, *76*, 2125–2154. [CrossRef]

21. Pani, S.K.; Wang, S.H.; Lin, N.H.; Tsay, S.C.; Lolli, S.; Chuang, M.T.; Lee, C.T.; Chantara, S.; Yu, J.Y. Assessment of aerosol optical property and radiative effect for the layer decoupling cases over the northern South China Sea during the 7-SEAS/Dongsha Experiment. *J. Geophys. Res. Atmos.* **2016**, *121*, 4894–4906. [CrossRef]

22. Tosca, M.G.; Campbell, J.; Garay, M.; Lolli, S.; Seidel, F.C.; Marquis, J.; Kalashnikova, O. Attributing accelerated summertime warming in the southeast united states to recent reductions in aerosol burden: Indications from vertically-resolved observations. *Remote Sens.* **2017**, *9*, 674. [CrossRef]

23. Lolli, S.; Madonna, F.; Rosoldi, M.; Campbell, J.R.; Welton, E.J.; Lewis, J.R.; Gu, Y.; Pappalardo, G. Impact of varying lidar measurement and data processing techniques in evaluating cirrus cloud and aerosol direct radiative effects. *Atmos. Meas. Tech.* **2018**, *11*, 1639. [CrossRef]

24. Campbell, J.R.; Lolli, S.; Lewis, J.R.; Gu, Y.; Welton, E.J. Daytime cirrus cloud top-of-the-atmosphere radiative forcing properties at a midlatitude site and their global consequences. *J. Appl. Meteorol. Climatol.* **2016**, *55*, 1667–1679. [CrossRef]

25. Lolli, S.; Campbell, J.R.; Lewis, J.R.; Gu, Y.; Marquis, J.W.; Chew, B.N.; Liew, S.C.; Salinas, S.V.; Welton, E.J. Daytime Top-of-the-Atmosphere Cirrus Cloud Radiative Forcing Properties at Singapore. *J. Appl. Meteorol. Climatol.* **2017**, *56*, 1249–1257. [CrossRef]

26. Holben, B.; Eck, T.; Slutsker, I.; Tanré, D.; Buis, J.; Setzer, A.; Vermote, E.; Reagan, J.; Kaufman, Y.; Nakajima, T.; et al. AERONET—A Federated Instrument Network and Data Archive for Aerosol Characterization. *Remote Sens. Environ.* **1998**, *66*, 1–16. [CrossRef]

27. Welton, E.J.; Campbell, J.R. Micropulse Lidar Signals: Uncertainty Analysis. *J. Atmos. Ocean. Technol.* **2002**, *19*, 2089–2094. [CrossRef]

28. Bissonnette, L.R.; Roy, G.; Fabry, F. Range–Height Scans of Lidar Depolarization for Characterizing Properties and Phase of Clouds and Precipitation. *J. Atmos. Ocean. Technol.* **2001**, *18*, 1429–1446. [CrossRef]

29. Campbell, J.R.; Hlavka, D.L.; Welton, E.J.; Flynn, C.J.; Turner, D.D.; Spinhirne, J.D.; Scott, V.S.; Hwang, I.H. Full-Time, Eye-Safe Cloud and Aerosol Lidar Observation at Atmospheric Radiation Measurement Program Sites: Instruments and Data Processing. *J. Atmos. Ocean. Technol.* **2002**, *19*, 431–442. [CrossRef]

30. Kay, S. *Fundamentals of Statistical Signal Processing: Detection Theory*; Fundamentals of Statistical Signal Processing; PTR Prentice-Hall: Upper Saddle River, NJ, USA, 1993.

31. Soille, P. *Morphological Image Analysis: Principles and Applications*; Springer-Verlag: Berlin/Heidelberg, Germany, 2003.

32. Löffler-Mang, M.; Joss, J. An Optical Disdrometer for Measuring Size and Velocity of Hydrometeors. *J. Atmos. Ocean. Technol.* **2000**, *17*, 130–139. [CrossRef]

33. Yum, S.S.; Cha, J.W. Suppression of very low intensity precipitation in Korea. *Atmos. Res.* **2010**, *98*, 118–124. [CrossRef]

34. Alparone, L.; Selva, M.; Aiazzi, B.; Baronti, S.; Butera, F.; Chiarantini, L. Signal-dependent noise modelling and estimation of new-generation imaging spectrometers. In Proceedings of the WHISPERS 2009, Grenoble, France, 26–28 August 2009

35. Gryspeerdt, E.; Stier, P.; White, B.A.; Kipling, Z. Wet scavenging limits the detection of aerosol effects on precipitation. *Atmos. Chem. Phys.* **2015**, *15*, 7557–7570. [CrossRef]

36. Remer, L.A.; Kaufman, Y.; Tanré, D.; Mattoo, S.; Chu, D.; Martins, J.V.; Li, R.R.; Ichoku, C.; Levy, R.; Kleidman, R.; et al. The MODIS aerosol algorithm, products, and validation. *J. Atmos. Sci.* **2005**, *62*, 947–973. [CrossRef]

37. Hunt, W.H.; Winker, D.M.; Vaughan, M.A.; Powell, K.A.; Lucker, P.L.; Weimer, C. CALIPSO lidar description and performance assessment. *J. Atmos. Ocean. Technol.* **2009**, *26*, 1214–1228. [CrossRef]

38. Pappalardo, G.; Amodeo, A.; Apituley, A.; Comeron, A.; Freudenthaler, V.; Linné, H.; Ansmann, A.; Bösenberg, J.; D'Amico, G.; Mattis, I.; et al. EARLINET. *Atmos. Meas. Tech.* **2014**, *7*, 2389–2409. [CrossRef]

39. Bösenberg, J.; Hoff, R. *Plan for the Implementation of the GAW Aerosol Lidar Observation Network GALION:(Hamburg, Germany, 27–29 March 2007)*; GAW Report 178; WMO: Geneva, Switzerland, 2007.

 *remote sensing*

Article

Aerosol Optical Radiation Properties in Kunming (the Low–Latitude Plateau of China) and Their Relationship to the Monsoon Circulation Index

Haoyue Wang [1,2], Chunyang Zhang [1,3], Ke Yu [4], Xiao Tang [5], Huizheng Che [6,*], Jianchun Bian [7], Shanshan Wang [2], Bin Zhou [2], Rui Liu [8], Xiaoguang Deng [1], Xunhao Ma [1], Zhe Yang [1], Xiaohang Cao [1], Yuehua Lu [1], Yuzhu Wang [1] and Weiguo Wang [1]

[1] Department of Atmospheric Sciences, Yunnan University, Kunming 650500, China; wanghaoyue22@ynu.edu.cn (H.W.); chunyang.zhang@tianjin-air.com (C.Z.); 12018000975@mail.ynu.edu.cn (X.D.); mxh@mail.ynu.edu.cn (X.M.); yangzhe@mail.ynu.edu.cn (Z.Y.); cxh@mail.ynu.edu.cn (X.C.); lu_yuehua@mail.ynu.edu.cn (Y.L.); wangyuzhu@mail.ynu.edu.cn (Y.W.); wangwg@ynu.edu.cn (W.W.)

[2] Shanghai Key Laboratory of Atmospheric Particle Pollution and Prevention (LAP3), Department of Environmental Science and Engineering, Fudan University, Shanghai 200433, China; shanshanwang@fudan.edu.cn (S.W.); binzhou@fudan.edu.cn (B.Z.)

[3] Tianjin Airlines Limited Liability Company (Tianjin Airlines Ltd.), Tianjin 300000, China

[4] Yunnan Meteorological Information Center, Kunming 650032, China; yk@ynu.edu.cn

[5] Institute of Atmospheric Physics, Chinese Academy of Sciences, State Key Laboratory of Atmospheric Boundary Layer Physics and Atmospheric Chemistry (LAPC), Beijing 100029, China; tangxiao@mail.iap.ac.cn

[6] State Key Laboratory of Severe Weather (LASW) and 1Key Laboratory of Atmospheric Chemistry (LAC), Chinese Academy of Meteorological Sciences (CAMS), Beijing 100081, China

[7] Key Laboratory of Middle Atmosphere and Global Environment Observation, Institute of Atmospheric Physics, Chinese Academy of Sciences, Beijing 100029, China; bjc@mail.iap.ac.cn

[8] Department of Geographic Information Science, Yunnan University, Kunming 650500, China; ruil529@ynu.edu.cn

* Correspondence: chehz@cma.gov.cn; Tel.: +86-10-58993116

Received: 10 October 2019; Accepted: 2 December 2019; Published: 5 December 2019

Abstract: Based on the Langley method and the EuroSkyRad (ESR) pack retrieval scheme, we carried out the retrieval of the aerosol properties for the CE–318 sunphotometer observation data from March 2012 to February 2014 in Kunming, China, and we explored the possible mechanisms of the seasonal variations. The seasonal variation of the aerosol optical depth (AOD) was unimodal and reached a maximum in summer. The retrieval analysis of the Angstrom exponent (α) showed the aerosol types were continental, biomass burning (BB), and urban/industrial (UI); the content of the desert dust (DD) was low, and it may have contained a sea–salt (SS) aerosol due to the influence of the summer monsoon. All the aerosol particle spectra in different seasons showed a bimodal structure. The maximum and submaximal values were located near 0.2 μm and 4 μm, respectively, and the concentration of the aerosol volume was the highest in summer. In summer, aerosol particles have a strong scattering power but a weak absorption power; this pattern is the opposite in winter. The synergistic effect of the East Asian monsoon and the South Asian monsoon seasonal oscillations can have an important impact on the variation of the aerosol properties. The oscillation variation characteristic of the total vertical columnar water vapor (CWV) and the monsoon index was completely consistent. The aerosol types and sources in the Yunnan–Kweichow Plateau and the optical radiation properties were closely related to the monsoon circulation activities during different seasons and were different from other regions in China.

Keywords: Yunnan–Kweichow Plateau; low–latitude plateau monsoon climate; aerosol type and source; aerosol properties; monsoon index; seasonal variation

---

1. Introduction

The complexity of aerosol sources (primary and aged aerosol) determines the diversity of the composition types and particle sizes, and it has important effects on the weather, climate, air quality, and human health. The direct and indirect effects of aerosols are some of the most poorly–understood variables affecting atmospheric and water circulation. It is necessary to combine aerosol measurements and radiation techniques with model simulations to accurately determine the aerosol effect [1–4]. Model simulation uses atmospheric radiation transfer models, such as the Santa Barbara DISORT(Discrete Ordinates Radiative Transfer) Atmospheric Radiative Transfer (SBDART), Global Atmospheric Model (GAME), Moderate Resolution Atmospheric Transmission (MODTRAN), and RSTAR. To estimate the accuracy of the models, the modeled global, diffuse, and Sun's direct radiation at ground level have been compared with experimental measurements. The Sun–sky radiometric measurements at ground level were also needed to characterize the aerosol properties. The CE–318 sunphotometer is used internationally. Systems such as Prede POM can be used to retrieve the optical properties of aerosols from solar and sky radiance measurements [5]. Aerosols with different particle sizes have different effects on the environment, transport, and human health [6,7]. However, due to the complex composition of aerosols and the uncertainty of their spatiotemporal distribution, the physicochemical and optical properties of aerosols vary [8], which makes environmental research and assessing the radiation effects of climate change challenging [9–13]. Therefore, quantitative remote sensing analysis of aerosol properties in different geographical locations has important theoretical and practical implications.

At present, a number of international and regional aerosol ground–based observation networks have been established. For example, the Aerosol Robotic Network (AERONET), established by the National Aeronautics and Space Administration (NASA), provides the most extensive aerosol database in the world [14]. The European Skynet Radiometer Network (ESR) is a new type of network in partnership with SKY–Radiometer NETwork–SKYNET (Skynet–Asia) in Japan [15,16]. There is also the Global Atmosphere Watch Precision Filter Radiometer Network (GAWPFR NET) [17]. The Chinese Sun Hazemeter Network (CSHNET), the China Aerosol Remote Sensing Network (CARSNET) [18], and the Campaign for the Atmospheric Aerosol Research Network of China (CARE–China) were established in China [19,20]. ground–based network observation can directly provide basic data for research; it can also provide a reference for the satellite detection data and numerical simulation results and provide an observational basis for the impact of environmental, weather and climate change [14,21–25].

The sunphotometer is one of the most widely used instruments for ground–based passive telemetry to accurately characterize aerosol properties, and its retrieval algorithm is an important component. Dubovik et al. evaluated the physical quantities, errors, and information from different sources. They gave different weights to the data in the retrieval process and applied advanced numerical optimization techniques to obtain the final statistical optimal solution. Through this algorithm, they obtained parameters such as the aerosol particle volume size distribution, complex refractive index, and single scattering albedo by using ground–based observation data [26]. Olmo et al. applied the Nakajima algorithm to incorporate the randomly distributed ellipsoid approximation to retrieve the aerosol parameters [27]. The ESR.pack retrieval scheme proposed by Estellés et al. [28] was based on the Nakajima algorithm and Skyrad.pack to improve and compile the applications of the CE–318, POM, and various sunphotometers. He et al. [29] compared the retrieval results of the aerosol properties generated by Nakajima et al. [30] and Dubovik et al. [26]. Huang et al. used the sun direct radiation data to retrieve the aerosol optical depth (AOD) and Angstrom exponent as well as the single scattering albedo (SSA) and scattering phase functions from the sky scattering data [31]. At present, the internationally–used, comprehensive aerosol retrieval results come from AERONET (for business retrieval) [26] and the ESR.pack retrieval scheme that covers ESR and Skynet–Asia [28]. Estellés et al. performed a comparison of the aerosol properties derived by the ESR.pack with AERONET and obtained good results [32].

The Yunnan–Kweichow Plateau (100–111°E, 22–30°N) is located in the southeastern part of the Qinghai–Tibet Plateau in southwestern China and is a typical low–latitude plateau monsoon climate zone. Its geographical location (it is adjacent to Southeast Asia and South Asia, and it is an important source of aerosols in southwest China due to the strong influence of the monsoon circulation) [33,34] and climate (the annual temperature difference is small and the daily range is large, the dry season and wet season are distinct, and the sunshine and ultraviolet radiation are strong) are unique. It is also an important path for water vapor transport of convection and advection in China. Research shows that aerosols have important effects on global and regional water vapor variation [35,36] and influence the regional climate and environment. However, there are few ground–based stations in the Yunnan–Kweichow Plateau, and there is a lack of systematic research on aerosols. Therefore, this paper explores the relationship between the seasonal and interannual variation characteristics of the East Asian monsoon and South Asian monsoon and the variation of the aerosol optical radiation properties, which were based on the combination of the aerosol properties retrieved from the CE–318 observation data from the Kunming Atmospheric Ozone Monitoring Station, No. 209 of the Global Ozone Observing System (GO3OS), and the monsoon circulation index. It is important to know and understand the aerosol variation and monsoon activities in the Yunnan–Kweichow Plateau and their impact on the environment and climate in specific areas.

2. Instruments and Data

2.1. Instruments

The Kunming atmospheric ozone monitoring station (25.03°N, 102.68°E; 1917 m above sea level (a.s.l.)) is located in the center of the city, and it is equipped with a CE–318 sunphotometer (CE318NTS8, France). It has eight channels with central wavelengths λ at 340, 380, 440, 500, 670, 870, 1020 and 1640 nm. The instrument can track the Sun for direct radiation observation, and it scans the sky for the scattered radiation of the Almucantar–azimuth angle, principal plane standard–scattering angle, and principal plane polarization–zenith angle.

The CE–318 sunphotometer can be used for atmospheric environmental monitoring and the radiometric calibration of remote sensing satellite sensors. Calibration is carried out every 6 months using the calibration facilities at the Chinese Academy of Meteorological Sciences [18,37]. We conducted sun direct radiation calibration by comparing the instrument with the master sunphotometers in Beijing. The master sunphotometers were calibrated using the Langley method at either Izaña (Spain, 28.31°N, 16.50°W; 2391.0 m a.s.l.) or Mauna Loa (HI, USA, 19.54°N, 55.58°W; 3397.0 m a.s.l.) [10]. The sky scattered radiation channel is calibrated by integrating the sphere radiation source. Tao et al. described the sphere calibration methods and protocols for CARSNET. The CARSNET sphere calibration results were compared with the original values provided by Cimel, the manufacturer; the linear interpolation method can be used to obtain the calibration coefficient for each period [37,38].

2.2. Data

We used the CE–318 observation data of the Kunming atmospheric ozone monitoring station from March 2012 to February 2014 (including eight bands of Sun direct radiation data and four bands of equal zenith angle scanning scattered radiation data; the minimum observation time point is in minutes, and there are missing measurements when it is cloudy) and the total ozone column data observed in real time by the Dobson ozone spectrophotometer (No. D003). The absorption coefficients of ozone and the water vapor at different wavelengths and temperatures were derived from the ESR.

We used ERA–Interim daily four times (00,06,12,18 UTC) over the same period of the wind field and relative humidity reanalysis data provided by the European Center for Medium–range Weather Forecasts (ECMWF). The data have a total vertical resolution of 37 layers, with a horizontal resolution of 0.25° × 0.25°. The daily surface meteorological data provided by the National Meteorological Base Station in Kunming included the hourly ground pressure, temperature, and relative humidity. The East Asian monsoon index (EAMI) and South Asian monsoon index (SAMI) were provided by the China National Climate Center.

NASA provided the water vapor data observed by the Atmospheric Infrared Detector (AIRS) carried by the Aqua satellite, which was compared with the vertical columnar water vapor (CWV) retrieved by the CE–318 observation data. The MODIS Fire Points Product Data (MCD14ML) were provided by the University of Maryland website (ftp://fuoco.geog.umd.edu), and the description and validation of the data are shown in the literature [39].

3. Retrieval Algorithms

3.1. Sun Direct Radiation

The direct solar irradiance of the ground–based measurement of a specific wavelength (λ) is based on the Beer–Bouguer–Lambert rule, which is expressed as the output voltage V on the CE–318 [40]:

$$V = V_0 R^{-2} T_g exp(-m\tau) \tag{1}$$

where V_0 is the calibration voltage constant and can be obtained by extrapolating from a series of observations to $m = 0$. $R = r_t \cdot r_m^{-1}$ is the Sun–Earth distance factor at the time of measurement, r_t is the distance between the Sun and the Earth at the time of measurement, and r_m is the mean distance between the Sun and the Earth. T_g is the gas absorption transmission rate (mainly considered ozone and water vapor). In the CE–318 channel, only the water vapor absorption band at 936 nm cannot be ignored, so T_g of the other channels is 1. $m = (cos\theta)^{-1}$ is the air mass factor, and θ is the solar zenith angle. τ is the total optical depth of the vertical atmosphere. Taking the logarithm of Equation (1) on both sides, the reciprocal of the slope of the line drawn by $\ln V + 2\ln R$ and m is τ, which is the Langley method [41,42]:

$$\tau = -\frac{1}{m}[ln(V/V_0) + 2lnR] \tag{2}$$

In the formula, $\tau = \tau_a + \tau_g + \tau_r$ is composed of aerosol scattering τ_a, gas absorption τ_g (such as ozone and water vapor), and Rayleigh scattering τ_r. Except for the 936 nm channel, τ_g of the other channels can be ignored, and τ_r can be calculated from the actual measured value of the ground pressure. Then the AOD is $\tau_a = \tau - \tau_r$.

Assuming that the aerosol particles follow the Junge volume size distribution [43], the calculation is:

$$n(r) = \frac{dN(r)}{dr} = c(z)r^{-(v+2)} \tag{3}$$

where r is the radius of the spherical particle, $N(r)$ is the total number of aerosol particles per unit area, v is the Junge parameter, and $c(z)$ is a function of height z. If λ is independent of the Junge spectrum type and the complex refractive index, Ångström generalizes the relationship between τ_a and λ [44]:

$$\tau_a(\lambda) = \beta\lambda^{-\alpha} \tag{4}$$

Here, α is the Angstrom exponent. Usually, $0 < \alpha < 4$ reflects the scale characteristic of the aerosol particle size and is inversely proportional to the particle size. When the coarse particles dominate, α tends to 0, and when the particle size is on the molecular scale, α is close to 4 [45]. β is the Angstrom turbidity coefficient, and the AOD at $\lambda = 1$ μm. When $\beta \geq 0.20$, the atmosphere is turbid; if $\beta \leq 0.10$, the atmosphere is relatively clean [46]. From Equation (4), the following can be obtained:

$$\alpha = -\frac{ln[\tau_a(\lambda_1)/\tau_a(\lambda_2)]}{ln(\lambda_1/\lambda_2)} \tag{5}$$

$$\beta = exp[ln\tau_a(\lambda) + \alpha ln\lambda] = \tau_a(\lambda)\lambda^\alpha \tag{6}$$

In the formula, if $\tau_a(\lambda_1)$ and $\tau_a(\lambda_2)$ of two wavelengths λ_1 and λ_2 are known, α and β are obtained. Thus, $\tau_a(\lambda)$ of arbitrary λ under the same conditions is calculated.

The formula of the water vapor transmission rate T_w on the channel is [47]

$$T_w = exp(-aw^b) \tag{7}$$

Here, w is the total amount of water vapor in the slant path, and a and b are constants. The Sun direct radiation response of CE–318 at the 936 nm water vapor absorption band is:

$$V = V_0 R^{-2} T_w exp(-m\tau_{ar}) \tag{8}$$

where $\tau_{ar} = \tau_a + \tau_r$, and τ_a is obtained by interpolating two channels at 870 and 1020 nm. At the same time, $w = m \cdot W_C$, and W_C is the total amount of the vertical water vapor column (CWV). By combining this with Equations (7) and (8), we obtain:

$$lnV + m\tau_{ar} = ln(V_0 R^{-2}) - am^b W_c^b \tag{9}$$

W_C (CWV) can be obtained, which is an improvement of the Langley method because it includes the influence of water vapor, making it more accurate and reliable [42].

3.2. Equal Zenith Angle Scattered Radiation

In the single channel of the radiation transmission model, the surface sun direct radiation is defined as E, and the surface scattered radiation is F [30]:

$$E = E_0 exp(-m_0\tau)$$
$$F(\theta, \phi) \equiv F(\Theta) = Em_0\Delta\Omega[\omega\tau P(\Theta) + q(\Theta)] \tag{10}$$

where E_0 is the sun direct radiation of the channel at the upper boundary of the atmosphere (unit: W·m^{-2}·μm^{-1}), and m_0 is the atmospheric optical mass. For $\theta \leq 75°$, $m_0 = (cos\theta)^{-1}$. $\Delta\Omega$ is the stereoscopic observation angle of the sunphotometer, calculated at 1.2°, ω is the SSA of the entire atmosphere [48], Θ is the scattering angle, $P(\Theta)$ is the total phase function when the scatter angle is Θ [49,50], and $q(\Theta)$ represents the contribution of multiple scattering (MS). The relationship between Θ and θ is observed in the actual equal zenith angle scan observation:

$$cos(\Theta) = cos^2\theta + sin^2\theta cos(\phi - \phi_0) \tag{11}$$

At this time, φ_0 is the solar azimuth angle, and $0 \leq \Theta \leq 2\theta_0$. We introduce $G(\Theta)$ and define it as:

$$G(\Theta) \equiv \frac{F(\Theta)}{Em_0\Delta\Omega} = \omega\tau P(\Theta) + q(\Theta) \equiv \beta(\Theta) + q(\Theta) \tag{12}$$

where $\beta(\Theta) = \omega\tau P(\Theta)$ is a single scattering equal to the total differential scattering coefficient, including Rayleigh scattering and Mie scattering. The AOD is defined as:

$$\tau_a(\lambda) = \int_{r_m}^{r_M} \pi r^2 Q_{ext}(x, \widetilde{m})n(r)dr \tag{13}$$

Here, Q_{ext} is the extinction effective factor of spherical particles, $x = (2\pi/\lambda)\cdot r$ is the Mie size parameter ($x \ll 1$ is Rayleigh scattering, $x \geq 1$ is Mie scattering), r_m and r_M are the minimum and maximum values of the aerosol particle radius, respectively, and $\widetilde{m} = m_R + im_I$ represents the complex refractive index of the aerosol. If Q_{ext} is only a scattering effective factor, the corresponding calculation result is the scattering optical depth τ_{as}, from which the formula of the SSA can be obtained:

$$\omega_a = \frac{\tau_{as}}{\tau_a}. \tag{14}$$

The differential scattering coefficient β_a of the entire layer of the atmospheric aerosol is defined as:

$$\beta_a(\Theta) = \frac{\lambda^2}{2\pi} \int_{r_m}^{r_M} [i_1(\Theta, x, \widetilde{m}) + i_2(\Theta, x, \widetilde{m})]n(r)dr \tag{15}$$

where i_1 and i_2 are the Mie scattering intensity functions, thereby defining the aerosol phase function as:

$$P_a(\Theta) = \frac{\beta_a(\Theta)}{\omega_a \tau_a} \tag{16}$$

The asymmetry factor g is the first moment of the phase function and is used to describe the relative intensity of the forward scattering:

$$g = \frac{1}{2} \int_{-1}^{+1} P(\Theta) \cos \Theta dcos\Theta \tag{17}$$

Here, g varies from -1 to 1. Usually, the Mie scattering has a peak $g > 0$ in the forward direction, and Rayleigh scattering has the same property in all directions $g = 0$. g is also an important parameter for discussing radiation transmission problems and aerosol properties.

The aerosol particle size spectrum $n(r)$ is the number of particles ($cm^{-2}\cdot cm^{-1}$) in a unit cross-section of the gas column and a unit radius interval [51]. The volume spectrum distribution $V(r)$ of the particles is defined as the volume of the aerosol ($cm^3\cdot cm^{-2}$) in a unit cross–section gas column and a unit logarithmic radius interval:

$$V(r) = \frac{4\pi}{3}r^4 n(r) \tag{18}$$

In the ESR retrieval scheme, τ_a and β_a can be summarized as:

$$\begin{aligned}
\tau_a(\lambda) &= \frac{2\pi}{\lambda} \int_{r_m}^{r_M} K_{ext}(x, \widetilde{m})v(r)d(lnr) \\
\beta_a(\Theta) &= \frac{2\pi}{\lambda} \int_{r_m}^{r_M} K(\Theta, x, \widetilde{m})v(r)d(lnr)
\end{aligned} \tag{19}$$

where K_{ext} and K are core functions:

$$K_{ext}(x) = \frac{3}{4}\frac{Q_{ext}(x)}{x} \quad K(\Theta, x, \widetilde{m}) = \frac{3}{2}\frac{i_1 + i_2}{x^3} \tag{20}$$

The values of the above two equations determine the radius of the aerosol particles, which in turn have a greater impact on the physical and optical properties of the aerosol [52,53].

3.3. Retrieval Scheme

The Sun direct radiation data are filtered by clouds, and the instrument retrieves the AOD of 8 channels (340, 380, 440, 500, 670, 87, 1020 and 1640 nm), α, β, CWV, and other properties. Smironv et al.'s [54] cloud filtering algorithm was applied to remove artifacts introduced by clouds. When retrieving $V(r)$, $P(\Theta)$, g, SSA, $\widetilde{m}$ and the other parameters using the sky scattering data observed by equal zenith angle scanning, it was necessary to eliminate cloud interference by examining the data symmetry. Holben et al.'s [14] quality control and cloud–removal scheme was applied to obtain high precision retrieval information: (1) The scattered radiation data used for retrieval had to satisfy the symmetry angle of more than 21 angles (the maximum quantity was 28); (2) the fitting error of the retrieval result was less than 5%; and (3) it needed to meet the condition of $\theta > 50°$ to perform the retrieval calculations.

4. The Characteristic Analysis of the Seasonal Variations

4.1. Optical Depth

Figure 1 shows the seasonal mean variation of the eight λ channel AOD year by year and the variation characteristics of AOD440 from March 2012 to February 2014. In Figure 1a, the AOD decreases with the increase of λ, and both have the same characteristics. The value for the AOD from June to August (JJA) in summer is greater than that of March to May (MAM) in spring, which is greater than that of autumn or September to November (SON), which is greater than that for winter or December to February (DJF). The seasonal mean is less than 0.80, but there are some differences in the seasonal mean variation year by year. This seasonal variation is related to the monsoon circulation activities (as shown in Figure A1) and environmental pollution emissions.

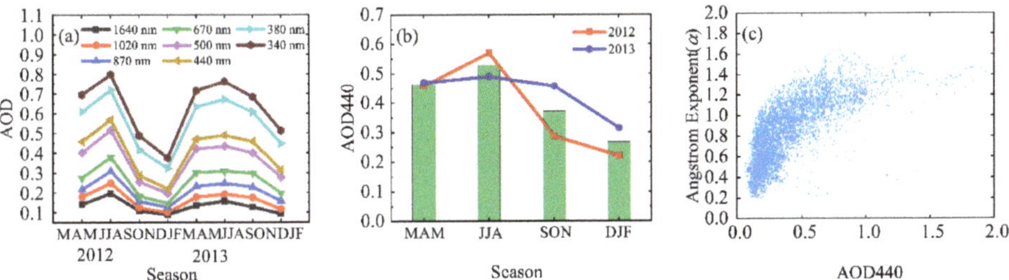

Figure 1. (a) Seasonal mean variation of the AOD in the 8 bands year by year. (b) Seasonal mean columnar distribution of AOD440 and comparison of the seasonal variations year by year. (c) Distribution of scatter data between AOD440 and the Angstrom Exponent (α).

The seasonal mean value of AOD440 from Figure 1b is between 0.2 and 0.6 and is less than 0.50, except for the summer of 2012. After analyzing the AOD of the eight channels, we found that they reach their maximum value in the summer of 2012; there are two reasons for this. First, combined with the seasonal variation of the 700 hPa level atmospheric circulation and relative humidity (RH) field in Figure A1, the summer monsoon circulation in 2012 brought more water vapor than that in 2013, and the summer RH around Kunming was 90% in 2012 and 80% in 2013. Second, it may be related to the increase of aerosol particles in the atmosphere caused by the large–scale municipal construction in the urban area of Kunming in the summer of 2012. However, we cannot rule out that the increase of aerosol particles was caused by other factors.

In Figure 1c, the value of AOD440 is concentrated between 0.10 and 1.0, and the Angstrom Exponent (α) is uniformly distributed in the range of 0.2–1.6. α increases with the increase of AOD440, indicating that it is mainly affected by fine particle aerosols generated by human activities, and the particle types and sources are different in different seasons (Table A1).

4.2. Angstrom Exponent and Turbidity Coefficient

Figure 2 shows a seasonal variation of the Angstrom Exponent (α) and frequency distributions at different intervals. The statistical results show the α value is distributed between 0.2 and 1.7. The highest frequency of 0.6–1.0 is 39.90%, and 1.4–1.8 only occurs 2.54% of the time. Figures 5 and 11 show radius r of the main control aerosol particles is more than 0.5 μm.

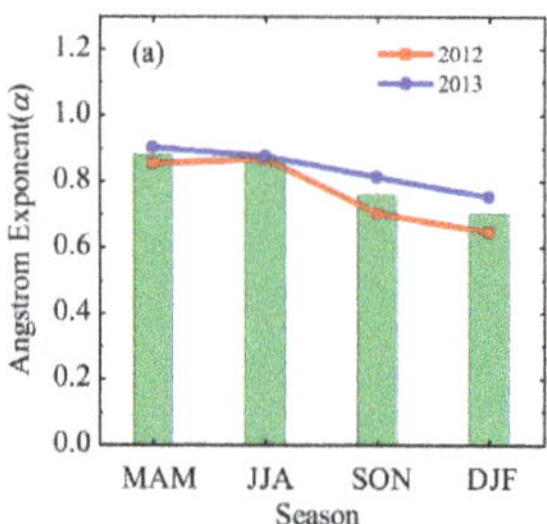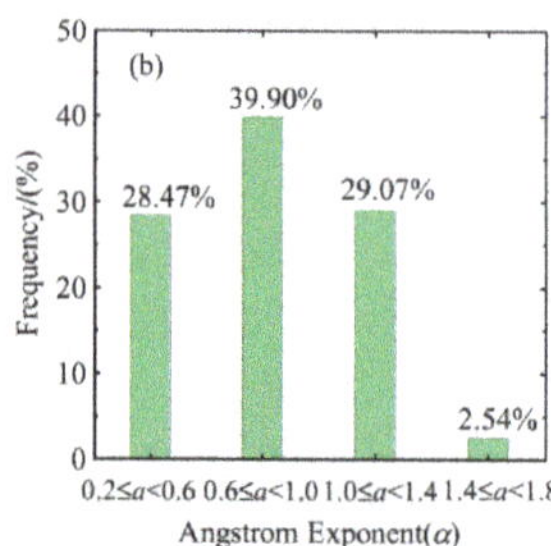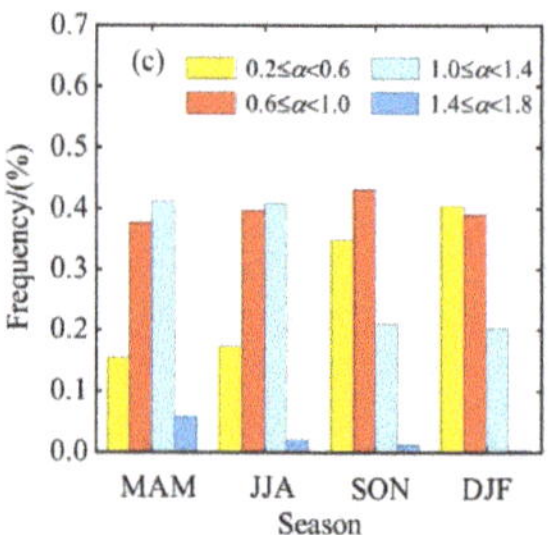

Figure 2. Retrieval results of the Angstrom Exponent (α). (**a**) Seasonal variation of α values. (**b**) The frequency distribution of α values at different intervals. (**c**) Frequency distribution of α in different seasons.

In the summer of 2012, the α value was slightly higher than that in the spring, while the autumn and winter values decreased in turn, reflecting that the mean aerosol particle size r in the spring and summer (autumn and winter) was relatively small (large). In 2013, the α value gradually decreased from spring to winter, indicating that there is a certain difference in the mean aerosol particle size r in different seasons. The grain size r is slightly larger (small) in autumn and winter (spring and summer). On the whole, the seasonal mean of the Angstrom Exponent (α) is between 0.6 and 0.9, which is slightly greater in spring and summer than in autumn and winter. It is noted that the mean value of α in the autumn and winter of 2012 decreased significantly, and the annual mean α value was slightly lower than that in 2013, mainly because the water vapor transport in 2012 was greater than that in 2013, so the hygroscopic growth effect was more significant.

For frequency statistics in different seasons, the distribution frequency of $0.6 \leq \alpha < 1.0$ varies with the season and is the most stable and highest in autumn. In spring and summer, $1.0 \leq \alpha < 1.4$ ($0.2 \leq \alpha < 0.6$) increases (decreases), while in autumn and winter, it decreases (increases), which shows the opposite frequency seasonal variation. The distribution frequency of $1.4 \leq \alpha < 1.8$ is the lowest and gradually decreases from spring to winter.

A further analysis of the data in Figure 2, Figure A2, and Figure A3 showed that the mean value of α is 0.88 in spring, and the frequency in the range of 0.6–1.4 is more than 75%. This result is mainly due to the frequent biomass burning around the Yunnan–Kweichow Plateau (Southeast Asia–South Asia) (Table A2), which produces a large amount of fine particle aerosols. It is also related to local anthropogenic emissions (industrial pollution, coal burning, motor vehicles, and other human activities). The mean value of α in summer is 0.87, and the frequency in the range of 0.6–1.4 is more than 80%; compared with the data from spring, the frequency of the small value range 0.6–1.0 increased, while that of the large value range 1.0–1.4 decreased slightly. These results are mainly because the summer monsoon circulation in East Asia and South Asia brings adequate water vapor to make fine particle aerosols grow hygroscopically, which leads to the large AOD in summer (Figure A1). The mean value of α in autumn is 0.76, and the frequency in the range of 0.6–1.4 is about 65%; the frequency in the small value range increased significantly. The mean value of α in winter is 0.70. The frequency between 0.6 and 1.4 is less than 60%, while the frequency between 0.2 and 0.6 is more than 40%, and the frequency between 1.4 and 1.8 is almost 0. In winter, Southeast Asia–South Asia also experienced more frequent biomass burning, resulting in the transport of aerosol particles to the Yunnan–Kweichow Plateau.

By comparing the values of the different intervals' α in Table A1, we found that there are main aerosols, such as UI (urban/industrial), BB (biomass burning), continental, and DD (desert dust), with some SS (sea–salt) aerosols being imported in the wet season. However, the most dominant content is that of the continental aerosol, followed by the BB and UI aerosols; the DD aerosol content is relatively low. From spring to winter, the dominant particles are coarse particles with a high frequency and fine particles with a low frequency.

Figure 3 shows the seasonal mean variation of the β coefficients for the 440 nm and 870 nm bands. The β values at the 2 channels differ little and vary substantially. The comparison shows that the β value is consistent with the seasonal variation of the AOD in Figure 1. The seasonal mean is between 0.10 and 0.30, which is the largest in summer, followed by the values for spring, autumn, and winter. The mean value of β in the summer of 2012 is the largest, indicating that the degree of atmospheric turbidity was the highest.

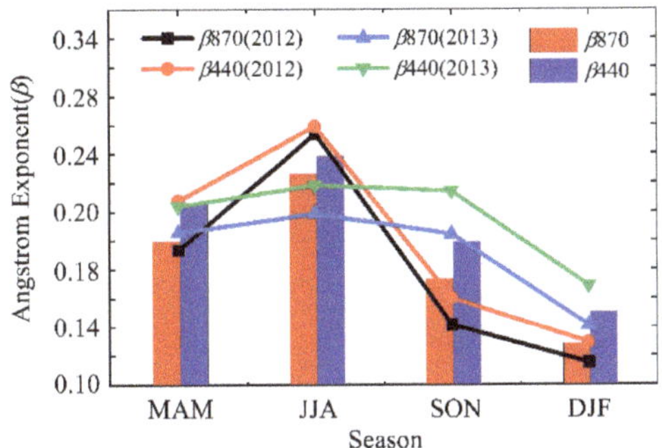

Figure 3. Seasonal mean variations in the Angstrom turbidity coefficient β at 440 nm and 870 nm and seasonal mean variations year by year.

Table A3 shows the division of different β values and atmospheric turbidity. Figure 3 and Table A3 illustrate that there is little atmospheric turbidity. The degree of atmospheric turbidity in spring and summer is significantly higher than that in autumn and winter; the β value in winter is slightly higher than 0.1, and the atmosphere is the cleanest.

4.3. Total Column Water Vapor

Figure 4 shows a comparison of the retrieval results of the CWV and observations with AIRS. The seasonal variation of the CWV is obvious, with an annual mean of about 1.0 g·cm^{-2} that reaches the maximum of more than 3.0 g·cm^{-2} in summer. Spring and autumn are similar, and the minimum is in winter. This is consistent with the climatic characteristics of the outbreak and end of summer monsoon over the Yunnan–Kweichow Plateau, and the retrieval results are consistent with the CWV from the AIRS detection.

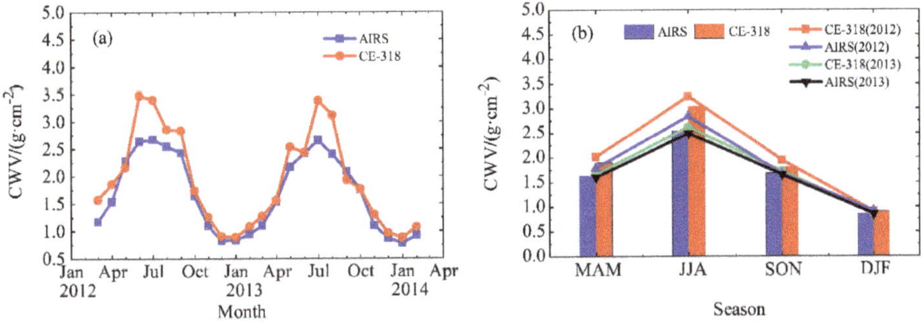

Figure 4. Retrieval results of column water vapor (CWV). (**a**) The comparison of the CWV inter-monthly variation in CE–318 and AIRS retrieval. (**b**) The comparison of CWV seasonal variation in CE–318 and AIRS retrieval.

It is not difficult to see that the seasonal variation of the AOD and CWV is similar, but slightly different. Overall, it reaches the maximum in summer and the minimum in winter, but the difference is that the AOD is significantly higher in spring than that in autumn, while the CWV has no significant size difference in spring and autumn.

4.4. Particle Spectrum Distribution and Complex Refractive Index

Figure 5 shows the mean annual aerosol particle spectrum distribution in different seasons. The distribution of the particle spectra varies in different seasons. However, the variation situation is basically the same, and it shows a bimodal shape. The maximum values of the fine mode and the coarse mode are located near 0.2 μm and 4 μm, respectively. The particle spectral structure is similar to the ratio of the continental and UI aerosol models in the Standard Chinese Radiation Atmosphere [55]. The seasonal variation analysis shows the volume concentration in summer, and the distribution of fine particles and coarse particles reaches the maximum. The maximum value of the fine particles is 0.07, and the maximum value of the coarse particles is 0.06. The main reason is the same as that for the seasonal variation of the AOD. In winter it is the smallest, with a maximum particle value of 0.05 and a maximum coarse particle value of 0.02. In winter, due to the influence of the dry inland winter monsoon (Figure A1), the aerosol content is the lowest, and the particle volume concentration is also the smallest.

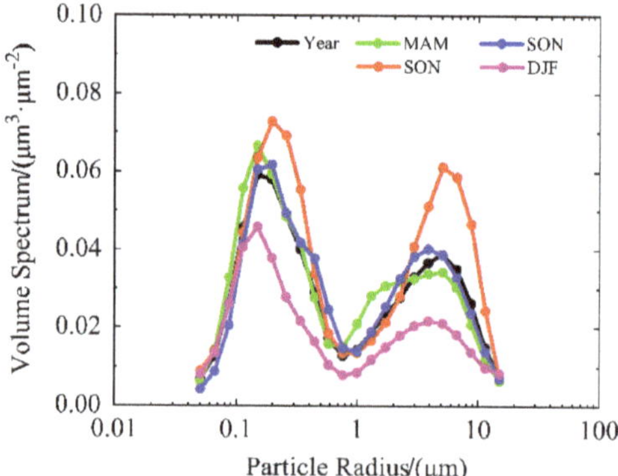

Figure 5. The mean annual aerosol particle spectrum distribution in different seasons.

In contrast, the retrieval results of the aerosol volume spectrum (the fine mode of 0.1–0.2 μm has a maximum value) in the Beijing area [56] are slightly smaller, reflecting the relatively light pollution in Kunming. Fine particles with a radius r below 1 μm occupy the main body and are mainly particles in the Aitken nuclei mold and the accumulation mode; the components of these particles are UI and continental aerosols.

Figure 6 shows the relationship between the annual mean of the FR and FI in the complex refractive index and λ, and only the seasonal mean variation of the 440 nm complex refractive index is extracted. The variation of the FR and FI is the opposite when they vary with λ, and its absolute value increases with the increase of λ. The variation of the FR is between 1.40 and 1.50, which is more obvious than the variation of FI, which is between 0.005 and 0.015. The variations in the FR and FI are not significant in the relatively short 440–670 nm band, while the FR and FI are significantly different in the relatively longer bands. The difference in the FR for different aerosol particles is not large, and the FI values can differ by several orders of magnitude.

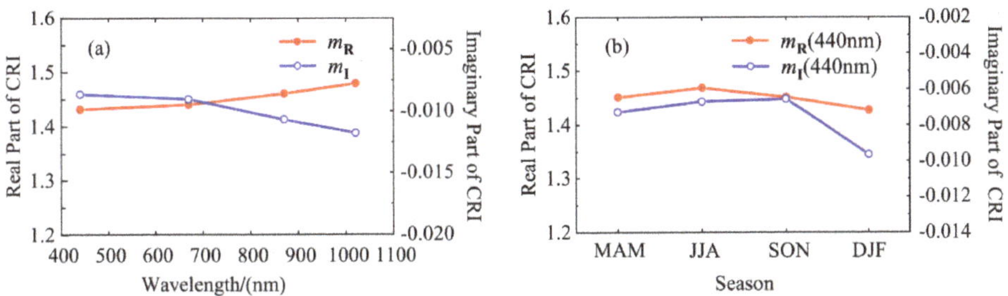

Figure 6. (**a**) Relationship between m_R and m_I of the annual average aerosol complex refractive index and wavelength (λ). (**b**) Seasonal variation of m_R and m_I of the aerosol complex refractive index at 440 nm.

The FR and FI characterize the scattering and absorption properties, respectively, of an aerosol; they offer a comprehensive reflection of the aerosol absorption properties of different components. Black carbon is the most absorbent component in aerosols, and minerals and dust are also important absorbent components [57]. Table A4 shows the statistical mean results of the values of the aerosol FR and FI for different components. Figure 6 illustrates that the content of water vapor and ammonium sulfate in the aerosol is relatively high; it also contains a very low amount of dust and the black carbon aerosol.

Studies [58] have shown that the value of the FR in urban/industrial areas is between 1.4 and 1.47, and if the area affected by the ocean is large, the value of the FR will be low. From the seasonal variation of the FR in Figure 6, it can be determined that Kunming is an urban/industrial area. The seasonal mean variation of the FI is below 0.01; it is the closest to 0.01 in winter, which is relatively large, while it is relatively small in summer, and thus the absorption aerosol content is low. However, there is also some uncertainty. Because the aerosol content in winter is low, the FI value is relatively large, so a more in–depth discussion of the retrieval results is needed to explain the variation of the FI. It may be worthwhile to use the refractive index to study the seasonal cycles of the aerosol types.

4.5. Single Scattering Albedo and Asymmetry Factor

Figure 7 shows the seasonal mean variation of the SSA in the 440 nm band and the g in the 4 bands. In Figure 7a, the seasonal variation of the SSA is a unimodal type; in summer (winter), it reaches a maximum (minimum) of about 0.96 (0.90). The water vapor content is high in summer, and the particle size and volume increase after the hygroscopic aerosol absorbs water, which enhances the scattering. The summer monsoon brings abundant water vapor and a little SS aerosol, and the non–absorbing sulfate aerosol enhances the scattering ability to some extent. The amount of wind and sand near the ground in spring is relatively large, which increases the amount of the flying dust and floating dust particles; the single scattering albedo (SSA) is slightly higher than that in autumn. The SSA is smaller in winter; the main reason is that the strong absorption of the aerosol scattering ability is relatively weak, which is consistent with the analysis of the complex refractive index in Figure 6.

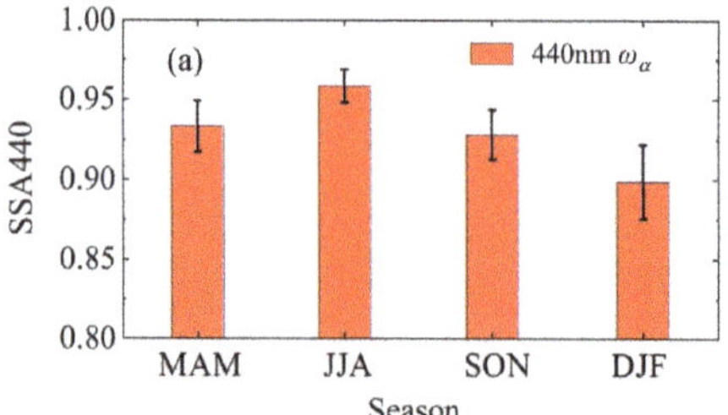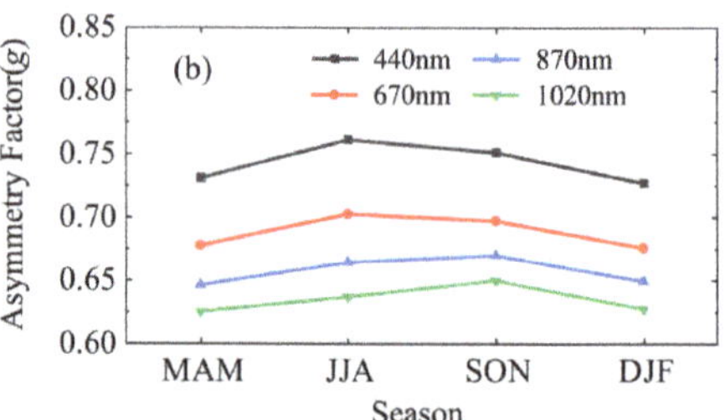

Figure 7. (**a**) Seasonal variation of the aerosol single scattering albedo (SSA) at 440 nm (the deviation line is the standard deviation). (**b**) Seasonal variation of the asymmetry factor (g) in 4 bands.

Figure 7b reflects the scattering rule of the aerosol particle $g > 0$, and it exhibits the variation that g decreases with the increase of λ. The shorter λ the stronger is aerosol forward scattering. The g value of the same λ in different seasons is not much different and is not affected by the seasonal variation. The result is reflected in that g has no obvious seasonal variation rule, but the inverse relationship between g and λ is particularly obvious.

5. Case Analysis

5.1. Retrieval of Direct Radiation

Figure 8 shows the daily variation of the AOD, Angstrom Exponent (α), and CWV from 9:00 to 18:00 (Beijing Time) on January 9, 2014. The AOD varies with λ and shows the typical Mie scattering characteristics. The Mie scattering process exists when the aerosol scale parameter $R_{mM} = 2\pi r \cdot \lambda^{-1}$ is 0.1–50 [59]. Particles with r in the range of 0.3–0.7 μm have the greatest influence on the extinction of visible light. This result shows that the shorter λ the larger AOD, and the stronger is extinction effect of the particles.

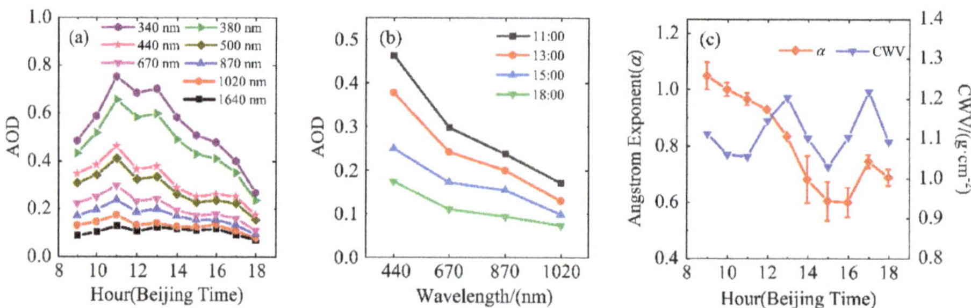

Figure 8. Daily variation of the AOD, Angstrom Exponent (α), and CWV from 9:00 to 18:00 (Beijing Time) on January 9, 2014. (**a**) The 8 wavelengths (λ) correspond to the hourly mean AOD. (**b**) The AOD at 440, 670, 870, and 1020 nm at 11, 13, 15, and 18 o'clock, respectively. (**c**) The hourly mean Angstrom Exponent (α) (the deviation line is the standard deviation) and the CWV (unit: g·cm^{-2}).

The daily variation of the hourly mean of the Angstrom Exponent (α) ranges from 0.6 to 1.2, which is significantly higher in the morning than in the afternoon; it decreases significantly in the afternoon and increases after 16:00. In the morning (afternoon), r of the aerosol's dominant particle is smaller (larger). By combining the AOD data and Table A5, we see that the AOD reaches the maximum at around 11:00, when the extinction effect of the aerosol is the strongest. At this time, the Angstrom Exponent (α) is 0.967. It is speculated that at the moment when the extinction effect is the strongest throughout the day, the dominant particle r should be around 1 μm. The CWV hourly mean is lower and varies from 1.0 g·cm^{-2} to 1.3 g·cm^{-2}, which is consistent with the climatic characteristics of the winter monsoon (dry and little rain).

5.2. Retrieval of Scattered Radiation

In order to comprehensively analyze the equal zenith angle observation data, the retrieval calculations were performed on the observation data at 10:00, 11:00, 12:00, and 13:00 (Beijing Time) on 9 January, 2014, and combined with the retrieval results of the direct radiation for analysis.

5.2.1. The Detection of Sky Radiation

Figure 9 shows the clear sky data after the equal zenith angle scan and the cloud detection; the abscissa indicates the difference angle with the solar azimuth (the positive and negative signs indicate the right–handed and left–handed scans, respectively, when the CE–318 is observing), and the ordinate indicates the response value corresponding to the amount of radiation received by the instrument when scanning through the azimuth. The data points scanned in 2 directions at different time points have good symmetry, and the data points in the curve satisfy the condition of $(A_1 - A_r) \cdot (A_1 + A_r)^{-1} \cdot 0.5$ < 10%, where Al and Ar represent the response values of the radiation received during the left–handed and right–handed scans of the instrument, respectively.

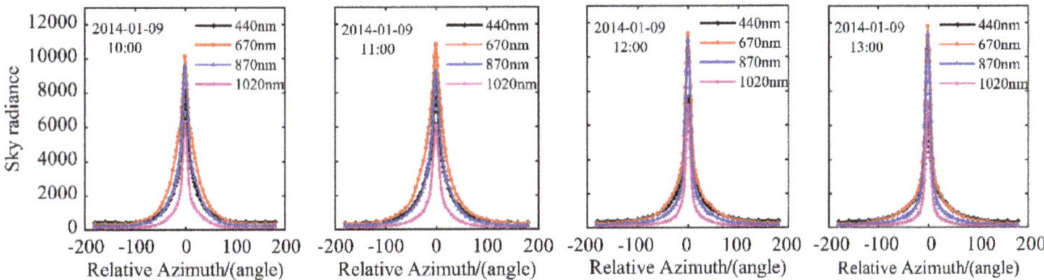

Figure 9. Data from the equal zenith angle scan observations at 4 times (Beijing Time) on 9 January, 2014.

5.2.2. SSA and Phase Function

Table 1 shows the variation of the SSA (ω_a) and g with λ at the 4 points. At 12:00, ω_a reaches a maximum value, and AOD440 is greater than 0.40. Table 1 also shows that ω_a is greater than 0.87 and shows a decrease with the increase of λ. Scattering plays a dominant role in the extinction effect of the aerosol on the radiation; λ is shorter, and the proportion of scattering is greater. By combining the CWV and Angstrom Exponent (a) data in Figure 1c, we find that the aerosol grows hygroscopically at 12:00 and 13:00, resulting in the enhanced scattering effect; therefore, the SSA reaches the maximum at 12:00. g decreases slightly with the increase of λ, but the difference at different times is small.

Table 1. Variation of the SSA (ω_a) and g with λ at 10:00, 11:00, 12:00, and 13:00 (Beijing Time) on 9 January 2014.

Wavelength (nm)	Time (Hour)							
	10:00		11:00		12:00		13:00	
	ω_a	g	ω_a	g	ω_a	g	ω_a	g
440	0.929926	0.741374	0.937808	0.74002	0.962463	0.737678	0.941437	0.740384
670	0.914820	0.696706	0.932619	0.688753	0.962108	0.692356	0.938427	0.688036
870	0.887368	0.675157	0.909843	0.665487	0.957335	0.662634	0.923731	0.661896
1020	0.871032	0.657125	0.900150	0.647351	0.954952	0.63962	0.914045	0.641120

Figure 10 shows the variation of $P(\Theta)$ with Θ for different λ. $P(\Theta)$ with different λ has a good consistency at each Θ angle and reaches the minimum near $\Theta = 120°$. The exponential curve of Θ and $P(\Theta)$ accords with the Junge model of the aerosol phase function.

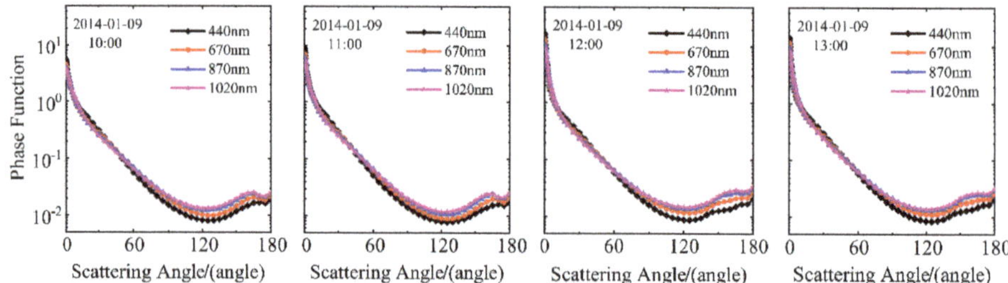

Figure 10. The variation of the scattering phase function $P(\Theta)$ with Θ at 4 times (Beijing Time) on 9 January, 2014.

5.2.3. Particle Spectrum and Turbidity

The range of particle r was determined to be 0.05–15 µm, and it contains 22 intervals in the ESR retrieval scheme with reference to the Skyrad scheme. The initial $\widetilde{m}$ of the particle is given during the retrieval process, and theoretical studies have shown that the effect on the retrieval results is small, so the empirical value of $\widetilde{m}$ is 1.500−0.005i. Figure 11 shows the relationship between the particle volume size distributions and the particle radius r. The measurement results at different times are from the bimodal spectrum and basically keep the variation in sync. The first peak mode r is located at 0.1–0.5 µm, and the second peak area r is located at 3–8 µm. For the 2 peak areas, the particle volume concentration is more concentrated on the fine particles with a smaller particle size r. The main reason is that anthropogenic aerosol particles are mostly fine particles such as those from the combustion processes, which produce a large number of submicron particles [60]. At 15 µm, the number of particles at 10:00 is greater than the number of particles at 11:00, 12:00, and 13:00; the radius of those particles can be more than 15 µm, which indicates that coarse aerosol particles may exist in the atmosphere. In addition, if the CWV content does not reach the saturation growth of the hygroscopic aerosol, it may also show an increase in the aerosol concentration of fine particles, resulting in a decrease in the effective radius of the particles.

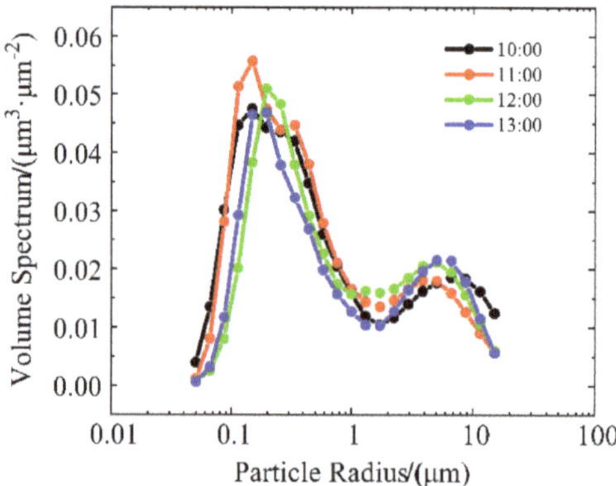

Figure 11. Aerosol particle spectrum at 4 times (Beijing Time) on 9 January, 2014.

Figure 12 shows a daily variation of the turbidity coefficient. The $\beta440$ and $\beta870$ coefficients calculated using the AOD of 440 nm and 870 nm, respectively, are very close. Since the β coefficient represents the AOD at the wavelength of 1 μm (Equation (4)), it is proved that the calculated β coefficient is highly reliable. It was found that β and the AOD maintain almost simultaneous characteristics. It is also verified by the physical meaning of β that the AOD in Figure 1a decreases with the increase of λ. Obviously, the β value can reflect both the degree of turbidity in the atmosphere and the degree of extinction of the atmosphere. Table A3 shows the division of different β values and atmospheric turbidity. In Figure 12, the daily variation range of the β value is $0.10 \leq \beta \leq 0.20$, the air quality is neither turbid nor clean, and the degree of turbidity at noon is relatively high.

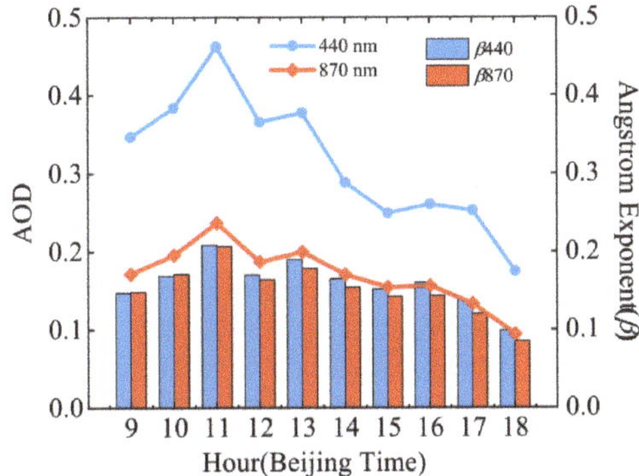

Figure 12. The Angstrom turbidity coefficient β and the daily variation of the AOD corresponding to λ at 440 nm and 870 nm at 9:00 to 18:00 on 9 January, 2014.

5.3. Atmospheric Circulation and Specific Humidity Field

Through the analysis of the weather map of the 600–700 hPa levels on 9 January, 2014 (figure omitted), it can be seen that the dry west wind and southwest wind from the Indian Peninsula-Bangladesh–Myanmar prevailed throughout the day over the Yunnan–Kweichow Plateau. Not only is it less humid, but the air is very dry. The 600 hPa in Kunming is the westerly circulation, and the atmosphere is very dry and has a specific humidity of about 2, the speed of southwest wind at 700 hPa gradually increases, and there is a significant increase at 14:00 compared with the value at 8:00; the specific humidity is significantly higher than 600 hPa but only 4–6. This shows that the dry weather with little rain in winter is also conducive to the long distance transport of aerosols from Southeast Asia–South Asia (aerosol pollutants in the atmosphere) to the southwestern area of China. However, the contribution and influence of the aerosol load in the Kunming area need to be further observed and verified by numerical simulation.

6. Analysis of the Causes of the Seasonal Variation

Kunming is located in the central part of the Yunnan–Kweichow Plateau and is affected by the East Asian monsoon and the South Asian monsoon. The annual sunshine is about 2200 h, and the ultraviolet radiation is strong. The monthly mean temperature is between 9.1 °C and 20.7 °C, and the monthly precipitation is between 11.3 mm and 204.0 mm. The precipitation in the rainy season from May to October accounts for 85% of the whole year and for more than 60% in summer. The southeast's warm and humid airflow from the western Pacific and the southwest's warm and humid airflow from the Indian Ocean meet over the Yunnan–Kweichow Plateau (which contains Yunnan), and a variety of aerosol species combine with water vapor to generate aerosol particles, which are transported over long distances (Southeast Asia–South Asia and the Iranian plateau to North Africa) [33,61], or the aerosol particles are hygroscopically grown. Under the conditions of a high temperature, high humidity, and strong ultraviolet radiation, chemical and photochemical reactions are favored to generate new aged aerosol particles. Therefore, due to the increase in the type and quantity of aerosol particles in summer (which makes the AOD reach its maximum), there is an increase in the volume concentration of the aerosol particles and the concentration of the coarse and fine particles. Diversification of the particle size (the Angström Exponent (α) is higher than the annual mean value) affects the degree of the atmospheric turbidity (the β coefficient reaches the largest value). Because the physical, chemical, and optical properties (absorption and scattering cross section) of the aerosol changed greatly, the extinction effect was enhanced. Moreover, the increase of the water vapor (the CWV reaches the maximum value) enhances the extinction of the solar radiation absorption and scattering, which leads to the enhancement of the secondary or multiple scattering of the aerosol particles and promotes the variation of the aerosol properties.

Precipitation in the dry season from November to April of the following year accounts for only 15% of the whole year. There is little rain and plenty of sunshine in winter. It is mainly influenced by the west wind circulation and cold air from the higher latitude continent. The source and the physicochemical properties of the aerosol are different from those in the summer, which leads to the characteristic parameters of the AOD, CWV, and β coefficient change with the seasons and the AOD, CWV, and β coefficient reach the minimum values in winter.

In the spring, there are few clouds and little rain, the air is dry, the evapotranspiration is strong, the solar ultraviolet radiation is enhanced, and the diurnal range of the temperature is large. The temperature drops rapidly in the autumn, which is about 2 °C lower than that in the spring, the precipitation is reduced and less than 30% of summer (but more than that in spring and winter), and the air is dry. The values of the aerosol properties in spring and autumn are between those of summer and winter, but they are slightly different in the two seasons. Comparing the spring and autumn, the aerosol properties are higher (larger) to different degrees. The main reason is that the wind speed in the near–surface layer in spring is much larger than that in the autumn, so that the flying dust and floating dust generated by the exposed surface can enter the atmosphere, which makes the amount of aerosol particles increase slightly. The increase of the desert dust aerosol in spring makes the average AOD higher than that in autumn, and the particle radius changes obviously make the turbidity stronger. Due to the influence of the summer monsoon circulation, the CWV values in spring and autumn are quite different (there is no obvious difference and regularity). The wet season starts at the end of spring and ends in autumn, but the variations of the CWV in the two seasons are slightly different. A more specific difference analysis is needed to combine the characteristics of the actual monsoon circulation evolution with different years, characteristics of water vapor transport, and variation in meteorological factors.

Figure 13 shows the characteristics of the seasonal variation and inter–annual activities of the EAMI and SAMI and aerosol properties normalized time series from March 2012 to February 2014. EAMI and SAMI are consistent with the variation in the aerosol properties, and the CWV and CWV (AIRS) are exactly the same as the seasonal variation of the monsoon index. The positive (negative) phase of the summer (winter) oscillation variation is significant, while those of spring and autumn are relatively weak, reflecting the difference in the aerosol type and source and the optical–radiation properties in different seasons. Except for the influence of other factors, it is mainly affected by the variation in the monsoon circulation activities (Figures A1 and A3). The interannual variation difference of the monsoon circulation can also be reflected in the variation of the aerosol properties. The positive phase of the EAMI and SAMI variations in the summer of 2012 is much longer than that in 2013, resulting in the difference of the variation of the aerosol properties in the summer during those two years.

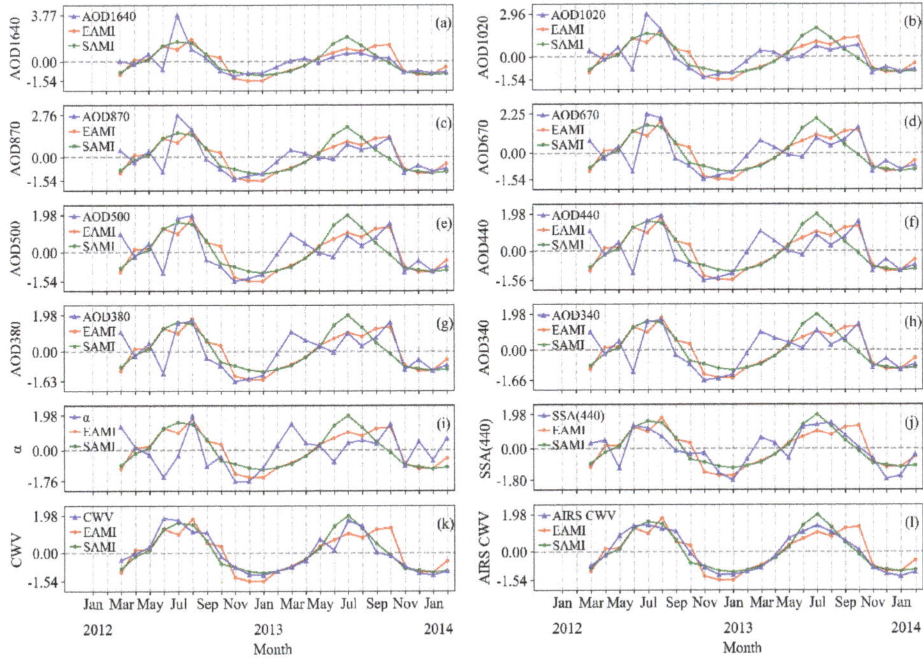

Figure 13. The characteristics of the seasonal variation and inter–annual activities of EAMI and SAMI and aerosol properties normalized time series from March 2012 to February 2014. The aerosol properties in sub-graphs are (**a**) AOD1640, (**b**) AOD1020, (**c**) AOD870, (**d**) AOD670, (**e**) AOD500, (**f**) AOD440, (**g**) AOD380, (**h**) AOD340, (**i**) Angstrom Exponent(α), (**j**) SSA(440), (**k**) CWV, (**l**) AIRS CWV.

The linear fitting regression analysis of the aerosol properties and the monsoon index normalized sequence shows that y (the response of the aerosol properties to the monsoon circulation) = kx (the variation of the monsoon circulation). Only the influence of the monsoon circulation variation is considered here. The physical meaning of the slope k is the sensitivity of the aerosol properties to the monsoon circulation variation, while $kx\times\%$ characterizes the relative influence rate of the monsoon circulation on the variation of the aerosol properties.

Figures 14 and 15 show the correlation analysis of the normalized sequence between EAMI and SAMI and aerosol properties containing CWV (AIRS CWV). The CWV and CWV (AIRS) have the best linear relationship with the monsoon index. The relationship not only reflects the transport of the water vapor from the atmospheric circulation but also verifies the reliability of the aerosol properties retrieval. The linear relationship between the AOD at 8 bands, SSA at 440 nm, CWV (AIRS CWV), EAMI, and SAMI all pass the significance test of more than 99%, but the linear relationship between the Angstrom Exponent (α) and the monsoon index is relatively insignificant.

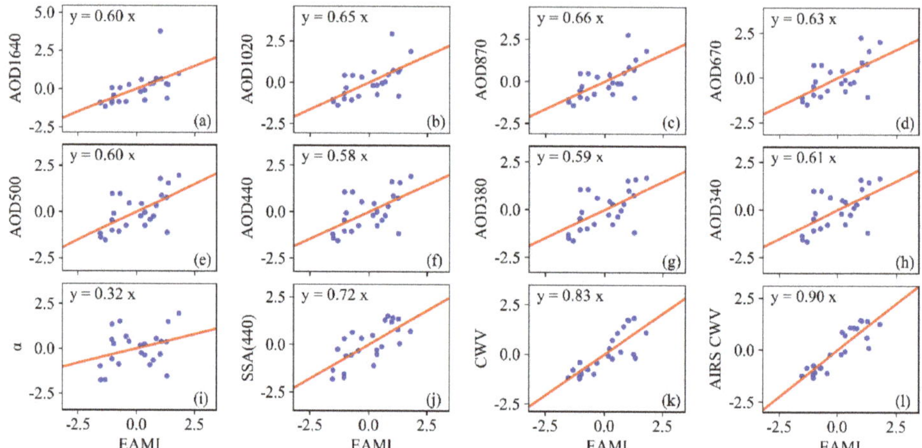

Figure 14. The correlation analysis between the standardized sequence of the EAMI and aerosol properties: (**a**) AOD1640, (**b**) AOD1020, (**c**) AOD870, (**d**) AOD670, (**e**) AOD500, (**f**) AOD440, (**g**) AOD380, (**h**) AOD340, (**i**) Angstrom Exponent(α), (**j**) SSA(440), (**k**) CWV, (**l**) AIRS CWV. Except for the Angstrom Exponent (α) (which passed 95% for the significance test), all passed the significance test with more than 99%.

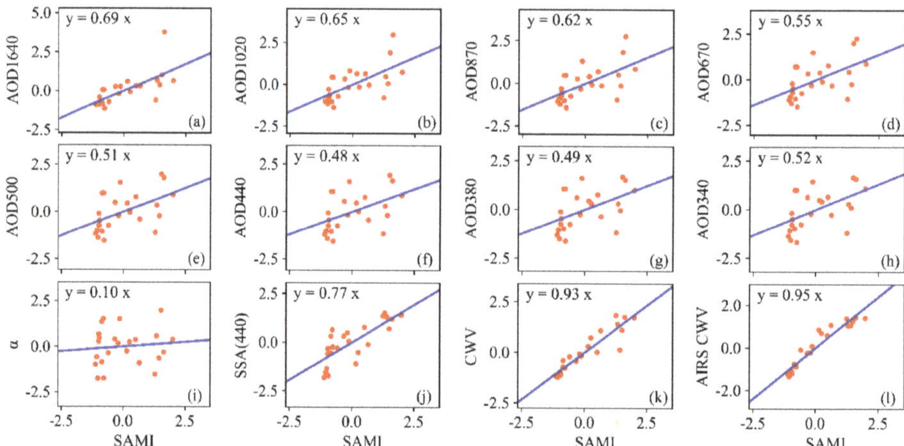

Figure 15. Correlation analysis between the normalized sequence of the SAMI and aerosol properties: (**a**) AOD1640, (**b**) AOD1020, (**c**) AOD870, (**d**) AOD670, (**e**) AOD500, (**f**) AOD440, (**g**) AOD380, (**h**) AOD340, (**i**) Angstrom Exponent(α), (**j**) SSA(440), (**k**) CWV, (**l**) AIRS CWV. The except for the Angstrom Exponent (α) (which only passed 69% of the significance test), all passed the significance test with more than 99%.

Table 2 shows the sensitivity k_E of the aerosol properties to the EAMI variation and the influence rate (%) of the variation of the East Asian monsoon circulation on the relative variation of the aerosol properties.

Table 2. Sensitivity k_E of the aerosol properties to the EAMI variation and the influence rate (%) of the East Asian monsoon circulation variation on the relative variation of the aerosol properties.

		East Asian Monsoon Relative Impact Rate %											
		Sensitivity k_E of Aerosol Properties to EAMI variation											
Year	Month	0.60	0.65	0.66	0.63	0.60	0.58	0.59	0.61	0.32	0.72	0.83	0.90
		AOD_{1640}	AOD_{1020}	AOD_{870}	AOD_{670}	AOD_{500}	AOD_{440}	AOD_{380}	AOD_{340}	α	SSA	CWV	AIRS CWV
	Mar	−62.3	−68.1	−68.2	−65.7	−62.7	−60.6	−61.6	−63.1	−33.5	−74.5	−86.1	−93.3
	Apr	9.5	10.4	10.4	10.0	9.6	9.2	9.4	9.6	5.1	11.4	13.1	14.2
	May	11.8	12.9	12.9	12.4	11.9	11.5	11.7	11.9	6.4	14.1	16.3	17.7
	Jun	76.1	83.2	83.4	80.4	76.6	74.1	75.3	77.1	41.0	91.1	105.2	114.0
2012	Jul	59.0	64.6	64.7	62.3	59.4	57.5	58.5	59.8	31.8	70.7	81.6	88.5
	Aug	107.1	117.1	117.4	113.1	107.8	104.2	106.0	108.5	57.7	128.2	148.1	160.5
	Sep	32.6	35.7	35.8	34.5	32.9	31.8	32.3	33.1	17.6	39.1	45.1	48.9
	Oct	19.9	21.7	21.8	21.0	20.0	19.3	19.7	20.1	10.7	23.8	27.5	29.8
	Nov	−79.4	−86.9	−87.0	−83.9	−80.0	−77.3	−78.6	−80.5	−42.8	−95.1	−109.8	−119.0
	Dec	−91.2	−99.7	−99.9	−96.3	−91.8	−88.7	−90.3	−92.4	−49.1	−109.1	−126.1	−136.6
	Jan	−92.2	−100.8	−101.0	−97.3	−92.8	−89.7	−91.2	−93.4	−49.7	−110.3	−127.4	−138.1
	Feb	−58.4	−63.8	−64.0	−61.6	−58.7	−56.8	−57.8	−59.1	−31.4	−69.8	−80.7	−87.4
	Mar	−42.2	−46.1	−46.2	−44.5	−42.4	−41.0	−41.7	−42.7	−22.7	−50.5	−58.3	−63.2
	Apr	−19.1	−20.9	−20.9	−20.2	−19.2	−18.6	−18.9	−19.4	−10.3	−22.9	−26.4	−28.6
	May	20.0	21.8	21.9	21.1	20.1	19.4	19.8	20.2	10.8	23.9	27.6	29.9
2013	Jun	41.3	45.2	45.2	43.6	41.6	40.2	40.9	41.8	22.2	49.4	57.1	61.9
	Jul	61.8	67.6	67.8	65.3	62.2	60.2	61.2	62.7	33.3	74.0	85.5	92.6
	Aug	48.7	53.3	53.4	51.4	49.0	47.4	48.2	49.4	26.2	58.3	67.3	73.0
	Sep	74.6	81.5	81.7	78.7	75.0	72.5	73.8	75.5	40.2	89.2	103.1	111.7
	Oct	79.4	86.9	87.1	83.9	80.0	77.3	78.6	80.5	42.8	95.1	109.8	119.0
	Nov	−44.8	−49.0	−49.1	−47.3	−45.1	−43.6	−44.3	−45.4	−24.1	−53.6	−61.9	−67.1
	Dec	−62.9	−68.8	−68.9	−66.4	−63.3	−61.2	−62.3	−63.7	−33.9	−75.3	−86.9	−94.2
2014	Jan	−62.6	−68.5	−68.6	−66.1	−63.0	−60.9	−62.0	−63.4	−33.7	−74.9	−86.5	−93.8
	Feb	−26.9	−29.4	−29.5	−28.4	−27.1	−26.2	−26.6	−27.3	−14.5	−32.2	−37.2	−40.3

During the dry season, from November to March (or April) of the following year, the aerosol characteristic parameter value decreases with the weakening of the East Asian monsoon circulation (negative phase) and reaches the minimum value in January. During the spring/summer transition from March (or April) to May, the positive phase of the East Asian monsoon circulation variation gradually increases. At this time, the variation of the aerosol characteristic parameter value under the influence of the monsoon circulation starts to change from decreasing to increasing. During the rainy season from May to October, the values of the aerosol properties increase rapidly in the period of the East Asian monsoon circulation enhancement (positive phase). It should be noted that the aerosol properties reached a maximum value in August 2012. However, the maximum value not only was delayed by two months but was relatively small. It shows that the interannual activity of the East Asian monsoon circulation is significantly different. The negative phase of the East Asian monsoon circulation variation gradually increases during the autumn–winter transition from October to November. At this time, the variation of the aerosol characteristic parameter value begins to change from increasing to decreasing due to the variation of the monsoon circulation. The transformation of the positive and negative (negative and positive) phases of the monsoon circulation affects the seasonal variation characteristics of the aerosol properties. However, the activity intensity of the monsoon circulation in different years and the difference in the transition period have different effects on different aerosol properties. Except for the CWV (AIRS CWV), the effect of the SSA on the 440 nm band is the most significant, and the effect on the AOD at different wavelengths is more consistent.

Table 3 shows the sensitivity k_S of the aerosol properties to the SAMI variation and influence rate (%) of the South Asian monsoon circulation on the relative variation of the aerosol properties. Compared with the East Asian monsoon circulation, the South Asian monsoon circulation weakens 1 month earlier. From October to April of the following year, the South Asian monsoon circulation weakens, and the aerosol characteristic parameter values decrease accordingly, also reaching a minimum value in January. During the transition period from April to May, the monsoon circulation has little effect on the variation of the aerosol properties due to the aerosol characteristic parameter values' change from decreasing to increasing.

Table 3. Sensitivity k_S of the aerosol properties to the SAMI variation and the influence rate (%) of the South Asian monsoon circulation on the relative variation of the aerosol properties.

| Year | Month | South Asian Monsoon Relative Impact Rate %
Sensitivity k_S of Aerosol Properties to EAMI Variation | | | | | | | | | | | |
		0.69 AOD_{1640}	0.65 AOD_{1020}	0.62 AOD_{870}	0.55 AOD_{670}	0.51 AOD_{500}	0.48 AOD_{440}	0.49 AOD_{380}	0.52 AOD_{340}	0.10 α	0.77 SSA	0.93 CWV	0.95 AIRS CWV
2012	Mar	−58.9	−55.4	−52.4	−47.2	−43.0	−40.7	−42.0	−44.6	−8.8	−65.9	−79.0	−81.1
	Apr	−13.7	−12.9	−12.2	−10.9	−10.0	−9.4	−9.7	−10.3	−2.0	−15.3	−18.3	−18.8
	May	9.9	9.3	8.8	7.9	7.2	6.8	7.0	7.5	1.5	11.0	13.2	13.6
	Jun	86.5	81.4	77.0	69.3	63.2	59.7	61.6	65.4	13.0	96.8	116.1	119.2
	Jul	111.1	104.6	98.9	89.0	81.1	76.7	79.1	84.1	16.6	124.3	149.1	153.1
	Aug	103.5	97.4	92.0	82.9	75.5	71.4	73.7	78.3	15.5	115.7	138.8	142.5
	Sep	43.7	41.1	38.9	35.0	31.9	30.2	31.1	33.0	6.5	48.9	58.6	60.2
	Oct	−41.0	−38.6	−36.5	−32.9	−30.0	−28.3	−29.2	−31.0	−6.1	−45.9	−55.1	−56.5
	Nov	−53.6	−50.4	−47.6	−42.9	−39.1	−37.0	−38.1	−40.5	−8.0	−59.9	−71.8	−73.8
	Dec	−69.8	−65.7	−62.1	−55.9	−51.0	−48.2	−49.7	−52.8	−10.5	−78.1	−93.7	−96.2
2013	Jan	−76.4	−71.9	−67.9	−61.2	−55.8	−52.7	−54.4	−57.8	−11.4	−85.4	−102.5	−105.2
	Feb	−67.3	−63.4	−59.9	−53.9	−49.2	−46.5	−48.0	−50.9	−10.1	−75.3	−90.3	−92.8
	Mar	−54.5	−51.3	−48.5	−43.6	−39.8	−37.6	−38.8	−41.2	−8.2	−60.9	−73.1	−75.0
	Apr	−23.2	−21.8	−20.6	−18.6	−16.9	−16.0	−16.5	−17.5	−3.5	−25.9	−31.1	−31.9
	May	16.8	15.8	15.0	13.5	12.3	11.6	12.0	12.7	2.5	18.8	22.6	23.2
	Jun	98.0	92.2	87.1	78.5	71.5	67.6	69.8	74.1	14.7	109.6	131.4	134.9
	Jul	137.3	129.2	122.1	109.9	100.2	94.8	97.8	103.8	20.6	153.6	184.2	189.1
	Aug	90.9	85.6	80.9	72.8	66.4	62.8	64.7	68.8	13.6	101.7	122.0	125.2
	Sep	33.5	31.6	29.8	26.9	24.5	23.2	23.9	25.4	5.0	37.5	45.0	46.2
	Oct	−9.3	−8.7	−8.2	−7.4	−6.8	−6.4	−6.6	−7.0	−1.4	−10.4	−12.4	−12.8
	Nov	−58.7	−55.2	−52.2	−47.0	−42.8	−40.5	−41.8	−44.4	−8.8	−65.6	−78.7	−80.8
	Dec	−65.3	−61.4	−58.1	−52.3	−47.6	−45.1	−46.5	−49.4	−9.8	−73.0	−87.6	−89.9
2014	Jan	−72.7	−68.4	−64.6	−58.2	−53.0	−50.2	−51.7	−55.0	−10.9	−81.3	−97.5	−100.1
	Feb	−66.9	−63.0	−59.5	−53.6	−48.9	−46.2	−47.7	−50.6	−10.0	−74.9	−89.8	−92.2

From May to September, the South Asian monsoon circulation increases, and the aerosol characteristic parameter values continued to increase, reaching a maximum value in August. During the transition period from September to October, the South Asian monsoon circulation begins to weaken, which changes the aerosol characteristic parameter value from increasing to decreasing and the impact of the variation of the monsoon circulation on the aerosol characteristic parameter values is also small.

By combining Tables 2 and 3, it is clear that the effects of the East Asian monsoon and the South Asian monsoon on the variation in aerosol properties are similar. Although there is a seasonal difference, the synergy between the East Asian monsoon and the South Asian monsoon is obvious, and the variation in the aerosol properties is consistent with their seasonal variations. The sensitivity of the CWV and CWV (AIRS) to the EAMI and SAMI variations is 0.83–0.90 and 0.93–0.95, respectively. The CWV (AIRS CWV) is more significantly affected by the South Asian monsoon than the East Asian monsoon, which indicates the contribution of the water vapor from the atmospheric circulation. The sensitivity of the SSA in the 440 nm band to the EAMI and SAMI is 0.72 and 0.77, while the sensitivity of the AOD for different λ to the EAMI and SAMI ranges from 0.58 to 0.66 (mean value is 0.62) and 0.48 to 0.69 (mean value is 0.56), which indicates that the extinction of the aerosol particles is deeply affected by the variation of the monsoon circulation activity. It is also noted that the sensitivity of the AOD to the EAMI is greater than that of the SAMI, and the influence of the inter–monthly variation of the EAMI on the aerosol properties is more drastic than that of the SAMI. This may be due to the complexity of the East Asian monsoon and the South Asian monsoon in the process of converging over the Yunnan–Kweichow Plateau, which needs to be studied in detail.

In summary, aerosol properties that vary with the seasons are affected and controlled by the variation in the monsoon circulation activity. For example, the AOD is closely related to the seasonal variation of the CWV. The increase of the water vapor content can affect the variation of the aerosol particle size and quantity, and the physical and chemical reactions of aerosol particles, which promotes the increase of the AOD caused by the extinction of the aerosol. In 2012 and 2013, the variation difference of the monsoon circulation can affect and control the difference in the aerosol sources and types, and the variation in the intensity of the seasonal and interannual activity may also change the number of aerosol particles and the formation difference of the aged aerosol. However, since only two years of data were used, it is necessary to verify whether it is credible through the evaluation of longer–term data. In addition, it should be noted that except for the influence of the monsoon circulation, there is an impact from the increased local emissions of fugitive dust on the aerosol properties. For example, the differences between the interannual and monthly variation in the aerosol properties in Kunming in 2012 and 2013 are not only related to the interannual differences in monsoon circulation but also related to the increase in the local emissions.

7. Conclusions

The relationship between the influence of the East Asian and South Asian monsoon index on the seasonal variation of the aerosol properties is discussed below based on the retrieval of the aerosol optical and radiation properties from the observational data of the CE–318 sunphotometer at the Kunming atmospheric ozone monitoring station. The conclusions are as follows.

The AOD decreases with the increase of λ and is consistent with the typical Mie scattering properties. The seasonal variation of the AOD, Angstrom turbidity coefficient β, and CWV is unimodal, the summer (winter) is the largest (smallest), and the spring is greater than the autumn. The CWV values in spring and autumn showed no significant difference, and the seasonal variation of the Angstrom Exponent (α) was not obvious. The turbidity in the atmosphere was relatively low. This indicates that the aerosol properties of the low–latitude plateaus in southwestern China are different from those in other parts of China. Although they are affected by anthropogenic emissions from the Southeast Asia–South Asia region and biomass burning aerosols, the correlation between the intensity of the seasonal activities of the monsoon circulation and the local anthropogenic emissions and meteorological factors will affect the aerosol properties, so further in–depth simulation studies are needed.

The seasonal and annual variations of the aerosol particle size distributions are bimodal. The complex refractive index and single scattering albedo showed that the summer aerosol particles had a stronger scattering power and a relatively weak absorption power; in winter, they exhibited the opposite. In summer, the water vapor content was high, and after the hygroscopic aerosol particles absorbed water, the particle size and volume increased, resulting in enhanced scattering. In order to explain the enhancement of the aerosol absorption capacity in winter, the retrieval result for the FI needs to be further discussed.

The variation of the aerosol properties with the season was significantly affected by the monsoon circulation activity, and the synergistic effect of the East Asian monsoon and South Asian monsoon made the variation of the aerosol properties have the same seasonal oscillation characteristics as that of the monsoon. In the transition period of the season (especially during the transition periods of the dry and wet seasons in Kunming), the monsoon circulation had little effect on the variation of the aerosol properties. The AOD was closely related to the seasonal variation of the CWV. The increase or decrease of the water vapor content did affect the variation of the aerosol particle size and quantity, thus affecting the physical and chemical reactions of the aerosol particles. Furthermore, the extinction of the aerosol was promoted to cause an increase in the AOD. Only in 2012 and 2013 did the variation difference of the monsoon circulation affect the source and type of aerosol. The variation in the intensity of the seasonal and interannual activity may have also changed the number of aerosol particles and the formation difference of the aged aerosol. It should also be noted that, in addition to the influence of the monsoon circulation, there was an impact from the increased local emissions of fugitive dust on the aerosol properties.

8. Discussion

This study only dealt with the retrieval and discussion of the aerosol properties in Kunming. Follow–up work needs to compare the aerosol properties of the retrieval between the satellite data and CE–318 observation data. We need to simulate the seasonal variation for the monsoon circulation activity variation and aerosol type and source in greater depth, as well as the direct radiative forcing effects of aerosols. This study investigated a limited time span of the observational data from the Kunming atmospheric ozone monitoring station, so the rule of seasonal change needs data from a longer period of time, and the long–term evolution trend of the interannual and interdecadal variations needs further research. Furthermore, a more in–depth statistical analysis of the relationship between the aerosol properties and meteorological elements is needed.

Author Contributions: Conceptualization, H.C., B.Z., H.W. and W.W.; Data curation, C.Z., X.T., J.B., S.W. and R.L.; Formal analysis, K.Y., Z.Y., X.C. and Y.L.; Methodology, Y.W.; Software, X.D. and X.M.; Original draft, H.W. and C.Z.

Funding: This work was supported by the "Second Tibetan Plateau Scientific Expedition and Research Program (STEP)" (grant no. 2019QZKK0604), the National Science Fund for Distinguished Young Scholars and the National Natural Science Foundation of China (grant nos. 41825011, 41807308, 21777026, 41641044, and 21477021), and the Open Project of Shanghai Key Laboratory of Atmospheric Particle Pollution and Prevention (LAP3) (grant no. FDLAP19009).

Acknowledgments: We want to thank ESR for the ESR.pack retrieval program, ECMWF for the ERA–Interim reanalysis data, the University of Maryland for the MODIS fire point product data, NASA for the AIRS water vapor observation data, the Kunming National Meteorological Base Station for the ground meteorological elements, and the China National Climate Center for the EAMI and SAMI data.

Conflicts of Interest: The authors declare no conflict of interest.

Appendix A

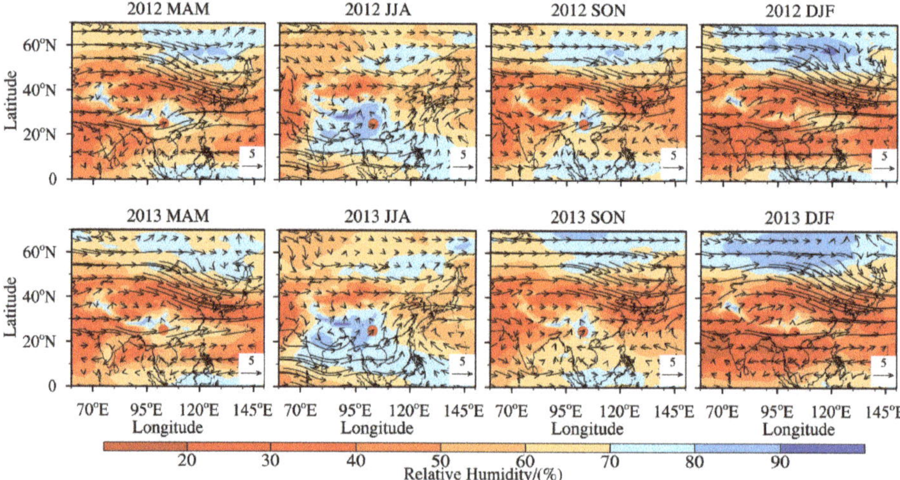

Figure A1. Seasonal variation of the 700 hPa wind field and relative humidity field from 2012 to 2013; the red point is the location of Kunming, the colored area is the Relative Humidity (%), and the reference wind vector is 5 m·s^{-1}.

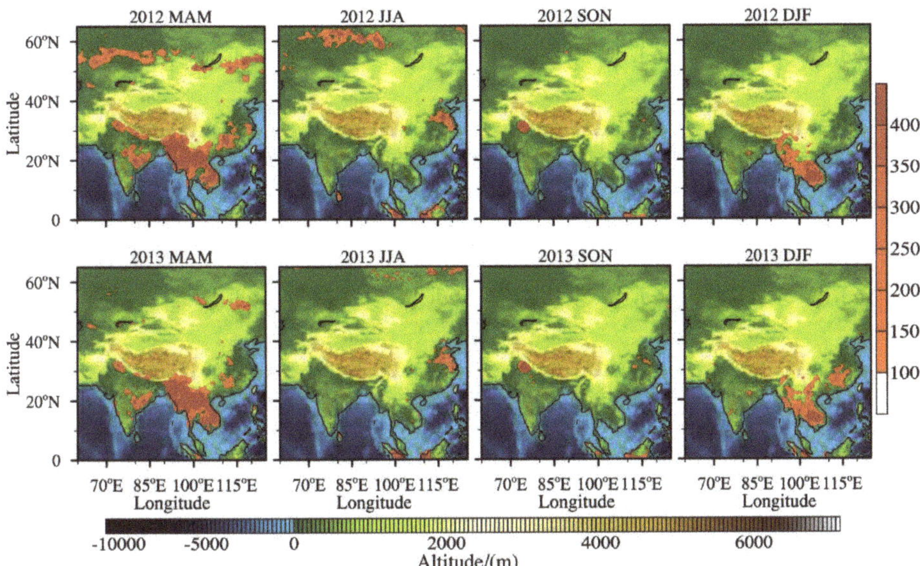

Figure A2. The emissions of the seasonal biomass burning (forest fires, straw burning, and burning) in South Asia–Southeast Asia from 2012 to 2013; the red part is the number of fire points for biomass burning. Refer to the color bar on the right side of the figure for more information.

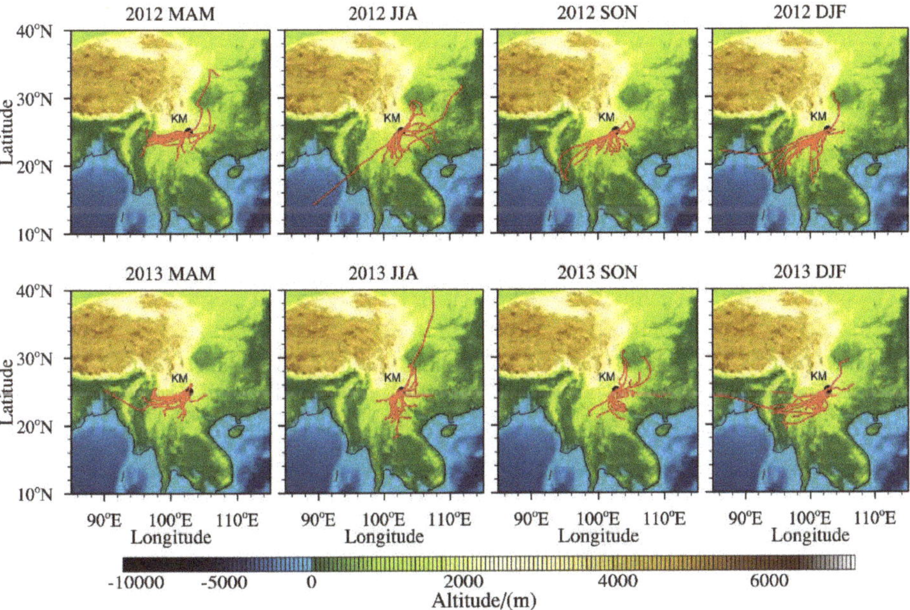

Figure A3. The 72-h backward trajectory of the Hybrid Single–Particle Lagrangian Integrated Trajectory (HYSPLIT) mode of Kunming at 750 hPa from March 2012 to February 2014.

Table A1. Characterization of the Angstrom Exponent (α) [45,62].

α	$-1 \leq \alpha \leq 0.5$	$0.5 \leq \alpha \leq 1.5$	$1.1 \leq \alpha \leq 1.8$	$1.1 \leq \alpha \leq 2.3$	$1.1 \leq \alpha \leq 2.4$
Aerosol type	DD	Continental	SS	BB	UI

Table A2. The frequency of the biomass burning fire points in different seasons in Southeast Asia and South Asia (0–43°N, 59–140°E) from 2012 to 2013.

Year	Spring (Frequency)	Summer (Frequency)	Autumn (Frequency)	Winter (Frequency)
2012	191,180	34,909	28,131	60,710
2013	170,232	36,465	32,871	68,031

Table A3. Characterization of Angstrom's Atmospheric Turbidity Coefficient β [63].

β	$\beta \geq 0.4$	$\beta \geq 0.2$	$\beta \leq 0.1$
Degree of turbidity	Unclean	Relatively clean	Clean

Table A4. Mean statistical results of the complex refractive index of aerosol in different compositions* [58,64–66].

Aerosol Component	Real (FR)				Imaginary (FI)			
	440 nm	670 nm	870 nm	1020 nm	440	670	870	1020
H_2O	1.33	1.33	1.33	1.33	—	—	—	—
Ammonium sulfate	1.53	1.53	1.53	1.53	—	—	—	—
sand and dust	1.57	1.57	1.57	1.57	0.01	0.004	0.004	0.004
Black carbon	1.95	1.95	1.95	1.95	0.66	0.66	0.66	0.66

* Combined with the spectral properties of the complex refractive index of the aerosol, it is assumed that the imaginary parts at 870 nm and 1020 nm are approximately equal to those at 670 nm.

Table A5. Statistical relationship between the Angstrom exponent (α) and mean radius r of the aerosol particles [59].

α	0	1.3	1.5	2.0	2.25	3.0	3.8~4.0
r (μm)	>2.0	0.6	0.5	0.22~0.25	0.15	0.062~0.10	≤0.02

References

1. Holben, B.; Tanré, D.; Smirnov, A.; Eck, T.; Slutsker, I.; Abuhassan, N.; Newcomb, W.; Schafer, J.; Chatenet, B.; Lavenu, F.; et al. An emerging ground–based aerosol climatology: Aerosol optical depth from AERONET. *J. Geophys. Res. Atmos.* **2001**, *106*, 12067–12097. [CrossRef]
2. Di, H.G.; Wang, Q.Y.; Hua, H.B.; Li, S.W.; Yan, Q.; Liu, J.J.; Song, Y.H.; Hua, D.X. Aerosol Microphysical Particle Parameter Inversion and Error Analysis Based on Remote Sensing Data. *Remote Sens.* **2018**, *10*, 1753. [CrossRef]
3. Stachlewska, I.S.; Samson, M.; Zawadzka, O.; Harenda, K.M.; Janicka, L.; Poczta, P.; Janicka, L.; Poczta, P.; Szczepanik, D.; Heese, B.; et al. Modification of Local Urban Aerosol Properties by Long–Range Transport of Biomass Burning Aerosol. *Remote Sens.* **2018**, *10*, 412. [CrossRef]
4. Zhang, Y.; Li, Z.Q.; Liu, Z.H.; Zhang, J.; Qie, L.L.; Xie, Y.S.; Hou, W.Z.; Wang, Y.Q.; Ye, Z.X. Retrieval of the Fine–Mode Aerosol Optical Depth over East China Using a Grouped Residual Error Sorting (GRES) Method from Multi–Angle and Polarized Satellite Data. *Remote Sens.* **2018**, *10*, 1838. [CrossRef]
5. Estellés, V.; Campanelli, M.; Expósito, F.; Utrillas, M.; Martínez–Lozano, J. *Aerosol Radiative Forcing in the European Skynet Radiometers Network*; European Geosciences Union: Munich, Germany, 2012; Available online: https://meetingorganizer.copernicus.org/EGU2012/EGU2012-10204.pdf (accessed on 30 December 2012).

6. Su, X.L.; Cao, J.J.; Li, Z.Q.; Li, K.T.; Xu, H.; Liu, S.X.; Fan, X.H. Multi–Year Analyses of Columnar Aerosol Opticaland Microphysical Properties in Xi'an, a Megacity inNorthwestern China. *Remote Sens.* **2018**, *10*, 1169. [CrossRef]

7. Li, Z.B.; Roy, D.P.; Zhang, H.K.; Vermote, E.F.; Huang, H.Y. Vermote and Haiyan Huang, Evaluation of Landsat–8 and Sentinel–2A Aerosol Optical Depth Retrievals across Chinese Cities and Implications for Medium Spatial Resolution Urban Aerosol Monitoring. *Remote Sens.* **2019**, *11*, 122. [CrossRef]

8. Zhang, L.; Li, J. Variability of Major Aerosol Types in China Classified Using AERONET Measurements. *Remote Sens.* **2019**, *11*, 2334. [CrossRef]

9. Tao, M.; Chen, L.; Wang, Z.; Tao, J.; Che, H.; Wang, X.; Wang, Y. Comparison and evaluation of the MODIS Collection 6 aerosol data in China. *J. Geophys. Res. Atmos.* **2015**, *120*, 6992–7005. [CrossRef]

10. Che, H.; Xia, X.; Zhu, J.; Li, Z.; Dubovik, O.; Holben, B.; Goloub, P.; Chen, H.; Estelles, V.; Cuevas-Agulló, E. Column aerosol optical properties and aerosol radiative forcing during a serious haze–fog month over North China Plain in 2013 based on ground–based sunphotometer measurements. *Atmos. Chem. Phys.* **2014**, *14*, 2125–2138. [CrossRef]

11. Qin, W.M.; Liu, Y.; Wang, L.C.; Lin, A.W.; Xia, X.G.; Che, H.Z.; Bilal, M.; Zhang, M. Characteristic and Driving Factors of Aerosol OpticalDepth over Mainland China during 1980–2017. *Remote Sens.* **2018**, *10*, 1064. [CrossRef]

12. McPhetres, A.; Aggarwal, S. An Evaluation of MODIS–Retrieved Aerosol Optical Depth over AERONET Sites in Alaska. *Remote Sens.* **2018**, *10*, 1384. [CrossRef]

13. Lopes, F.J.S.; Silva, J.J.; Marrero, J.C.A.; Taha, G.; Landulfo, E. Synergetic Aerosol Layer Observation After the 2015 Calbuco Volcanic Eruption Event. *Remote Sens.* **2019**, *11*, 195. [CrossRef]

14. Holben, B.; Eck, T.; Slutsker, I.; Tanré, D.; Buis, J.; Setzer, A.; Vermote, E.; Reagan, J.; Kaufman, Y.; Nakajima, T.; et al. AERONET—A Federated Instrument Network and Data Archive for Aerosol Characterization. *Remote Sens. Environ.* **1998**, *66*, 1–16. [CrossRef]

15. Uchiyama, A.; Yamazaki, A.; Togawa, H.; Asano, J. Characteristics of Aeolian dust observed by SKY–Radiometer in the Intensive Observation Period 1 (IOP1). *J. Meteorol. Soc. Jpn.* **2005**, *83*, 291–305. [CrossRef]

16. Takamura, T.; Takenaka, H.; Cui, Y.; Nakajima, T.; Higurashi, A.; Fukuda, S.; Kikuchi, N.; Nakajima, T.; Sano, I.; Pinker, R. Aerosol and Cloud Validation System Based on SKYNET Observations: Estimation of Shortwave Radiation Budget Using ADEOS–II/GLI Data. *J. Remote Sens. Soc. Jpn.* **2009**, *29*, 40–53.

17. Wehrli, C. GAWPFR: A Network of Aerosol Optical Depth Observations with Precision Filter Radiometers. WMO/GAW Experts Workshop on a Global Surface–Based Network for Long Term Observations of Column Aerosol Optical Properties. 2005. Available online: https://library.wmo.int/index.php?lvl=notice_display& id=11094#.XeYm625uKM9 (accessed on 12 December 2006).

18. Che, H.; Zhang, X.; Chen, H.; Damiri, B.; Goloub, P.; Li, Z.; Zhang, X.; Wei, Y.; Zhou, H.; Dong, F.; et al. Instrument calibration and aerosol optical depth validation of the China Aerosol Remote Sensing Network. *J. Geophys. Res. Atmos.* **2009**, *114*, D03206. [CrossRef]

19. Xin, J.; Wang, Y.; Li, Z.; Wang, P.; Wang, S.; Wen, T.; Sun, Y. Introduction and Calibration of the Chinese Sun Hazemeter Network. *Chin. J. Environ. Sci.* **2006**, *27*, 1697–1702.

20. Xin, J.; Wang, Y.; Pan, Y.; Ji, D.; Liu, Z.; Wen, T.; Wang, Y.; Li, X.; Sun, Y.; Sun, J.; et al. The Campaign on Atmospheric Aerosol Research Network of China: CARE–China. *Bull. Am. Meteorol. Soc.* **2015**, *96*, 1137–1155. [CrossRef]

21. Mangold, A.; Backer, H.D.; Paepe, B.D.; Dewitte, S.; Chiapello, I.; Derimian, Y.; Kacenelenbogen, M.; Léon, J.-F.; Huneeus, N.; Schulz, M.; et al. Aerosol analysis and forecast in the European Centre for Medium–Range Weather Forecasts Integrated Forecast System: 3. Evaluation by means of case studies. *J. Geophys. Res.* **2011**, *116*, D03302.

22. Zhang, M.; Wang, L.C.; Bilal, M.; Gong, W.; Zhang, Z.Y.; Guo, G.M. The Characteristics of the Aerosol Optical Depth within the Lowest Aerosol Layer over the TibetanPlateau from 2007 to 2014. *Remote Sens.* **2018**, *10*, 696. [CrossRef]

23. Kim, M.; Kim, J.; Torres, O.; Ahn, C.; Kim, W.; Jeong, U.; Go, S.J.; Liu, X.; Moon, K.J.; Kim, D.-R. Optimal Estimation–Based Algorithm to RetrieveAerosol Optical Properties for GEMS Measurementsover Asia. *Remote Sens.* **2018**, *10*, 162. [CrossRef]

24. Zhang, L.; Zhang, M.; Yao, Y.B. Multi–Time Scale Analysis ofRegional AerosolOptical Depth Changes in National–Level UrbanAgglomerations in China Using Modis Collection6.1 Datasets from 2001 to 2017. *Remote Sens.* **2019**, *11*, 201. [CrossRef]

25. Lin, C.A.; Chen, Y.C.; Liu, C.Y.; Chen, W.T.; Seinfeld, J.H.; Chou, C.C.-K. Satellite–Derived Correlation of SO2, NO2, andAerosol Optical Depth with Meteorological Conditions over East Asia from 2005 to 2015. *Remote Sens.* **2019**, *11*, 1738. [CrossRef]

26. Dubovik, O.; King, M. A flexible inversion algorithm for retrieval of aerosol optical properties from Sun and sky radiance measurements. *J. Geophys. Res. Atmos.* **2000**, *105*, 20673–20696. [CrossRef]

27. Olmo, F.; Quirantes, A.; Alcántara, A.; Lyamani, H.; Alados-Arboledas, L. Preliminary results of a non–spherical aerosol method for the retrieval of the atmospheric aerosol optical properties. *J. Quant. Spectrosc. Radiat. Transf.* **2006**, *100*, 305–314. [CrossRef]

28. Estellés, V.; Smyth, T.; Campanelli, M.; Utrillas, M. Development of an open source package for the processing of Sun–sky photometric data in the European Skyrad Users network (ESR). In Proceedings of the EGU General Assembly, Vienna, Austria, 19–24 April 2009; Volume 11, pp. EGU2009–EGU10952. Available online: https://meetingorganizer.copernicus.org/EGU2009/EGU2009-10952.pdf (accessed on 30 December 2009).

29. He, Q.; Yang, Y.; Geng, F.; Zhou, G.; Liu, D.; Wang, W. Algorithm comparison of aerosol parameter retrieval from sunphotometer measurements. *Acta Meteorol. Sin.* **2010**, *68*, 428–438.

30. Nakajima, T.; Tonna, G.; Rao, R.; Boi, P.; Kaufman, Y.; Holben, B. Use of sky brightness measurements from ground for remote sensing of particulate polydispersions. *Appl. Opt.* **1996**, *35*, 2672–2686. [CrossRef]

31. Huang, H.; Yi, W.; Qiao, Y. A method for retrieving aerosol optical properties based on Sun photometer. *J. Atmos. Environ. Opt.* **2012**, *7*, 175–180.

32. Estellés, V.; Campanelli, M.; Utrillas, M.; Expósito, F.; Martínez–Lozano, J. Comparison of AERONET and SKYRAD4.2 inversion products retrieved from a Cimel CE318 sunphotometer. *Atmos. Meas. Tech.* **2012**, *5*, 569–579. [CrossRef]

33. Streets, D.; Yarber, K.; Woo, J.; Carmichael, G. Biomass burning in Asia: Annual and seasonal estimates and atmospheric emissions. *Glob. Biogeochem. Cycles* **2003**, *17*. [CrossRef]

34. Sahu, L.; Saxena, P. High time and mass resolved PTR–TOF–MS measurements of VOCs at an urban site of India during winter: Role of anthropogenic, biomass burning, biogenic and photochemical sources. *Atmos. Res.* **2015**, *64*, 84–94. [CrossRef]

35. Gui, K.; Che, H.; Chen, Q.; Zeng, Z.; Zheng, Y.; Long, Q.; Sun, T.; Liu, X.; Wang, Y.; Liao, T.; et al. Water vapor variation and the effect of aerosols in China. *Atmos. Environ.* **2017**, *165*, 322–335. [CrossRef]

36. Zhu, J.; Che, H.; Xia, X.; Yu, X.; Wang, J. Analysis of water vapor effects on aerosol properties and direct radiative forcing in China. *Sci. Total Environ.* **2019**, *650*, 257–266. [CrossRef] [PubMed]

37. Tao, R.; Che, H.; Chen, Q.; Wang, Y.; Sun, J.; Zhang, X.; Lu, S.; Guo, J.; Wang, H.; Zhang, X. Development of an integrating sphere calibration method for Cimel sunphotometers in China aerosol remote sensing network. *Particuology* **2014**, *13*, 88–99. [CrossRef]

38. Yang, Z.; Zhang, X.; Che, H.; Zhang, X.; Hu, X.; Zhang, L. An Introductory Study on the Calibration of CE318 Sunphotometer. *J. Appl. Meteorol. Sci.* **2008**, *19*, 297–306.

39. Giglio, L. MODIS Collection 5 Active Fire Product User's Guide Version 2.5. 2013. Available online: http://modis-fire.umd.edu/files/MODIS_Fire_Users_Guide_2.5.pdf (accessed on 30 December 2013).

40. Liu, J.; Niu, S.J.; Zheng, Y.F. Optical Depth Characteristics of Yinchuan Atmospheric Aerosols Based on the CE–318 Sun Tracking Spectrophotometer Data. *J. Nanjing Inst. Meteorol.* **2004**, *27*, 615–622.

41. Zhu, X.S.; Zhou, J. Determination of Clear–Sky Columnar Water Vapor Using Solar Radiometer. *Sci. Atmos. Sin.* **1998**, *22*, 40–46.

42. Schmechtig, C.; Santer, R.P.; Roger, J.C.; Meygret, A. Automatic ground–based station for vicarious calibration. *Proc. SPIE-Int. Soc. Opt. Eng.* **1997**, *3221*, 309–317.

43. Junge, C. The Size Distribution and Aging of Natural Aerosols as Determined from Electrical and Optical Data on the Atmosphere. *J. Atmos. Sci.* **1955**, *12*, 13–25. [CrossRef]

44. Ångström, A. On the atmospheric transmission of Sun radiation and on dust in the air. *Geografiska Annaler* **1929**, *11*, 156–166.

45. O'Neill, N.; Dubovik, O.; Eck, T. Modified Ångström exponent for the characterization of submicrometer aerosols. *Appl. Opt.* **2001**, *40*, 2368–2375. [CrossRef]

46. Tanré, D.; Kaufman, Y.J.; Holben, B.N.; Chatenet, B.; Karnieli, A.; Lavenu, F.; Blarel, L.; Dubovik, O.; Remer, L.A. Climatology of dust aerosol size distribution and optical properties derived from remotely sensed data in the solar spectrum. *J. Geophys. Res.* **2001**, *106*, 18205–18217. [CrossRef]

47. Bruegge, C.; Halthore, R.; Markham, B.; Spanner, M.; Wrlgley, R. Aerosol optical depth retrievals over the Konza Prairie. *J. Geophys. Res. Atmos.* **1992**, *97*, 18743–18758. [CrossRef]

48. Mitchell, J.M., Jr. The Effect of Atmospheric Aerosols on Climate with Special Reference to Temperature near the Earth's Surface. *J. Appl. Meterol.* **1971**, *10*, 703–714. [CrossRef]

49. Liou, K.N. Animprint of Elsevier Science. In *An Introduction to Atmospheric Radiation*, 2nd ed.; Academic Press: San Diego, CA, USA, 2002; Available online: https://www.elsevier.com/books/an-introduction-to-atmospheric-radiation/liou/978-0-12-451451-5 (accessed on 29 April 2002).

50. Mishchenko, M.I.; Travis, L.D.; Kahn, R.A.; West, R.A. Modeling phase functions for dustlike tropospheric aerosols using a shape mixture of randomly oriented polydisperse spheroids. *J. Geophys. Res. Atmos.* **1997**, *102*, 16831–16847. [CrossRef]

51. Mathai, C.V.; Harrison, A.W.; Mathews, T. Aerosol Particle Size Distribution (0.1–1.0 μm) during the Chinooks of 1979 over Calgary, Canada. *J. Appl. Meteorol.* **1980**, *19*, 1101–1125. [CrossRef]

52. Viera, G.; Box, M. Information content analysis of aerosol remote–sensing experiments using singular function theory 1: Extinction measurements. *Appl. Opt.* **1987**, *26*, 1312–1339. [CrossRef]

53. Viera, G.; Box, M. Information content analysis of aerosol remote–sensing experiments using singular function theory 2: Scattering measurements. *Appl. Opt.* **1988**, *27*, 3262–3274. [CrossRef]

54. Smironv, A.; Holben, B.; Eck, T.; Dubovik, O.; Slutsker, I. Cloud–screening and quality control algorithms for the AERONET database. *Remote Sens. Environ.* **2000**, *73*, 337–349. [CrossRef]

55. Yin, H. *Atmospheric Radiology Foundation*; China Meteorological Press: Beijing, China, 1993.

56. Shi, C.; Yu, X.; Zhou, B.; Xiang, L.; Nie, H. Aerosol Optical Properties during Different Air–Pollution Episodes over Beijing. *Environ. Sci.* **2013**, *34*, 4139–4145.

57. D'Almeida, G.; Koepke, P.; Shettle, E. Atmospheric Aerosols: Global Climatology and Radiative Characteristics. *J. Med. Microbiol.* **1991**, *54*, 55–61.

58. Kubilay, N.; Cokacar, T.; Oguz, T. Optical properties of mineral dust outbreaks over the northeastern Mediterranean. *J. Geophys. Res. Atmos.* **2003**, *108*, 1981–1990. [CrossRef]

59. Sheng, P.; Mao, J.; Li, J. *Atmospheric Physic*; Peking University Press: Beijing, China, 2003.

60. Tang, X.; Zhang, Y.; Shao, M. *Atmospheric Environmental Chemistry*; China Higher Education Press: Beijing, China, 2006.

61. Hao, W.; Liu, M. Spatial and temporal distribution of tropical biomass burning. *Glob. Biogeochem.* **1994**, *8*, 495–503.

62. O'Neill, N.; Eck, T.; Holben, B.; Smirnov, A.; Royer, A.; Li, Z. Optical properties of boreal forest fire smoke derived from Sun photometry. *J. Geophys. Res. Atmos.* **2002**, *107*, AAC 6-1–AAC 6-19. [CrossRef]

63. Ångström, A. The parameters of atmospheric turbidity. *Tellus* **1964**, *16*, 64–75. [CrossRef]

64. Seinfeld, J.; Pandis, S. *Atmospheric Chemistry and Physics: From Air Pollution to Climate Change*, 2nd ed.; WILEY: New York, NY, USA, 2012.

65. Lesins, G.; Chylek, P.; Lohmann, U. A study of internal and external mixing scenarios and its effect on aerosol optical properties and direct radiative forcing. *J. Geophys. Res. Atmos.* **2002**, *107*, AAC 5-1–AAC 5-12. [CrossRef]

66. Hess, W.; Herd, C. Influence of carbon black morphology and surface activity on vulcanizate properties. *Rubber World* **1993**, *208*, 26.

Article

New Global View of Above-Cloud Absorbing Aerosol Distribution Based on CALIPSO Measurements

Wenzhong Zhang [1,2], Shumei Deng [3], Tao Luo [1,*], Yang Wu [1,4], Nana Liu [1,4], Xuebin Li [1], Yinbo Huang [1] and Wenyue Zhu [1]

[1] Key Laboratory of Atmospheric Optics, Anhui Institute of Optics and Fine Mechanics, Chinese Academy of Sciences, Hefei 230031, China; zwzzwz@mail.ustc.edu.cn (W.Z.); wy0608@mail.ustc.edu.cn (Y.W.); liunana@mail.ustc.edu.cn (N.L.); xbli@aiofm.ac.cn (X.L.); ybhuang@aiofm.ac.cn (Y.H.); zhuwenyue@aiofm.ac.cn (W.Z.)

[2] Science Island Branch of Graduate School, University of Science and Technology of China, Hefei 230031, China

[3] School of Environment and Energy Engineering, Anhui Jianzhu University, Hefei 230031, China; dsm_790510@ustc.edu.cn

[4] School of Environment Science and Optoelectronic Technology, University of Science and Technology of China, Hefei 230031, China

[*] Correspondence: luotao@aiofm.ac.cn; Tel.: +86-0551-65592979

Received: 3 September 2019; Accepted: 12 October 2019; Published: 16 October 2019

Abstract: Above-low-level-cloud aerosols (ACAs) have gradually gained more interest in recent years; however, the combined aerosol–cloud radiation effects are not well understood. The uncertainty about the radiative effects of aerosols above cloud mainly stems from the lack of comprehensive and accurate retrieval of aerosols and clouds for ACA scenes. In this study, an improved ACA identification and retrieval methodology was developed to provide a new global view of the ACA distribution by combining three-channel CALIOP (The Cloud–Aerosol Lidar with Orthogonal Polarization) observations. The new method can reliably identify and retrieve both thin and dense ACA layers, providing consistent results between the day- and night-time retrieval of ACAs. Then, new four-year (2007 to 2010) global ACA datasets were built, and new seasonal mean views of global ACA occurrence, optical depth, and geometrical thickness were presented and analyzed. Further discussion on the relative position of ACAs to low clouds showed that the mean distance between the ACA layer and the low cloud deck over the tropical Atlantic region is less than 0.2 km. This indicates that the ACAs over this region are more likely to be mixed with low-level clouds, thereby possibly influencing the cloud microphysics over this region, contrary to findings reported from previous studies. The results not only help us better understand global aerosol transportation and aerosol–cloud interactions but also provide useful information for model evaluation and improvements.

Keywords: above-cloud aerosol; low-level cloud; CALIPSO

1. Introduction

The long-range transport of aerosols plays an important role in several regions of the world, having a potential impact on aerosol–cloud interactions, atmospheric chemistry, and air quality [1–8]. In particular, aerosols often overlay lower level clouds [9], for example, biomass burning aerosols and wind-blown dust overlay low-level cloud deck over the Atlantic. The above-low-level cloud aerosols (ACAs) occupy about 25% of the mean aerosol optical depth (fine mode) at a global scale [9], and this fraction could be much higher regionally and seasonally [10]. Current models experience significant inter-model discrepancies in aerosol forcing assessments, especially over the aerosol–cloud overlap regions [11], that result from inter-model differences in both aerosol and cloud properties [11,12].

A recent evaluation showed that most models cannot reproduce the observed large aerosol load episodes [13]. Therefore, improved observations are needed to better understand ACA–cloud interactions and to constrain the aerosol–cloud radiative processes in models.

The ACA has gained increasing attention in recent years because of its important, but not well-understood, radiative and microphysical effects on clouds [14–19]. Different from clear-sky aerosols, which are usually associated with a negative (cooling) direct radiative effect [15,19–23], the absorption effect of the ACA is significantly amplified due to the strong reflection light of low clouds, resulting in a less negative or even positive (warming) direct radiative effect. By warming the free troposphere and cooling the surface below, the ACA can increase the low cloud cover by enhancing atmospheric stability [24]. However, if the ACA mixes directly with the cloud layer, warming of the ACA could reduce the relative humidity, dissipating the cloud [24]. Therefore, the radiative and microphysical effects of ACAs on clouds depend on both their loadings and their positions relative to the clouds [25].

Despite its importance, quantifying the effect of the ACA on the clouds from satellite observations is still challenging. For passive remote sensing, the ACA has usually been neglected, and only clear-sky aerosol concentrations can be derived from passive satellites [26,27]. This is because passive satellites employ the reflection of natural sunlight to retrieve aerosol and cloud properties separately. The neglect of the ACA in passive remote sensing can result in uncertainties in the retrieval of cloud micro- and macro-properties, such as the liquid water path, cloud optical thickness, and effective radius of cloud droplets [26,27]. Until recently, based on different absorption effects of the ACA in the visible and near-infrared channels, a "color ratio" (CR) method was used to retrieve the ACA optical depth (ACAOD) and aerosol-corrected cloud optical depth (COD) [26] and this was further improved to a multi-channel method to simultaneously retrieve ACAOD, COD, and cloud effective radius data [27]. The CR between a pair of wavelengths is a function of both the aerosol and cloud optical thicknesses, and the measured reflectance can be related to pairs of aerosol and cloud optical thicknesses. The results indicate that the mean liquid cloud optical thickness can be increased by roughly 6%, and the mean liquid effective radius can be increased by roughly 2.6% after correcting for the effect of the ACA [22,28–32]. An inter-satellite comparison of the ACAOD retrieved from NASA's A-train sensors revealed a good level of agreement between the passive sensors over homogeneous cloud fields [33]. However, passive satellites can only provide daytime ACAOD and are unable to determine the spatial position of aerosols with respect to the clouds.

Other than passive sensors, the active lidar Cloud–Aerosol Lidar with Orthogonal Polarization (CALIOP) onboard the Cloud–Aerosol Lidar and Infrared Pathfinder Satellite Observations (CALIPSO) can provide valuable vertically resolved information about aerosols and clouds. CALIPSO Level-2 data have already been employed in studies of ACA climate effects [19,22,23,34–36]. It was shown that the vertical structure of ACA and clouds was critical in determining the aerosol–cloud radiative forcing. However, inter-satellite comparisons of ACAOD [33,37,38] showed that the CALIPSO-derived ACAOD data was consistently lower than those derived from other satellite sensors in the A-Train, by a factor of four to six, as compared with passive retrieval. The main reason for this is that the current CALIPSO Level-2 aerosol retrieval algorithm employs the 532 nm channel, which cannot detect the true aerosol layer base because of the strong attenuation by the ACA in this channel [33,38]. Based on the CALIPSO Level-2 product, most of the ACA resides above the cloud deck at a distance of about 1 km over the tropical Atlantic region where both the ACA and low cloud frequently occur, indicating that weak cloud microphysical effects occur due to the aerosols not mixing with the cloud [24]. However, this could be misleading because the detected ACA layer base in the CALIPSO Level-2 product has been suggested to be higher than the real position [33,38]. Some other issues also limit the use of CALIPSO Level-2 data in ACA studies, such as the poor consistency between day- and night-time retrieval due to the poor signal-to-noise-ratio (SNR) performance of CALIPSO daytime observations.

This paper aims to provide a new global view of the ACA distribution based on an improved ACA detection method by combining the CALIPSO 532 and 1064 nm channels. Furthermore, efforts to address the low SNR issue in CALIPSO daytime observations were made to minimize the difference between day- and night-time retrieval. The data sources used in this study are introduced in Section 2. The new ACA identification and retrieval method is detailed in Section 3. Section 4 presents the results and discussions, and the conclusions are summarized in Section 5.

2. Data

CALIPSO, CloudSat and operational meteorology datasets were used to perform the study. CloudSat has not been able to operate at night-time and has operated in daylight-only mode since a spacecraft battery anomaly occurred in 2011. Therefore, the 2007–2010 observations were employed in this study to build both day- and night-time ACA datasets.

The CALIPSO Level-1B Version 4 product provides 532 nm total attenuated backscatter (β'_{532}), a perpendicular polarization component (β'_{532p}), and 1064 nm total attenuated backscatter (β'_{1064}) [39], with a 30 m vertical resolution below 8.2 km and a 1/3 km resolution in the horizontal direction along the ground track. The photomultiplier tube exhibited a nonideal recovery at 532 nm after encountering a strong backscattering objective, which was corrected following Li et al. [40]. The CALIPSO Level-2 Vertical Feature Mask (VFM) and Aerosol Profile Products (APro) files were used for comparison in this study.

The cloud type, cloud top height, and cloud bottom height were provided by the 2B-CLDCLASS-Lidar product [41]. The 2B-CLDCLASS-Lidar product used combines the CloudSat Cloud Profiling Radar (CPR), which is a 94 GHz microwave radar, and CALIPSO lidar measurements and thus can provide more accurate and complete cloud mask information than radar-only or lidar-only measurements [42].

The meteorological reanalysis data MERRA2 (The Modern Era Retrospective-analysis for Research and Application, Version 2 [43]) assimilated the meteorological data using a modern satellite database, which was released by the Global Modelling and Assimilation Systems of Goddard Space Flight Centre of NASA. It was used to provide temperature and pressure profiles for molecular backscattering estimation in this study.

3. Methodology

Previous studies have shown that the CALIPSO operational aerosol product tends to miss the bottom of dense ACA layers and underestimates the 532 nm aerosol optical depth because of the relatively strong aerosol attenuation in the 532 nm channel [33,37,44]. In contrast, both the aerosol attenuation and molecular backscattering in the 1064 nm channel are smaller than those in the 532 nm channel. Therefore, CALIPSO 532 and 1064 nm channel lidar observations were combined to develop an ACA retrieval methodology in this study, as detailed below. A flow chart of the methodology is shown in Figure 1.

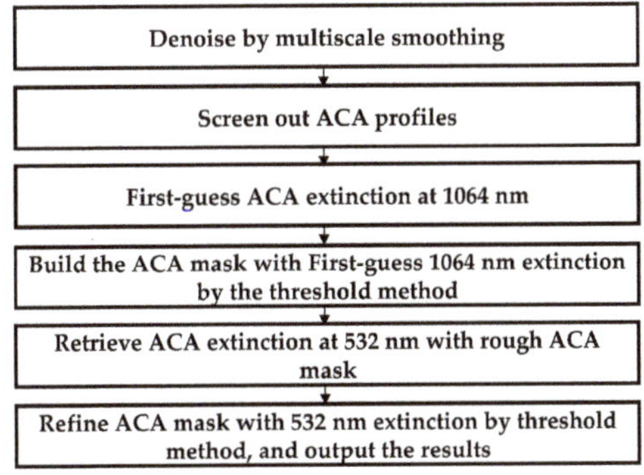

Figure 1. Flow chart of the new above-low-level cloud aerosol (ACA) identification and retrieval methodology.

Firstly, daytime CALIPSO observations experience lower SNRs than night-time observations, resulting in a strong apparent day- and night-time difference in the ACA global distribution (refer to Figure 3 in Alfaro-Contreras et al. [45]). To overcome the poor SNR issue in daytime CALIOP signals, different smoothing scales were selected for day- and night-time observations to provide consistent retrieval. For the β'_{532} and β'_{1064} night-time observations, 20 km horizontal averaging was adopted. For the β'_{532} and β'_{1064} daytime observations, multiscale averaging in different signal-strength bins was selected according to the night-time noise level. As shown in Figure 2, daytime noise was calculated with horizontal averaging of 20, 40, 60, 80, and 100 km and compared to the 20 km night-time results. According to Figure 2, β'_{532} daytime signals of higher than 2.3×10^{-2} km^{-1}sr^{-1}, between 2.3×10^{-2} and 2.8×10^{-3} km^{-1}sr^{-1}, and lower than 2.8×10^{-3} km^{-1}sr^{-1} were exhibited at 20, 40, and 60 km, respectively. The maximum averaging scale was selected to be 60 km, after which the SNR performance improved a little. To avoid involving cloud signals in the moving-smoothed aerosol signals, only non-cloud data were smoothed using the cloud mask from the CloudSat 2B-CLDCLASS-Lidar product.

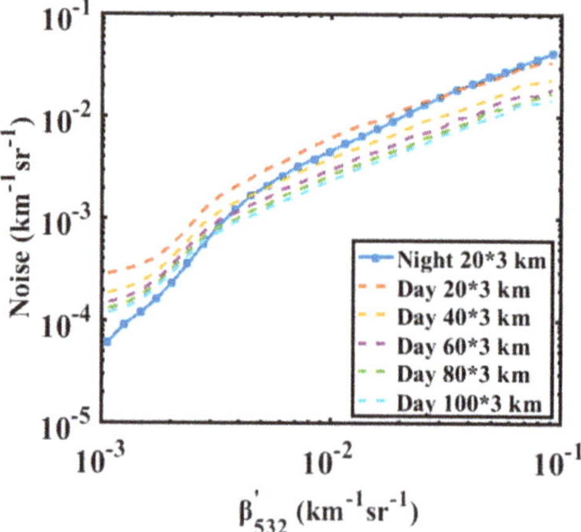

Figure 2. Comparison of noise between the 20 km moving–smoothing in the night-time and multiscale moving–smoothing in the daytime.

Secondly, after multiscale smoothing, ACA profiles were screened out using the CloudSat 2B-CLDCLASS-Lidar product and the estimated two-way transmittance at 1064 nm. Low cloud was defined as a cloud top lower than 3 km above ground level (AGL). Only single-layer low cloud or multi-layer thin cirrus cloud above the low cloud cases were considered in this study. Low-level cloud sample quantity is presented for each 2.5° × 2.5° grid box and averaged for winter (DJF, i.e., December, January, and February), spring (MAM, i.e., March, April, and May), summer (JJA, i.e., June, July, and August), and autumn (SON, i.e., September, October and November) in Figure 3. For each low cloud profile, the two-way transmittance at 1064 nm within 6 km above the cloud top was estimated by ignoring the molecular backscattering, as

$$x = e^{-2\zeta} = 1 - 2S \int \beta'_{1064} dr \tag{1}$$

Here, x is the two-way transmittance, and $\int \beta' dr$ is the integrated attenuated backscatter at 1064 nm within 6 km above the cloud top, and S is the lidar ratio at 1064 nm. For cirrus cloud, S was chosen to be 19 sr. For aerosol, the S at 1064 nm has a smaller range than that in the 532 nm channel [46] and was assumed to be 40 sr in this step. By considering the noise, $e^{-2*0.015} = 0.97$ was chosen as the threshold, and values smaller than that were identified as possible ACA cases.

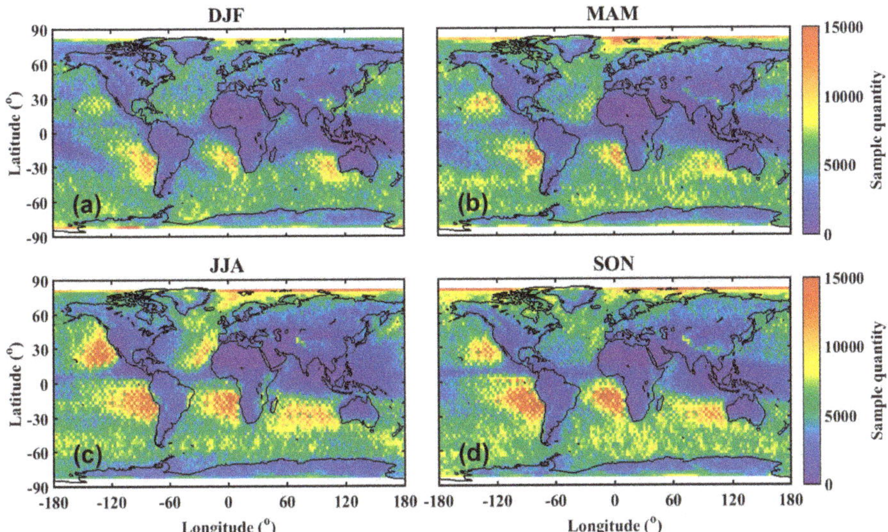

Figure 3. Low cloud samples quantity in (**a**) winter (DJF, i.e., December, January, and February), (**b**) spring (MAM, i.e., March, April, and May), (**c**) summer (JJA, i.e., June, July, and August), and (**d**) autumn (SON, i.e., September, October, and November).

Thirdly, following the screening-out of possible ACA cases, the first-guess ACA extinction at 1064 nm was retrieved with the forward integration scheme [47] to build the ACA mask. In this step, the 1064 nm S was chosen to be 40 sr. The first-guess ACA mask was set to unity when the first-estimate ACA extinctions were larger than four times the measured noise, indicating the possible presence of aerosols. Then, the first-guess ACA top and bottom were identified as the highest and lowest points above the cloud top where the ACA mask equals 1.

Finally, using the first-guess ACA top and bottom from the last step, the ACA extinctions from the 532 nm channel were retrieved, with the lidar ratio determined according to the aerosol type classification, identical to the method used by Omar et al. [46]. The ACA mask was thus further refined according to the 532 nm extinctions using a threshold of four times of the measured noise.

Figure 4 presents comparisons of an ACA case derived from the new method and the CALIPSO level-2 V4 aerosol and cloud product. This is an outflowed smoke case immediately above the top of low marine clouds, as can be seen from the smoothed β'_{532} and data in Figures 4b and 4c, respectively. It should be noted that the same case was demonstrated in Jethva et al. [33]. The retrieval from the new method (Figure 4e) was close to that from passive methods and the CALIOSP "DR" and "CR" methods (refer to Figure 2 in Jethva et al. [33]). In contrast, the ACAOD from the CALIPSO Level-2 product was much smaller than that from the new method, with a strong underestimation by a factor of up to about 2. This is because the standard CALIOP Level-2 aerosol and cloud products use the 532 nm signal to detect respective layers, which, in the presence of thick smoke layers, can be incorrectly assigned due to strong attenuation at 532 nm. Figure 4h shows that only the upper portion of the layer with dense aerosols was detected, while most of the lower portion of the ACA was missed. We tested the projection of a new extinction (Figure 4f) to the CALIOP Level-2 aerosol mask in Figure 4h, and this produced a similar ACAOD to L2. Benefitting from the aid of the 1064 nm extinction coefficient, the new method was shown to provide a complete aerosol mask. Therefore, the new method can provide more accurate ACA detection and retrieval results than the CALIPSO Level-2 product.

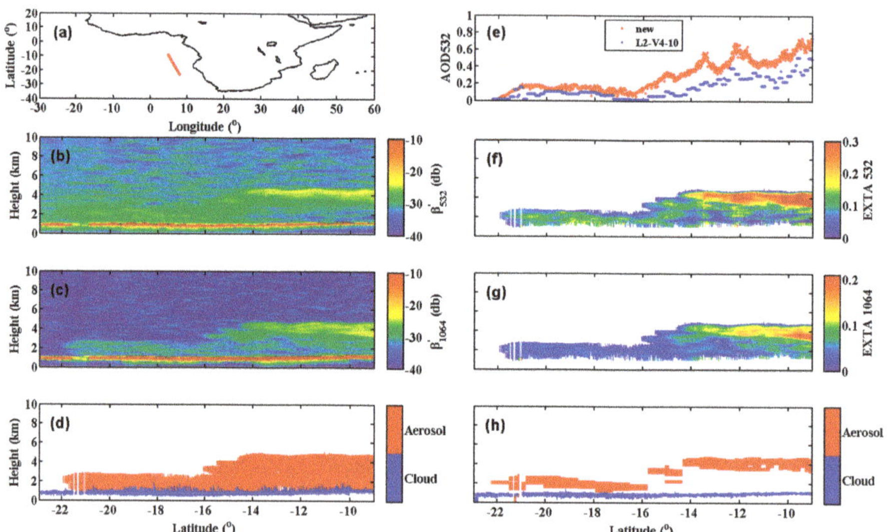

Figure 4. ACA case on 2 August 2007, 13:12:11 UTC: (**a**) geo-location; (**b**) $\beta\prime_{532}$; (**c**) $\beta\prime_{1064}$; (**d**) ACA mask from the new method in this study; (**e**) aerosol optical depth at 532 nm determined using the new method (red) and CALIPSO Level-2 V4 APro files (blue); (**f**) extinction coefficient of ACA at 532 nm; (**g**) extinction coefficient of ACA at 1064 nm; (**h**) ACA and cloud layer from the CALIPSO Level-2 VFM product.

4. Results

The new methodology was applied to the 2007–2010 observations to build a new global ACA dataset. Statistical results are also presented for each 2.5° × 2.5° grid box and averaged for DJF, MAM, JJA, and SON. Section 4.1 presents a comparison of day- and night-time cloudy-sky ACA occurrences (ACAO$_C$) and ACAOD, and Section 4.2 analyses the global distribution of the seasonal-mean ACAO$_C$ and ACAOD with different aerosol types. The ACAO$_C$ is defined as the number of ACA profiles divided by the number of low cloud profiles in each 2.5° grid box.

4.1. Day- and Night-Time Comparison of the New ACA Dataset

Figure 5 presents a comparison between night- and day-time global annual mean distributions of low-level cloud fractions (left column), ACAO$_C$ (middle column), and ACAOD (right column) derived from the new method using new four-year global ACA datasets. Table 1 summarizes the night- and day-time statistics of the annual mean ACAO$_C$ and ACAOD in different seasons globally and over the dust region (5–30° N, 60–16° W), smoke region (22°S–5° N, 18° W–15° E), and Eastern Asia (19–40° N, 100–140° W), respectively. These places have also been documented in previous studies [28,48–51].

As shown in Figure 5 and Table 1, the ACAO$_C$ and ACAOD show very similar patterns and values between day- and night-time retrieval. The global mean day- and night-time ACAO$_C$ are 0.125 and 0.108 respectively, and the global mean ACAOD values are 0.146 and 0.143. The correlation coefficient between day- and night-time is 0.938 for ACAO$_C$ and 0.796 for ACAOD. This shows that the new method developed in this study can produce reasonable day- and night- time results, better than the prior results derived from the CALIPSO level-2 product (refer to Figure 3 in Alfaro-Contreras et al. [45]). The night-time results from the CALIPSO level-2 product are quite similar to the results produced in this study. However, in the day-time, the CALIPSO level-2 products missed lots of weak aerosol layers due to their poor SNRs and thus gave quite different global distributions of ACAO$_C$ and ACAOD from those of night-time CALIPSO level-2 product and our method. The new method described in this study can detect more ACA cases over both source regions and long-range transport regions. Furthermore, the ACAOD values derived by the new method are larger than those from the CALIPSO Level-2 product. The CALIPSO Level-2 product retrieves ACAOD values that are too low over the tropical Atlantic region, a region where other passive methods suggest that high above-cloud aerosol loading could exist (i.e., refer to Figure 2 in Devasthale et al. [52]).

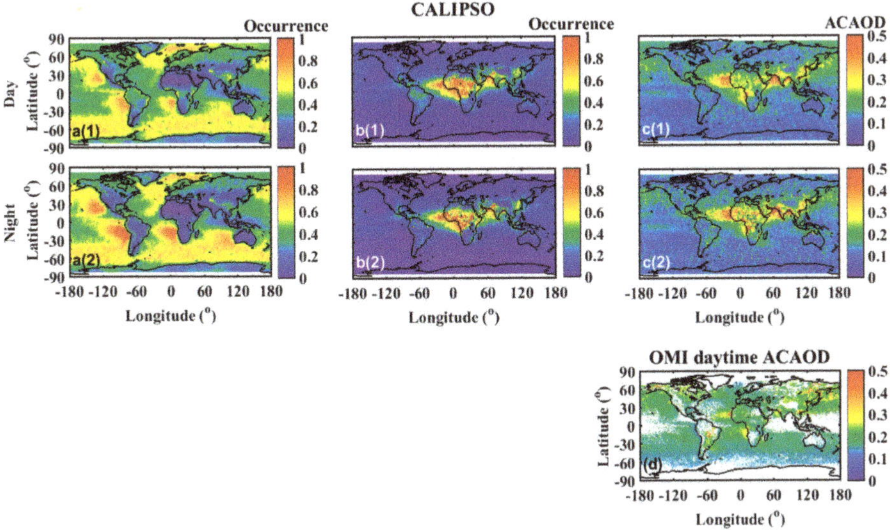

Figure 5. Day- and night- time comparison of the global annual mean distribution of low-level cloud fractions (**a**(1)-(2)), cloudy-sky ACA occurrences (ACAO$_C$, **b**(1)-(2)), ACA optical depth (ACA optical depth (ACAOD), **c**(1)-(2)), and (**d**) is the global annual mean distribution of OMI daytime ACAOD.

Table 1. ACAO$_C$ and ACAOD data from different regions in the night- and day-time.

			Global	Eastern Asia	Smoke Region	Dust Region
ACAO$_C$	Day	DJF	0.098	0.326	0.496	0.237
		MAM	0.148	0.581	0.412	0.395
		JJA	0.147	0.299	0.575	0.646
		SON	0.116	0.233	0.516	0.349
	Night	DJF	0.117	0.377	0.520	0.240
		MAM	0.174	0.650	0.471	0.413
		JJA	0.173	0.350	0.588	0.677
		SON	0.139	0.275	0.583	0.407
ACAOD	Day	DJF	0.118	0.218	0.238	0.185
		MAM	0.143	0.311	0.199	0.264
		JJA	0.149	0.234	0.244	0.353
		SON	0.140	0.212	0.255	0.237
	Night	DJF	0.117	0.211	0.242	0.178
		MAM	0.140	0.308	0.212	0.260
		JJA	0.151	0.234	0.219	0.357
		SON	0.139	0.212	0.236	0.220

A preliminary comparison between our ACA dataset and passive satellite product was also done. Figure 5d shows the global distribution of annual-mean daytime 500 nm ACAOD derived from OMACA product version 3, which retrieves ACAOD from OMI's two near-UV observations (354 and 388 nm) [17]. The data processing of Figure 5d follows the same way as Figure 8 in Jethva et al. [10], except that only 2007–2010 data were used in this paper. Comparing to the daytime ACAOD result of this paper (Figure 5c(1)), it shows that both products have quite similar pattern of global distribution to each other, while with regional differences in values. Our ACA dataset retrieves larger ACAOD than OMACA product over some major aerosol source regions, such the main dust band region including Sahara and Middle-Eastern dust regions and related long-range transport regions (tropic Atlantic, Arabian Sea, and Southern Asia), central Africa wildfire region and eastern Asia region except southernmost part of China. In contrast, OMACA product retrieves larger ACAOD than our ACA dataset over some weak aerosol source regions, such as Siberia, Alaska and Southern Oceans. Further detailed evaluation of the ACA dataset developed in this paper with passive satellite products and its possible implements in improving passive ACA and cloud retrievals will be our study interest in the near future.

4.2. Global Distribution of the Seasonal Mean ACA Properties

Figures 6 and 7 illustrate the global distribution of the seasonal mean $ACAO_C$ and ACAOD data for all aerosol types (left column), dust aerosols (pure and pollute dust, middle column), and smoke and polluted continental aerosols (right column). Smoke and polluted continental aerosols were analyzed together in these figures, because these two types of aerosol have similar optical properties and are quite difficult to discriminate each other for the ACA case [46]. Furthermore, the existing classification algorithm is not suitable for the detection of above-cloud marine aerosols. As can be seen in those figures, above-cloud marine aerosol occurs over the western tropical Pacific region but is misclassified as smoke or polluted continental aerosols. The global mean $ACAO_C$ in each season was found to be 13% (DJF), 20% (MAM), 18% (JJA), and 15% (SON), and the global mean ACAOD was 0.12 (DJF), 0.14 (MAM), 0.15 (JJA), and 0.14 (SON), respectively.

As shown in Figures 6 and 7, ACA frequently occurs near the source regions, such as over the Sahara and Middle-Eastern dust regions (main dust band region), the Africa smoke region, and the Eastern Asia region. Those aerosols are mobilized far from the sources and travel along transoceanic pathways, such as dust transport over the tropical Atlantic, region and smoke transport over the southeast Atlantic region, resulting high ACAOc values there. For convenience, the ACA and its long-distance transport are discussed with regard to its main source as in the following text.

(1) Main dust band region: Above-cloud dust aerosols frequently occur along the main dust band, including in the Saharan to Middle-Eastern dust source regions and transport regions such as the Atlantic and Indian Oceans and India. Over these source regions, dust activity is strongest in MAM and JJA, resulting in more than 80% of low cloud having dust above it (Figure 6). This dust transports furthest, from the west to north Atlantic, in JJA, and it transports most widely in MAM. The above-cloud dust AOD values over the northern Atlantic were found to be 0.26 (MAM) and 0.35 (JJA) (Figure 7b2,b3). The easterly long-range transport of Saharan and Middle-Eastern dust can result in an above cloud occurrence of 65% with an AOD of 0.5 over the Indian Ocean and India in JJA, as shown in Figures 6b3 and 7b3. In SON, the period associated with the weakest dust activities, the above-cloud dust occurrence was found to decrease to ~60% over the source region and ~20–50% over the long-range transport regions. Correspondingly, the above-cloud dust AOD reduced to ~0.15–0.3 over these regions.

(2) Africa smoke region: Smoke aerosol frequently occurs as a result of biomass events in central Africa and is transported across the cloud deck over southeastern the Atlantic Ocean. As shown in Figure 6c(1)–c(4), the above-cloud smoke aerosol occurs most in JJA (about 52%), second-most in SON (about 49%) and third-most in DJF (about 33%), whereas little ACA occurs in MAM. The mean above-cloud smoke AOD values in each season were found to be 0.28 (DJF), 0.17 (MAM), 0.23 (JJA), and 0.24 (SON).

(3) Eastern Asia region: In MAM, the period when Asian dust activity is strongest, above-cloud dust aerosol is present for most of the year (about 29%), and this period is associated with the strongest AOD value of about 0.31. The dust from Asia will be transported long-range across the Pacific to northern America, resulting in about 20% of ACA occurrences and an ACAOD value of 0.25

over the Pacific Ocean. The above-cloud smoke and polluted continental aerosols over Eastern Asia are also most active in MAM, representing ~32% of occurrences and having an AOD of 0.31 above low clouds. These aerosols can also be transported over the Pacific and contribute to the ACA presence there. The occurrence of above-cloud smoke and polluted continental aerosols is ~25% and the ACAOD value is 0.23 over the Pacific. In other seasons, the ACA occurrence is about 7–25%, and this is associated with very weak long-range transport.

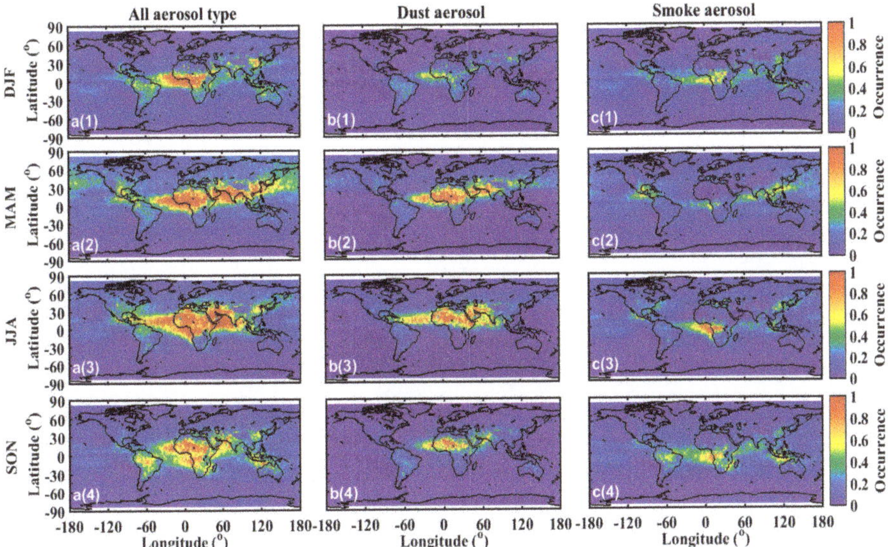

Figure 6. Global distribution of the seasonal mean $ACAO_C$ for all aerosol types (left column), pure dust and polluted dust aerosols (middle column) and smoke and polluted continental aerosols (right column).

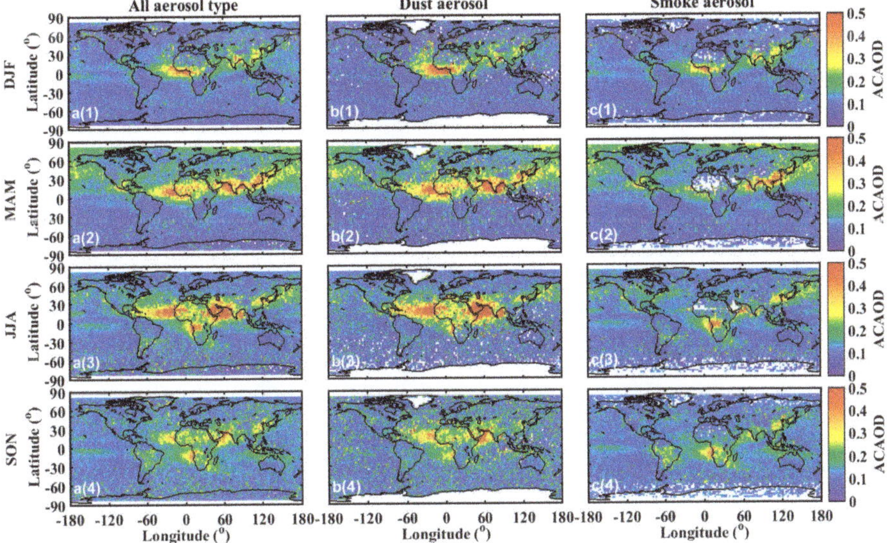

Figure 7. Global distribution of the seasonal mean ACAOD for all aerosol types (left column), pure dust and polluted dust aerosols (middle column) and smoke and polluted continental aerosols (right column).

Other than these major aerosol source regions, our results also indicate some weak aerosol source regions, such as South America, Central America, and Timor Sea, where ACA frequently occurs in some seasons, but this has rarely been studied in previous studies. South America and the Timor Sea can be considered to be anthropogenic source regions due to the biomass burning emissions [53,54]. During SON, the above-cloud smoke aerosol occurrence over South America is about 24% with an AOD value of about 0.16. Trade winds will carry the smoke aerosol produced from Northern Australian savannah fires and agricultural/peat burning fires in Indonesia over the Timor Sea [54], resulting in an above-cloud smoke aerosol occurrence of 57% with an ACAOD value of 0.20. The ACA over Central America mostly occurs in MAM as a result of both anthropogenic sources due to human activity and biomass burning activities [55]. The occurrence of above-cloud polluted continental and above-cloud smoke aerosol in Central America is about 35% with an AOD value of about 0.21 in MAM. Our results also indicate that smoke from southeastern Africa transports in an easterly direction across the Indian Ocean in SON, and may reach the west coast of Australia.

5. Discussion

The relative positions of the ACA and low clouds determine the response of low clouds to the ACA [24]. Figure 8 shows the 4-year zonal mean $ACAO_C$ (Figure 8a), ACAOD (Figure 8b), vertical distance between the ACA and cloud (VDAC, Figure 8c), and ACA geometrical thickness (AGT, Figure 8d). The $ACAO_C$ and ACAOD values have quite large inter-season variations at low latitudes, which are related to the dust and smoke activity. At low latitudes (between −40° and 40°N), the $ACAO_C$ can reach about 40%, and the ACAOD can reach 0.2–0.4 during high ACA activity seasons (i.e., MAM in the northern tropical region and SON in the southern tropical region). Additionally, the VDAC is smaller than about 0.5 km and the AGT can reach 1–2 km, because of the ACA being close to the main source region (Saharan dust and African smoke). When transported away from the source to high latitudes (<−40° and >40°), the $ACAO_C$ decreases as the latitude increases. However, the VDAC increases from 0.5 to about 1–1.5 km during long-range transport, associated with the decrease in AGT, except in MAM. In MAM, the AGT in the northern hemisphere is generally about 2 km, as a result of the long-range transport of Asian dust at mid-latitudes [56]. The different behaviors that occur at low and high latitudes indicate different cloud responses to the ACA. The stabilization effect of the ACA dominates the cloud response at the mid-to-high-latitude regions, because the ACA generally resides away from the low clouds [24]. In contrast, at low latitudes, the ACA is more likely to directly mix with low clouds and influence the cloud microphysics.

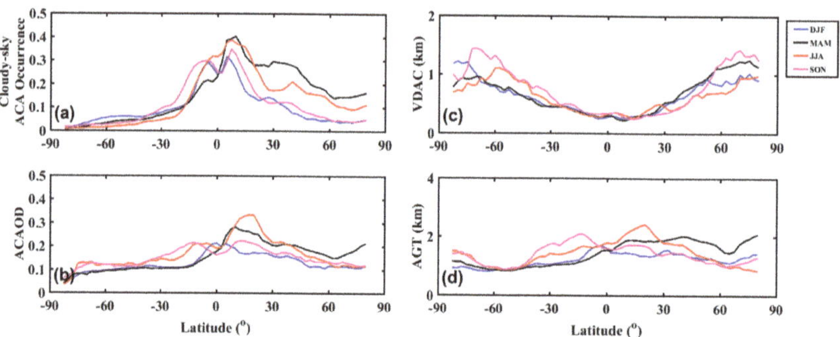

Figure 8. (**a**) $ACAO_C$ line chart with latitude. (**b**) ACAOD line chart with latitude. (**c**) Vertical distance between the ACA and cloud (VDAC) line chart with latitude. (**d**) ACA geometrical thickness (AGT) line chart with latitude.

A more insightful view of the relative positions of the ACA and low clouds at low altitudes is shown in Figure 9 by the 4-year mean longitude–altitude distribution of cloud (gray contour line) and ACA (color shade) occurrences over the dust (left column) and smoke regions (right column). The latitude ranges of the dust (tropical Atlantic) and smoke (southeast Atlantic) regions are the same as those stated in Section 4.1. Similar results by the CALIPSO level-2 product can be found in Figure 1 in Zuidema et al. [24], in which most of the ACA layer was shown to reside above the cloud deck at a mean distance of about 1 km [24]. According to the CALIPSO level-2 product, a semi-direct effect (atmospheric stabilization, reference) due to absorption in the aerosol layers above the cloud deck could dominate the ACA–cloud interaction over these regions. However, our results show that the bottom of the high ACA occurrence layer resides immediately above the most frequent cloud height (Figure 9). Our findings are different to those derived from the CALIPSO level-2 product. The main reason for this is the misdetection of the ACA layer bottom in the CALIPSO level-2 product (as stated in Section 3), which was also reported by Rajapakshe et al. (2017) using two seasons of night-time CATS 1064 nm observations over the southeast Atlantic region [44]. Therefore, as well as the semi-effect, smoke and dust layers can mix into the low cloud over the transport region of the Atlantic and influence the cloud microphysics (indirect effect) due to its proximity to the cloud layer [34].

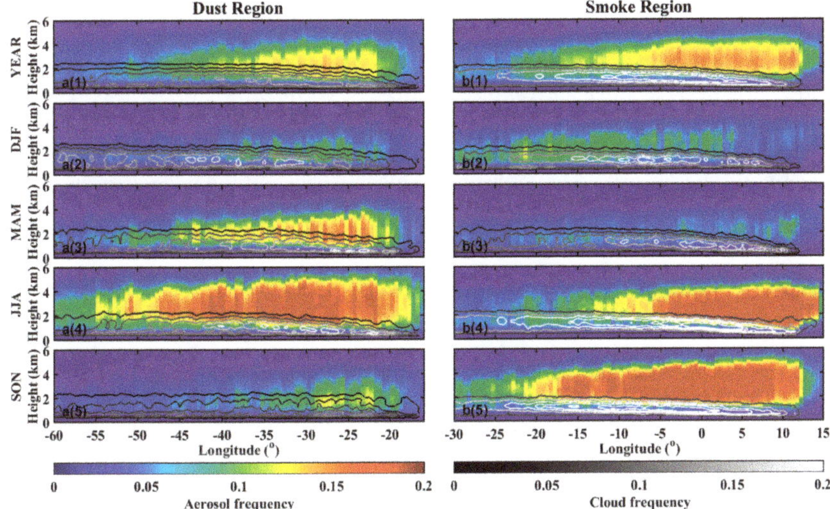

Figure 9. Positional relationship between the aerosol and cloud layers in the dust and smoke area with longitude.

6. Conclusions

This paper presented an improved ACA identification and retrieval methodology that combines CALIOP 532 and 1064 nm observations. The ACA was mainly detected with the 1064 nm channel by taking advantage of the weaker aerosol attenuation compared with that of the 532 nm channel. The selection of the 1064 nm signal for ACA identification allowed the detection of the full column of aerosols above the clouds, which the 532 nm signal misses due to its stronger attenuation. Effort was made to address the low SNR issue in the CALIOP day-time observation to provide consistent results between day- and night-time retrieval. Another feature of this methodology is the reliable cloud–aerosol distinction that occurs with the combination of CALIPSO and CloudSat observations.

Then, new four-year (2007 to 2010) global ACA datasets were built, and new global seasonal-mean views of ACA properties were presented and analyzed, including the occurrence of ACA and its optical properties. The results indicate that the new method can not only capture the main ACA occurrence regions, such as the tropical Atlantic region which is associated with outflows of African dust and smoke, but also weak ACA occurrence regions, such as Eastern Asia, and long-range transport regions, such as the Pacific, using both day- and night-time observations. Compared with the CALIPSO L evel-2 product, the newly derived day- and night-time global distributions of ACA properties show good agreement, benefitted from multiple moving-smoothing processes.

The relative vertical positions of the ACA and low clouds were presented and discussed. The results indicate that at high latitudes ($<-40°$ and $>\sim40°$), the ACA mainly resides above the low clouds at a distance greater than 0.5 km, indicating the semi-direct influence of the ACA on the cloud by stabilizing the atmospheric temperature profile. At low latitudes, the distance between the ACA and low clouds is smaller than 0.5 km. Especially over the Atlantic near the Saharan dust and African smoke source regions, the ACA mainly resides immediately above the clouds; this contrasts with the results derived from the CALIPSO operational product in previous studies. The ACA over the tropical Atlantic (dust region) and southeast Atlantic (smoke region) can directly mix with the cloud, influencing the lower cloud deck through both semi-effect and indirective effects. This highlights the complexity of the aerosol–cloud interactions over these regions, where the model still has difficulty reproducing the vertical aerosol structure and aerosol–cloud radiative effect [11–13].

To qualify the roles of these two effects, improve the passive cloud property retrieval, and constrain the relative model process, a height-resolved ACA database like the one developed in this study is needed. The new method developed in this study can provide a more complete and accurate global view of the ACA. Together with other satellite measurements such as MODIS [57] and CERES [58] and cloud property retrieval methods such as that described by Luo et al. [59], the radiative effects of the ACA and its influences on the macro- and micro-physics of low clouds will be further studied in future work. These results not only help us better understand global aerosol transportation and aerosol–cloud interactions but also provide useful information for model evaluation and improvements.

Author Contributions: Conceptualization, S.D.; Data curation, Y.W.; Formal analysis, W.Z. (Wenzhong Zhang); Funding acquisition, T.L.; Investigation, N.L.; Methodology, W.Z. (Wenzhong Zhang); Project administration, X.L.; Software, W.Z. (Wenzhong Zhang); Supervision, Y.H. and W.Z. (Wenyue Zhu); Validation, W.Z. (Wenzhong Zhang); Visualization, W.Z. (Wenzhong Zhang); Writing—original draft, W.Z. (Wenzhong Zhang); Writing—review and editing, T.L.

Funding: This research was funded by the Strategic Priority Research Program of Chinese Academy of Sciences, grant number XDA17010104. Tao Luo's work was also supported by the National Key Technology Support Program of China, grant number 2017YFC0209801 and Natural Science Foundation of China, grant number 41875041.

Acknowledgments: The authors thank the CALIPSO Team for providing data products, which were obtained from the NASA Langley Research Center Atmospheric Science Data Center. The authors are also grateful to the CloudSat team for making the data available; these data were obtained from the CloudSat Data Processing Center (http: //www.cloudsat.cira.colostate.edu). The authors would also like to thank the MERRA team for providing the data obtained from the Global Modeling and Assimilation Systems. The authors thank the OMACA team for providing data products, which were obtained from https://avdc.gsfc.nasa.gov/pub/data/satellite/Aura/OMI/V03/L2/OMACA. The authors acknowledge the anonymous reviewers and the editor for their valuable comments and suggestions.

Conflicts of Interest: The authors declare no conflict of interest.

References

1. Overpeck, J.; Rind, D.; Lacis, A.; Healy, R. Possible role of dust-induced regional warming in abrupt climate change during the last glacial period. *Nature* **1996**, *384*, 447–449. [CrossRef]
2. Tegen, I.; Lacis, A.A.; Fung, I. The influence on climate forcing of mineral aerosols from disturbed soils. *Nature* **1996**, *380*, 419–422. [CrossRef]
3. Alpert, P.; Kaufman, Y.J.; Shay-El, Y.; Tanre, D.; da Silva, A.; Schubert, S.; Joseph, J.H. Quantification of dust-forced heating of the lower troposphere. *Nature* **1998**, *395*, 367–370. [CrossRef]
4. Kinne, S.; Schulz, M.; Textor, C.; Guibert, S.; Balkanski, Y.; Bauer, S.E.; Berntsen, T.; Berglen, T.F.; Boucher, O.; Chin, M.; et al. An AeroCom initial assessment—Optical properties in aerosol component modules of global models. *Atmos. Chem. Phys.* **2006**, *6*, 1815–1834. [CrossRef]
5. Solomon, S.; Qin, D.; Manning, M.; Alley, R.B.; Berntsen, T.; Bindoff, N.L.; Chen, Z.; Chidthaisong, A.; Gregory, J.M.; Hegerl, G.C.; et al. *Technical Summary, Climate Change 2007: The Physical Science Basis, Contribution of Working Group I to the Fourth Assessment Report of the Intergovernmental Panel on Climate Change;*

Solomon, S., Qin, D., Manning, M., Marquis, M., Averyt, K.B., Tignor, M., Miller, H.L., Chen, Z., Eds.; Cambridge University Press: Cambridge, UK; New York, NY, USA, 2007; pp. 19–91.

6. Chaboureau, J.P.; Richard, E.; Pinty, J.P.; Flamant, C.; Girolamo, P.D.; Kiemle, C.; Behrendt, A.; Chepfer, H.; Chiriaco, M.; Wulfmeyer, V. Long-range transport of Saharan dust and its radiative impact on precipitation forecast: A case study during the convective and or graphically-induced precipitation study (COPS). *Q. J. R. Meteorol. Soc.* **2011**, *137*, 236–251. [CrossRef]

7. Zhang, X.Y.; Wang, Y.Q.; Niu, T.; Zhang, X.C.; Gong, S.L.; Zhang, Y.M.; Sun, J.Y. Atmospheric aerosol compositions in China: Spatial/temporal variability, chemical signature, regional haze distribution and comparisons with global aerosols. *Atmos. Chem. Phys.* **2012**, *12*, 779–799. [CrossRef]

8. Fan, J.; Leung, L.R.; DeMott, P.J.; Comstock, J.M.; Singh, B.; Rosenfeld, D.; Tomlinson, J.M.; White, A.; Prather, K.A.; Minnis, P.; et al. Aerosol impacts on California winter clouds and precipitation during CalWater 2011: Local pollution versus long-range transported dust. *Atmos. Chem. Phys.* **2014**, *14*, 81–101. [CrossRef]

9. Waquet, F.; Peers, F.; Ducos, F.; Goloub, P.; Platnick, S.; Riedi, J.; Tanré, D.; Thieuleux, F. Global analysis of aerosolproperties above clouds. *Geophys. Res. Lett.* **2013**, *40*, 5809–5814. [CrossRef]

10. Jethva, H.; Torres, O.; Ahn, C. A 12-year long global record of optical depth of absorbing aerosols above the clouds derived from the OMI/OMACA algorithm. *Atmos. Meas. Tech.* **2018**, *11*, 5837–5864. [CrossRef]

11. Schulz, M.; Textor, C.; Kinne, S.; Balkanski, Y.; Bauer, S.; Berntsen, T.; Berglen, T.; Boucher, O.; Dentener, F.; Guibert, S.; et al. Radiative forcing by aerosols as derived from the AeroCom present-day and pre-industrial simulations. *Atmos. Chem. Phys.* **2006**, *6*, 5225–5246. [CrossRef]

12. Stier, P.; Schutgens, N.A.J.; Bellouin, N.; Bian, H.; Boucher, O.; Chin, M.; Ghan, S.; Huneeus, N.; Kinne, S.; Lin, G.; et al. Host model uncertainties in aerosol radiative forcing estimates: Results from the Aero-Com Prescribed intercomparison study. *Atmos. Chem. Phys.* **2013**, *13*, 3245–3270. [CrossRef]

13. Peers, F.; Waquet, F.; Cornet, C.; Dubuisson, P.; Ducos, F.; Goloub, P.; Szczap, F.; Tanré, D.; Thieuleux, F. Absorption of aerosols above clouds from POLDER/PARASOL measurements and estimation of their direct radiative effect. *Atmos. Chem. Phys.* **2015**, *15*, 4179–4196. [CrossRef]

14. Chand, D.A.; Anderson, T.L.; Wood, R.; Charlson, R.J.; Hu, Y.; Liu, Z.; Vaughan, M. Quantifying above-cloud aerosol using spaceborne lidar for improved understanding of cloudy-sky direct climate forcing. *J. Geophys. Res.* **2008**, *113*, D13206. [CrossRef]

15. Chand, D.A.; Wood, R.; Anderson, T.L.; Satheesh, S.K.; Charlson, R.J. Satellite-derived direct radiative effect of aerosols dependent on cloud cover. *Nat. Geosci.* **2009**, *2*, 181–184. [CrossRef]

16. Waquet, F.; Riedi, J.; Labonnote, L.C.; Goloub, P.; Cairns, B.; Deuzé, J.-L.; Tanré, D. Aerosol remote sensing over clouds using A-train observations. *J. Atmos. Sci.* **2009**, *66*, 2468–2480. [CrossRef]

17. Torres, O.; Jethva, H.; Bhartia, P.K. Retrieval of aerosol optical depth above clouds from OMI observations: Sensitivity analysis and case studies. *J. Atmos. Sci.* **2012**, *69*, 1037–1053. [CrossRef]

18. Tsay, S.-C.; Hsu, N.-C.; Lau, W.K.-M.; Li, C.; Gabriel, P.M.; Ji, Q.; Holben, B.N.; Welton, E.J.; Nguyen, A.X.; Janjai, S.; et al. From BASE-ASIA toward 7-SEAS: A satellite-surface perspective of boreal spring biomass-burning aerosols and clouds in Southeast Asia. *Atmos. Environ.* **2013**, *78*, 20–34. [CrossRef]

19. Yu, H.; Zhang, Z. New Directions: Emerging satellite observations of above-cloud aerosols and direct radiative forcing. *Atmos. Environ.* **2013**, *72*, 36–40. [CrossRef]

20. Abel, S.J.; Highwood, E.J.; Haywood, J.M.; Stringer, M.A. The direct radiative effect of biomass burning aerosols over southern Africa. *Atmos. Chem. Phys.* **2005**, *5*, 1999–2018. [CrossRef]

21. Keil, A.; Haywood, J.M. Solar radiative forcing by biomass burning aerosol particles during SAFARI 2000: A case study based on measured aerosol and cloud properties. *J. Geophys. Res.* **2003**, *108*, 8467. [CrossRef]

22. Meyer, K.; Platnick, S.; Oreopoulos, L.; Lee, D. Estimating the direct radiative effect of absorbing aerosols overlying marine boundary layer clouds in the southeast Atlantic using MODIS and CALIOP. *J. Geophys. Res. Atmos.* **2013**, *118*, 4801–4815. [CrossRef]

23. Zhang, Z.; Meyer, K.; Platnick, S.; Oreopoulos, L.; Lee, D.; Yu, H. A novel method for estimating shortwave direct radiative effect of above-cloud aerosols using CALIOP and MODIS data. *Atmos. Meas. Tech.* **2014**, *7*, 1777–1789. [CrossRef]

24. Zuidema, P.; Redemann, J.; Haywood, J.; Wood, R.; Piketh, S.; Hipondoka, M.; Formenti, P. Smoke and clouds above the southeast Atlantic: Upcoming field campaigns probe absorbing aerosol's impact on climate. *BAMS* **2016**, *97*, 1131–1135. [CrossRef]

25. Shrestha, R.K.; Connolly, P.J.; Gallagher, M.W. Sensitivity of WRF cloud microphysics to simulations of a convective strom over the Nepal Himalayas. *Open Atmos. Sci. J.* **2017**, *11*, 29–43. [CrossRef]

26. Jethva, H.; Torres, O.; Remer, L.A.; Bhartia, P.K. A color ratio method for simultaneous retrieval of aerosol and cloud optical thickness of above-cloud absorbing aerosols from passive sensors: Application to MODIS measurements. *IEEE Trans. Geosci. Remote Sens.* **2013**, *51*, 3862–3870. [CrossRef]

27. Meyer, K.S.; Platnick, S.; Zhang, Z. Simultaneously inferring above-cloud absorbing aerosol optical thickness and underlying liquid phase cloud optical and microphysical properties using MODIS. *J. Geophys. Res.-Atmos.* **2015**, *120*, 5524–5547. [CrossRef]

28. Haywood, J.M.; Osborne, S.R.; Abel, S.J. The effect of overlying absorbing aerosol layers on remote sensing retrievals of cloud effective radius and cloud optical depth. *Q. J. R. Meteorol. Soc.* **2004**, *130*, 779–800. [CrossRef]

29. Wilcox, E.M.; Harshvardian; Platnick, S. Estimate of the impact of absorbing aerosol over cloud on the MODIS retrievals of cloud optical thickness and effective radius using two independent retrievals of liquid water path. *J. Geophys. Res.* **2009**, *114*, D05210. [CrossRef]

30. Coddington, O.M.; Pilewskie, P.; Redemann, J.; Platnick, S.; Russell, P.B.; Schmidt, K.S.; Gore, W.J.; Livingston, J.; Wind, G.; Vukicevic, T.L. Examining the impacts of overlying aerosols on the retrieval of cloud optical properties from passive remote sensing. *J. Geophys. Res.* **2010**, *115*, D10211. [CrossRef]

31. Alfaro-Contreras, R.; Zhang, J.; Campbell, J.R.; Holz, R.E.; Reid, J.S. Evaluating the impact of aerosol particles above cloud on cloud optical depth retrievals from MODIS. *J. Geophys. Res. Atmos.* **2014**, *119*, 5410–5423. [CrossRef]

32. Li, Z.; Zhao, F.; Liu, J.; Jiang, M.; Zhao, C.; Cribb, M. Opposite effects of absorbing aerosols on the retrievals of cloud optical depth from space-borne and ground-based measurements. *J. Geophys. Res. Atmos.* **2014**, *119*, 5104–5114. [CrossRef]

33. Jethva, H.; Torres, O.; Waquet, F.; Chand, D.; Hu, Y. How do A-train sensors intercompare in the retrieval of above-cloud aerosol optical depth? A case study-based assessment. *Geophys. Res. Lett.* **2014**, *41*, 186–192. [CrossRef]

34. Costantino, L.; Bréon, F.-M. Aerosol indirect effect on warm clouds over South-East Atlantic, from co-located MODIS and CALIPSO observations. *Atmos. Chem. Phys* **2013**, *13*, 69–88. [CrossRef]

35. Min, M.; Zhang, Z. On the influence of cloud fraction diurnal cycle and sub-grid cloud optical thickness variability on all-sky direct aerosol radiative forcing. *J. Quant. Spectrosc. Ra* **2014**, *142*, 25–36. [CrossRef]

36. Zhang, Z.; Meyer, K.; Yu, H.; Platnick, S.; Colarco, P.; Liu, Z.; Oreopoulos, L. Shortwave direct radiative effects of above-cloud aerosols over global oceans derived from 8 years of CALIOP and MODIS observations. *Atmos. Chem. Phys* **2016**, *16*, 2877–2900. [CrossRef]

37. Kacenelenbogen, M.; Redemann, J.; Vaughan, M.A.; Omar, A.H.; Russell, P.B.; Burton, S.L.; Rogers, R.R.; Ferrare, R.; Hostetler, C.A. An evaluation of CALIOP/CALIPSO's aerosol-above-cloud detection and retrieval capability over North America. *J. Geophys. Res. Atmos.* **2014**, *119*, 230–244. [CrossRef]

38. Liu, Z.; Winker, D.; Omar, A.; Vaughan, M.; Kar, J.; Trepte, C.; Hu, Y.; Schuster, G. Evaluation of CALIOP 532 nm aerosoloptical depth over opaque water clouds. *Atmos. Chem. Phys.* **2015**, *15*, 1265–1288. [CrossRef]

39. Winker, D.M.; Vaughan, M.A.; Omar, A.; Hu, Y.; Powell, K.A.; Liu, Z.; Hunt, W.H.; Young, S.A. Overview of the CALIPSO mission and CALIOP data processing algorithms. *J. Atmos. Ocean. Tech.* **2009**, *26*, 2310–2323. [CrossRef]

40. Li, J.; Hu, Y.; Huang, J.; Stamnes, K.; Yi, Y.; Stamnes, S. A New Method for Retrieval of the Extinction Coefficient of Water Clouds by using the Tail of the CALIOP Signal. *Atmos. Chem. Phys.* **2011**, *11*, 2903–2916. [CrossRef]

41. Wang, Z.; Vane, D.; Stephens, G.; Reinke, D. Level 2 Combined Radar and Lidar Cloud Scenario Classification Product Process Description and Interface Control Document. JPL Rep, 22. Available online: http://www.cloudsat.cira.colostate.edu/sites/default/files/products/files/2B-CLDCLASS-LIDAR_PDICD.P_R04.20120522.pdf (accessed on 18 February 2019).

42. Li, X.; Zheng, X.; Zhang, D.; Zhang, W.; Wang, F.; Deng, Y.; Zhu, W. Clouds over East Asia Observed with CollocatedCloudSat and CALIPSO Measurements: Occurrence and Macrophysical Properties. *Amosphere* **2018**, *9*, 168. [CrossRef]

43. Draper, C.S.; Reichle, R.H.; Koster, R.D. Assessment of MERRA-2 land surface energy flux estimates. *J. Clim.* **2018**, *31*, 671–691. [CrossRef]

44. Rajapakshe, C.; Zhang, Z.; Yorks, J.E.; Yu, H.; Tan, Q.; Meyer, K.; Platnick, S.; Winker, D.M. Seasonally transported aerosol layers over southeast Atlantic are closer to underlying clouds than previously reported. *Geophys. Res. Lett.* **2017**, *44*, 5818–5825. [CrossRef]

45. Alfaro-Contreras, R.; Zhang, J.; Campbell, J.R.; Reid, J.S. Investigating the frequency and interannual variability in global above-cloud aerosol characteristics with CALIOP and OMI. *Atmos. Chem. Phys.* **2016**, *16*, 47–69. [CrossRef]

46. Omar, A.H.; Winker, D.M.; Vaughan, M.A.; Hu, Y.; Trepte, C.R.; Ferrare, R.A.; Lee, K.-P.; Hostetler, C.A.; Kittaka, C.; Rogers, R.R.; et al. The CALIPSO automated aerosol classification and lidar ratio selection algorithm. *J. Atmos. Ocean. Technol.* **2009**, *26*, 1994–2014. [CrossRef]

47. Fernald, F.G. Analysis of atmospheric lidar observations: Some comments. *Appl. Opt.* **1984**, *23*, 652–653. [CrossRef] [PubMed]

48. Carlson, T.N.; Prospero, J.M.; Hanson, K.J. *Attenuation of Solar Radiation by Windborne Saharan Dust Off the West Coast of Africa*; NOAA Technical Memorandum ERL WMPO-7; U.S. Department of Commerce: Boulder, CO, USA, 1973.

49. Karyampudi, V.M.; Palm, S.P.; Reagen, J.A.; Fang, H.; Grant, W.B.; Hoff, R.M.; Moulin, C.; Pierce, H.F.; Torres, O.; Browell, E.V.; et al. Validation of the Saharan dust plume conceptual model using lidar, meteosat, and ECMWF data. *Bull. Am. Meteorol. Soc.* **1999**, *80*, 1045–1076. [CrossRef]

50. Campbell, J.R.; Welton, E.J.; Spinhirne, J.D.; Ji, Q.; Tsay, S.-C.; Piketh, S.J.; Barenbrug, M.; Holben, B.N. Micropulse lidar observations of tropospheric aerosols over northeastern South Africa during the ARREX and SAFARI-2000 dry season experiments. *J. Geophys. Res.* **2003**, *108*, 8497. [CrossRef]

51. Yu, H.; Zhang, Y.; Chin, M.; Liu, Z.; Omar, A.; Remer, L.A.; Yang, Y.; Yuan, T.; Zhang, J. An integrated analysis of aerosol above clouds from A-Train multi-sensor measurements. *Remote Sens. Environ.* **2012**, *121*, 125–131. [CrossRef]

52. Devasthale, A.; Thomas, M.A. A global survey of aerosol-liquid water cloud overlap based on four years of CALIPSO-CALIOP data. *Atmos. Chem. Phys.* **2011**, *11*, 1143–1154. [CrossRef]

53. Saulo, R.F.; Karla, M.L.; Maria, A.F.S.D.; Pedro, L.S.D.; Robert, C.; Elaine, P.; Paulo, A.; Georg, A.G.; Fernando, S.R. Monitoring the Transport of Biomass Burning Emissions in South America. *Environ. Fluid Mech.* **2005**, *5*, 135–167.

54. Mallet, M.; Cravigan, L.; Milic, A.; Alroe, J.; Ristovski, Z.; Ward, J.; Keywood, M.; Williams, L.; Miljevic, B. Composition, size and cloud condensation nuclei activity of biomass burning aerosol from north Australian savannah fires. *Atmos. Chem. Phys.* **2017**, *17*, 3605–3617. [CrossRef]

55. Torres, O.; Bhartia, P. OMI measurements of aerosol absorption over Central America. *AGU Spring Meeting*. 2007. Available online: https://www.researchgate.net/publication/253399702_OMI_measurements_of_aerosol_absorption_over_Central_America (accessed on 13 October 2019).

56. Han, Y.; Fang, X.; Zhao, T.; Kang, S. Long range trans-Pacific transport and deposition of Asian dust aerosols. *J. Environ. Sci.* **2008**, *20*, 424–428. [CrossRef]

57. Platnick, S.; King, M.D.; Ackerman, S.A.; Menzel, W.P.; Baum, B.A.; Riedi, J.C.; Frey, R.A. The MODIS cloud products: Algorithms and examples from Terra. *IEEE Trans. Geosci. Remote Sens.* **2003**, *41*, 459–473. [CrossRef]

58. Minnis, P.; Sun-Mack, S.; Young, D.F.; Heck, P.W. CERES Edition-2 Cloud Property retrievals using TRMM VIRS and Terra and Aqua MODIS data. Part I: Algorithms. Part II: Examples of average results and comparison with other data. *IEEE Trans. Geosci. Remote Sens.* **2011**, *49*, 4374–4430. [CrossRef]

59. Luo, T.; Wang, Z.; Li, X.; Deng, S.; Huang, Y.; Wang, Y. Retrieving the polar mixed-phase cloud liquid water path by combining CALIOP and IIR measurements. *J. Geophys. Res.* **2018**, *123*, 1755–1770. [CrossRef]

Article

Vertical Structures of Dust Aerosols over East Asia Based on CALIPSO Retrievals

Di Liu [1], Tianliang Zhao [1,*], Richard Boiyo [1,2], Siyu Chen [3,*], Zhengqi Lu [1], Yan Wu [4] and Yang Zhao [5,6]

[1] Collaborative Innovation Centre on Forecast and Evaluation of Meteorological Disasters, Key Laboratory of Meteorological Disaster, Ministry of Education (KLME), International Joint Laboratory on Climate and Environment Change (ILCEC), Key Laboratory for Aerosol-Cloud-Precipitation of China Meteorological Administration, School of Atmospheric Physics, Nanjing University of Information Science and Technology, Nanjing 210044, China; 20161102059@nuist.edu.cn (D.L.); rkipkemboi@must.ac.ke (R.B.); 20181103042@nuist.edu.cn (Z.L.)

[2] Department of Physical Sciences, Meru University of Science and Technology, P.O. Box 972-60200 Meru, Kenya

[3] Key Laboratory for Semi-Arid Climate Change of the Ministry of Education, College of Atmospheric Sciences, Lanzhou University, Lanzhou 730000, China

[4] Heilongjiang Meteorological Bureau, Harbin 150030, China; lsd@nuist.edu.cn

[5] State Key Laboratory of Severe Weather, Chinese Academy of Meteorological Sciences, Beijing 100081, China; 20161101009@nuist.edu.cn

[6] Nanjing University of Information Science and Technology, Nanjing 210044, China

*   Correspondence: tlzhao@nuist.edu.cn (T.Z.); chensiyu@lzu.edu.cn (S.C.)

Received: 1 February 2019; Accepted: 21 March 2019; Published: 23 March 2019

Abstract: The spatiotemporal and especially the vertical distributions of dust aerosols play crucial roles in the climatic effect of dust aerosol. In the present study, the spatial-temporal distribution of dust aerosols over East Asia was investigated using Cloud-Aerosol Lidar and Infrared Pathfinder Satellite Observations (CALIPSO) retrievals (01/2007–12/2011) from the perspective of the frequency of dust occurrence (FDO), dust top layer height (TH) and profile of aerosol subtypes. The results showed that a typical dust belt was generated from the dust source regions (the Taklimakan and Gobi Deserts), in the latitude range of 25°N~45°N and reaching eastern China, Japan and Korea and, eventually, the Pacific Ocean. High dust frequencies were found over the dust source regions, with a seasonal sequence from high to low as follows: spring, summer, autumn and winter. Vertically, FDOs peaked at about 2 km over the dust source regions. In contrast, FDOs decreased with altitude over the downwind regions. On the dust belt from dust source regions to downwind regions, the dust top height (TH) was getting higher and higher. The dust TH varied in the range of 1.9–3.1 km above surface elevation (a.s.e.), with high values over the dust source regions and low values in the downwind areas, and a seasonally descending sequence of summer, spring, autumn and winter in accord with the seasonal variation of the boundary layer height. The annual AOD (Aerosol Optical Depth) was generally characterized by two high and two low AOD centers over East Asia. The percent contribution of the Dust Aerosol Optical Depth to the total AOD showed a seasonal variation from high to low as follows: spring, winter, autumn and summer. The vertical profile of the extinction coefficient revealed the predominance of pure dust particles in the dust source regions and a mixture of dust particles and pollutants in the downwind regions. The dust extinction coefficients over the Taklimakan Desert had a seasonal pattern from high to low as follows: spring, winter, summer and autumn. The results of the present study offered an understanding of the horizontal and vertical structures of dust aerosols over East Asia and can be used to evaluate the performance aerosol transport models.

Keywords: CALIPSO; dust top height; frequency of dust occurrence; pure dust; polluted dust; extinction coefficient

1. Introduction

Dust aerosol, one of the most important aerosol species, modifies the energy balance and the hydrologic cycle directly by absorbing and scattering solar radiation and indirectly by altering cloud microphysical properties [1–3]. East Asia, the second largest contributor of dust aerosols in the world, emits nearly 600 tons of dust particles into the atmosphere annually through the erosion of soil mainly from natural conditions and partly from human activities [4,5]. Approximately 30% of the dust particles redeposit back into emission regions, 20% are transported to eastern China, and the rest are transported to the Pacific Ocean and beyond by westerly jets which can significantly affect the Asian monsoon system [6,7]. The deposition of dust particles serves as a key mineral supplement for the marine biopshere and for remote rainforests, thus altering the global carbon cycle [8].

It is well known that the vertical structure of dust aerosols plays a crucial role in the atmospheric thermal structure and the aerosol radiative forcing [9,10]. The vertical distribution of dust aerosols largely determines their residence and transport time [11]. Dust aerosols can also uplift to the upper troposphere and travel around the world via westerlies [12]. However, previous work using aircraft, upper-air sounding balloons and ground-based platforms are limited in space and time, offering little information on the vertical structure of dust aerosols. These limitations have been partially addressed with the launch of the Cloud-Aerosol Lidar and Infrared Pathfinder Satellite Observations (CALIPSO) satellite [13,14]. Huang et al. provided an analysis of the distribution of aerosols of different types on a global scale on the basis of five-year CALIPSO data [15]. Guo et al. used CALIPSO, MODIS (Moderate Resolution Imaging Spectroradiometer) and OMI (Ozone Monitoring Instrument) measurements to build three-dimensional (3D) patterns of the occurrence frequency of aerosol, dust and smoke in China [16]. Proestakis et al. described the 3D distribution of dust aerosols over southeastern Asia [17]. Nan et al. investigated the vertical distribution of dust particles for the Taklimakan Desert and for the downwind regions using CALIPSO measurement [18].

In this study, we used five years (January 2007 to December 2011) of CALIPSO Level 3 data to gain further insights into the generation, emission, transport, distribution and speciation of aerosols over East Asia. East Asia was divided into five regions to achieve a stronger understanding of the spatial distribution of dust aerosols in relation to emission sources and transport processes. For each region, the analysis was made of the frequency of dust occurrence (FDO), dust Top Height (TH), Aerosol Optical Depth (AOD), percent contribution of dusts to the total AOD (D_AOD), light extinction coefficient, speciation (pure dust versus pollution aerosols) and horizontal dust transport flux.

A previous study conducted over our study domain concentrated mainly on individual dust storm events involving pure dust [19]. However, human health concerns and policy on pollution controls require detailed knowledge of the speciation of dust (pure dust versus dust mixed with biomass burning smoke) and the mean behaviors of total dust (sum of pure and polluted dust). Another related study documented the seasonal variations of the aerosol extinction profile and occurrence frequency using five-year Level 3 CALIPSO products, focusing on global patterns [15]. In the present study, we also used the Level 3 products, but focused on regional patterns and specifically on the contrast between pure dusts generated from dust source regions in western China and those mixed with smoke from pollution source regions in eastern China. By analyzing spatial patterns of AOD, D_AOD and profiles of extinction coefficients of various aerosol types, we further investigated into long-distance dust transport, which is known to impact the public health of populated centers in eastern China, Korea and Japan.

Section 2 outlines the study domain, data and methodology related to the present work. The results and discussion are given in Section 3. Section 4 summarizes the main findings of the present study.

2. Data and Methods

CALIPSO was launched into a sun-synchronous orbit in April, 2006 with a repeating cycle of 16 days. The main instruments aboard CALIPSO include a Cloud-Aerosol Lidar with Orthogonal Polarization (CALIOP), a Wide Field Camera (WFC) and an Infrared Imager Radiometer (IIR).

These instruments can not only retrieve the vertical profiles of clouds and aerosols, but also provide information concerning cloud-aerosol interactions and surface radiative budgets [20]. CALIPSO continuously retrieves profiles of attenuated backscatter at 532 and 1064 nm and of polarized backscatter at 532 nm in the latitude range of 82°N–82°S with high horizontal and vertical resolutions of 333 m and 30–60 m, respectively [21].

Three-level products are included in CALIPSO data. Level 1 (L1) products offer a large volume of raw signals of high spatial resolutions. They are classified with a cloud-aerosol discrimination (CAD) algorithm into clouds and aerosols products, with negative and positive values representing aerosols and clouds, respectively. Level 2 (L2) products consist of six aerosol subtypes classified using a scene classification algorithm [22]. The six aerosol subtypes are pure dust, polluted dust, smoke, clean continental, polluted continental and clean marine aerosols with the corresponding Lidar ratios for the retrieval of aerosol light extinction coefficient. Level 3 (L3) products provide specific monthly variables, such as monthly mean AOD and extinction coefficients, from the L2 products. Relative to data products from other space-borne sensors, the CALIPSO data offer three main advantages for the investigation of the vertical distribution and transport of dust particles. First, CALIPSO can retrieve profiles of backscatter in different atmospheric layers to estimate the vertical profiles of aerosols and clouds, can distinguish between dust and other types of aerosols based on the observed depolarization ratios, and can minimize the interference associated with prevailing surface conditions, and hence, are able to retrieve aerosol profiles where other space-borne sensors may not be able to. Second, being an active remote sensor, it can retrieve information on aerosols and clouds in both daylight and night hours [23]. Third, CALIPSO can retrieve the vertical distributions of different aerosol subtypes; such information is crucial for understanding the long-range transport of aerosols at the regional and the global scales.

L3 is a globally monthly product with a $2° \times 5°$ latitude-longitude grid, and a vertical resolution of 60 m from the ground to 12 km with a total of 208 layers. Uncertainties exist when CALIPSO retrieves optical properties [24]. Data on extinction below the height of 180 m were excluded as a precautionary measure to avoid biases resulting from the physical interpretation of near-ground aerosols [25]. Following [15], we also only considered extinction coefficients of >0.001 km^{-1} to ensure high levels of accuracy.

To better understand the distribution of dust aerosols over different surface types, five homogenous regions over East Asia were analyzed in the present study (Figure 1). They include the Taklimakan Desert (TD; dust source region), the Gobi Desert (GD; dust source region), northern China (NC; a region affected by anthropogenic aerosols), southern China (SC; also under the influence of anthropogenic aerosols) and Korea-Japan (KJ; under the influence of marine aerosols).

The FDO is defined as the ratio of the number of CALIPSO overpasses with dust observations to the total number of CALIPSO overpasses (including conditions of clear air and aerosols) [17]:

$$\text{FDO} = \frac{P_{dust}}{P_{all}} \tag{1}$$

where P_{dust} represents the number of CALIPSO dust overpasses and P_{all} denotes the total number of CALIPSO overpasses.

The seasonally averaged CALIPSO L3 dust profile Top Height (TH) is by definition the height at which dust AOD contribution (D_AOD) aggregates to 98% [17].

The dust column concentration (M_{du}) and D_AOD (τ_{du}) are calculated with Equations (2) and (3) [26];

$$M_{du} = \left(\frac{\rho 4\pi}{3}\right) \int r^3 n(r) dr \tag{2}$$

$$\tau_{du}\tau_{du} = \pi \int Q(r) r^2 n(r) dr \tag{3}$$

where $n(r)$ represents particle size distribution of dust aerosol, ρ is dust aerosol density and Q is extinction index. Following other researchers [26], we obtain:

$$M_{du} = 2.7 \times \tau_{du} \text{ g m}^{-2} \tag{4}$$

according to Kaufman et al. (2005), the horizontal dust transport flux (F) is calculated with Equation (5) [26]:

$$F = M_{du} \times W \times L \text{ g s}^{-1} \tag{5}$$

where W represents the monthly mean west wind speed (m s^{-1}), M_{du} is the monthly mean column dust concentration (g m^{-2}), and L is longitudinal length (m). Following other researchers [26–28], we adopted wind fields of 500 hPa over the TD and GD and of 850 hPa over the NC and SC.

For the purpose of validation, the CALIPSO AOD and pure dust D_AOD were compared with those from the Modern-Era Retrospective analysis for Research and Applications, version 2 (MERRA-2), over the same time period. MERRA-2 calculation of these parameters is driven by anthropogenic emission inventories and natural dust emission parameterization and is bias-corrected by assimilation of MODIS and AVHRR satellite AOD products [29].

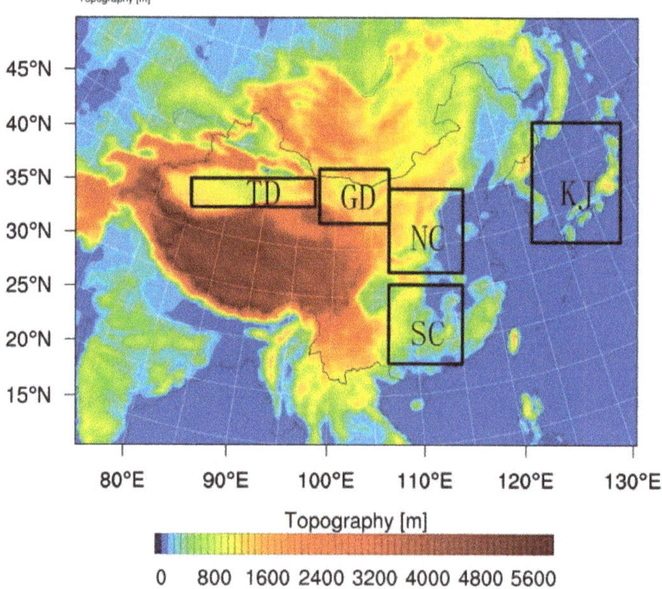

Figure 1. Topographical map of the study domain. Homogenous regions highlighted in black boxes include: (1) the Taklimakan Desert (TD), (2) the Gobi Desert (GD), (3) Northern China (NC), (4) Southern China (SC) and (5) Korea-Japan (KJ).

3. Results and Discussion

3.1. Frequency of Dust Occurrence (FDO)

The TD is the largest desert in China and the second largest shifting desert in the world. It emits large volumes of dust particles into the atmosphere annually. Most of the emitted particles are deposited back to the ground primarily as the result of weak tropospheric winds and complex topography: the center of the Tarim Basin is surrounded by mountains toward the south, north and west [30]. Dust episodes are less frequent but more severe over the GD. Most dust particles emitted by the GD are easily uplifted to the troposphere and transported downwind primarily due to the plateau topography (910–1520 m) of the GD. Additionally, as the main desert region of East Asia, the GD includes stationary deserts, such as the Tengger Desert and Badain Jaran Desert, and shifting deserts, such as the Mu Us Sandland. Over the NC and SC, abundant anthropogenic aerosols are mixed with pure dust through long-range transport, forming polluted dusts [31]. The KJ region suffers not only from polluted dust transported by westerlies from upwind regions (e.g., the NC and SC), but also from locally-formed marine aerosols. To understand the generation and transport of dust aerosols over East Asia, we divided the study period into four seasons, March–April–May (MAM), June–July–August (JJA), September–October–November (SON) and December–January–February (DJF), based on prevailing climatological conditions observed over the study domain.

FDO values show significant levels of spatiotemporal heterogeneity, in order from high to low as follows: MAM, JJA, SON and DJF (Figure 2). The highest FDOs of 18.4% were observed over the TD in MAM and decreased in JJA with the highest values of 17.1%. Distributions of FDOs observed during SON corresponded to those observed in JJA but with a lower value of 14.6% over the TD and higher values observed over the NC. The FDOs observed during DJF were the lowest relative to those of other seasons with the highest value of 11% found over the NC. These results were consistent with those reported previously for East Asia [17]. The spatial pattern of seasonally averaged FDOs was characterized by two high FDO centers over dust source regions (the TD and GD). There existed a dust belt from the dust source regions (the TD and GD) to eastern China, Japan, Korea and beyond, in the latitudinal range of 25°N to 45°N. In MAM, FDOs over KJ reached as high as 11 %, which was close to the value for the GD dust source region, indicating efficient long-distance transport in this season. That two distinct FDO patterns exited over the TD and GD implied that the diffusion and long-range transport of dust aerosols from the TD were limited, or to put it differently, most dust aerosols emitted from the TD could not be transported to downwind regions since the TD was mainly surrounded by rugged mountains [3]. Since the whole study domain is under the control of westerlies and obstacles from the Himalayan Mountains, dust aerosols from South Asia and India enter continental China from southwestern China, affecting the NC, SC and KJ in MAM and DJF [32–34]. In the NC, the highest value of FDOs of 16.1% was observed during MAM.

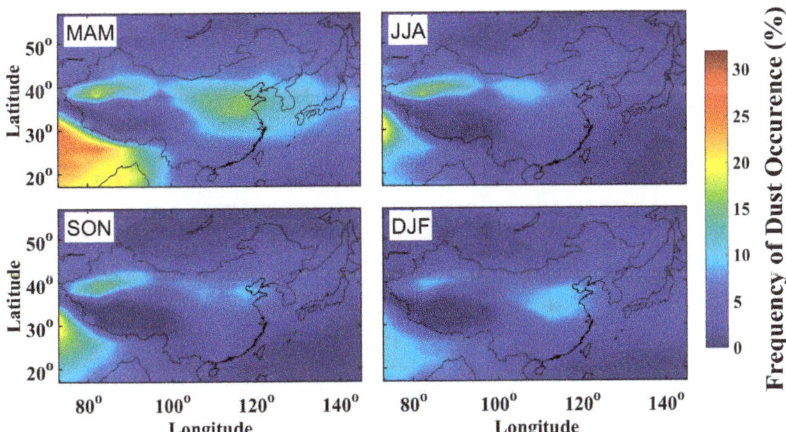

Figure 2. The seasonal variations in the frequency of dust occurrence over East Asia based on Cloud-Aerosol Lidar and Infrared Pathfinder Satellite Observations (CALIPSO) data.

The vertical distributions of the FDO over the dust source regions (TD and GD) follow a comparable pattern, with significant seasonality (Figure 3). The profile peaked at a height of about 2 km above the surface; beyond this height, the FDO decreased with increasing height. In other words, large volumes of dust aerosols were uplifted into the lower troposphere from the dust source regions, where they could undergo mixing and long-range transport (to eastern China and even to the western Pacific Ocean). The peak FDO values were 52.0% and 40.0% in MAM and 43.5% and 28.5% in JJA over the TD and GD, respectively.

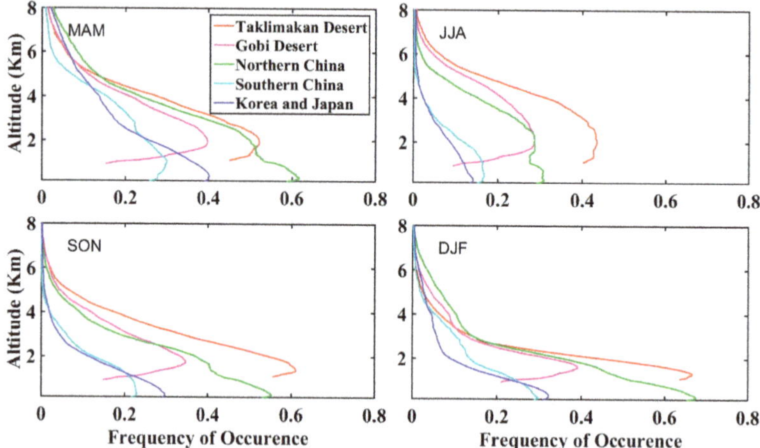

Figure 3. Vertical distribution of the frequency of dust occurrence over: (1) the Taklimakan Desert (red line), (2) the Gobi Desert (magenta line), (3) Northern China (green line), (4) Southern China (light blue line) and (5) Korea-Japan (dark blue) based on CALIPSO data.

Over the regions downwind of these sources (the NC, SC and KJ), FDO decreased with increasing height from the ground to the upper troposphere. The closer to the dust source regions, the higher the FDO: of these three regions, the overall FDO profile showed the highest values over the NC and lowest values over KJ, with the profile over SC generally falling in between. At the NC, the near surface FDO was greater than 60% in MAM, SON and DJF with the highest FDO of 67.1% recorded during DJF. This high value could be attributed to the proximity to the dust regions and to the dominance of aerosols generated by use of fossil fuels [35]. FDOs over the SC ranged from 27.0% to 29.9%, with peak values of 29.9% observed during MAM. The highest FDO observed over the KJ (40.2%) occurred during MAM, a pattern that resembled that of the SC. These profiles confirmed the presence of a dust belt in the latitude range of 25°N to 45°N that transported significant volumes of dust from the dust source regions.

3.2. Seasonal Distribution of Dust Top Height

The information on the dust TH can help elucidate mechanisms of long-range transport of the dust aerosols. The dust TH, which is defined as the height above surface elevation (a.s.e.), shows significant seasonal variations over East Asia (Figure 4). Over the TD and GD, the dust TH was the largest in MAM with an average value of 3.1 km (a.s.e.) and the lowest in DJF with an average value of 1.9 km (a.s.e.). Note that the variations here resembled those of the boundary layer height [36]. In comparison, over the downwind regions (the NC, SC and KJ), the dust TH values were still the highest in MAM with a range of 3.9~5.0 km, and the lowest in SON with a range of 2.7~3.2 km. The dust belts were evident in all four seasons. During MAM, the dust TH increased along the dust belt from the dust source regions to the downwind regions, implying progressive vertical expansion of the dust layer as the airmass absorbed dust particles from the ground along its transport pathway. Although not shown in Figure 4, other studies have demonstrated that, aided by the ascending movement of the East Asian and the North American troughs, the vertical expansion of the dust layer can continue all the way to North America [37]. Thus, these patterns of dust storm explain why the highest dust TH values were recorded during MAM in the downwind regions.

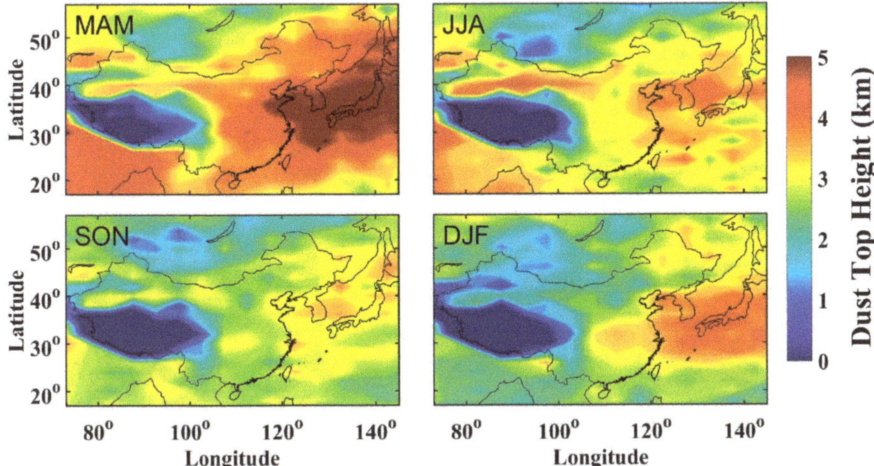

Figure 4. The seasonal distribution of dust Top Height (km) over East Asia based on CALIPSO data.

3.3. Seasonal Distribution of AOD and the Percentage of D_AOD to the Total AOD

To develop a stronger understanding of the contributions of dust aerosols to atmospheric aerosols, it is necessary to further analyze dust aerosols generated from anthropogenic activities by investigating the variations of AOD and the percent contribution of dust aerosols to the total AOD (D_AOD). The seasonally averaged AOD is generally characterized by two high AOD centers and two low AOD centers over East Asia (Figure 5). Economically developed and industrialized areas of eastern China and areas under the influence of natural dust sources constituted the high AOD centers. In contrast, less developed areas with smaller populations across the Tibetan Plateau and eastern Inner Mongolia constituted the low AOD centers. In addition to natural dust aerosols, aerosols emitted by anthropogenic activities, such as smoke particles from burning of agricultural biomass, sulphate and black and organic carbon aerosols from industrial activities could be responsible for enhanced levels of aerosol loadings during DJF over the SC with a high AOD value of 0.88 [38–40].

The data on D_AOD reveal once again a distinct dust belt, especially in MAM and DJF. In these two seasons, high percentages (>70%) were found along the belt that extends from the dust source regions to the coastal line of eastern China. The large percentage of greater than 95% observed over the dust source regions indicated that natural dust aerosols served as the most important component of atmospheric aerosols. Meanwhile, obvious seasonal variations were observed. The highest D_AOD during MAM was closely related to frequent and intensive dust storm events in the dust source regions. The distribution of D_AOD in SON was similar to that observed in MAM but with lower percentage values [41]. In JJA, the dust belt was much smaller in extent, limited mostly to the dust source regions (the TD and GD) but with abnormally high percentages (>85%). The small spatial extent of the dust belt in JJA can be partially explained by the prevailing climate: JJA is the rainy season in Eastern China, Japan and Korea, and most of the dust aerosols are removed by wet deposition.

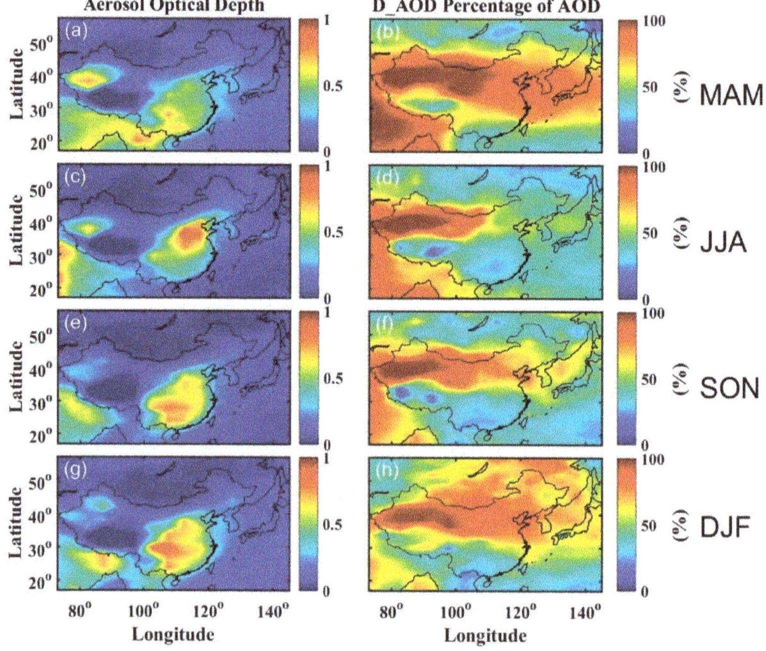

Figure 5. Seasonal distribution of Aerosol Optical Depths and percentages of D_AOD to the total AOD over East Asia based on CALIPSO data.

3.4. Extinction Coefficient of Pure and Polluted Dust

The vertical profiles of extinction coefficients for the three main aerosol types (total aerosol, pure dust and polluted dust) are presented in Figure 6. Here, polluted dust is defined as the mixture of pure dust and smoke particles generated from biomass burning, and total aerosol includes clean marine, pure dust, polluted dust, polluted continental, clean continental, polluted dust and smoke components [42]. Except for a small proportion of polluted dust in DJF, pure dust was the only component of the total aerosols over the dust source regions (TD and GD), showing that pure dust dominated throughout the year. In these two regions, high dust extinction coefficients (TD: 0.17 km^{-1}; GD: 0.06 km^{-1}) were observed at an altitude of 1~2 km during MAM.

As evidenced in Figure 6s,t, extinction coefficients of polluted dust (<0.04 km^{-1}) were observed over the dust source regions (the TD and GD) in the winter, largely as the result of fossil fuel burning in the winter heating period. Over the NC, the region downwind of the deserts, extinction coefficients of pure dust were much lower than those of polluted dust, indicating the dominate role of anthropogenic activities in this region across all four seasons. Another notable feature about NC is that extinction coefficients of pure dust were higher than those of polluted dust above the height of 1.1 km in MAM (Figure 6c), again confirming that emission and transport of dusts from the deserts were large contributors to air quality problems during this time of the year.

Over the SC and KJ, aerosol extinction profiles did not show considerable seasonal variations, and major aerosol subtypes could be listed in descending order as follows: total aerosols, polluted dust and pure dust. The extinction coefficients of pure dues were less than 0.04 km^{-1} year-round while polluted dust accounted for 40~60% of total aerosols, illustrating that local pollution sources dominated over the role of dust long-distance transport.

To further understand the dust transport pattern, we present the vertical distribution of the dust extinction coefficient along the W-E transact at 40°N (Figure 7). The highest dust extinction coefficients were observed over the TD (80–97.5°E). If we used a threshold value of 0.1 km^{-1}, the dust layer in this longitudinal range extended to a height of 4 km in MAM and JJA, 3 km in SON, and roughly 2.5 km in DJF. Fewer dust storm events occur in JJA than in MAM. However, it appears that such dust devils were able to generate large volumes of dust aerosols through localized disturbances, and because of strong convections, the emitted dust particles were uplifted to up to 4 km beyond the boundary layer height (BLH), where strong wind caused some dust aerosols to be transported to the downwind regions [36,37,43]. On the other hand, due to prevailing topographic conditions, only some particles were successfully transported to the downwind regions on account of topographic obstacles, especially in SON and DJF [3].

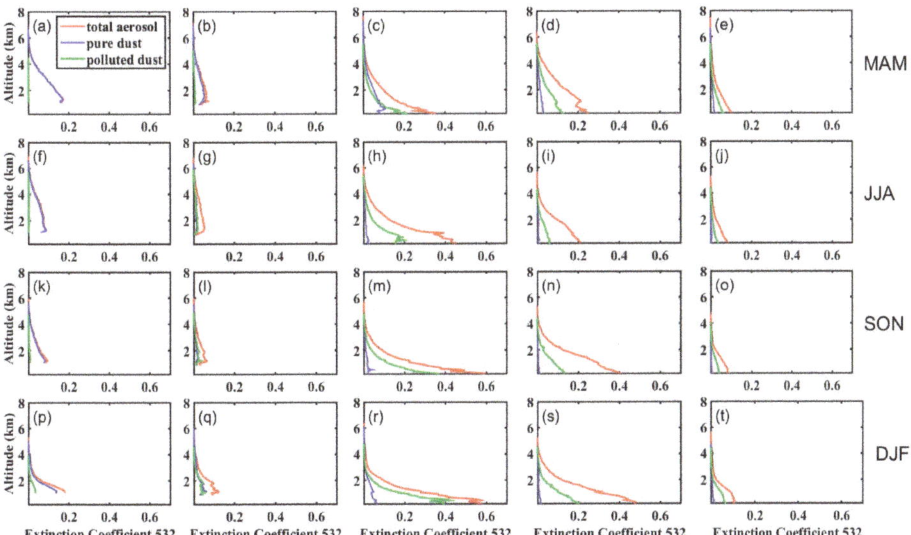

Figure 6. Profiles of the seasonal average aerosol extinction coefficient (km^{-1}) derived from CALIPSO observations over 1. The Taklimakan Desert (**a,f,k,p**), 2. The Gobi Desert (**b,g,l,q**), 3. Northern China (**c,h,m,r**), 4. Southern China (**d,i,n,s**) and 5. Korea-Japan (**e,j,o,t**).

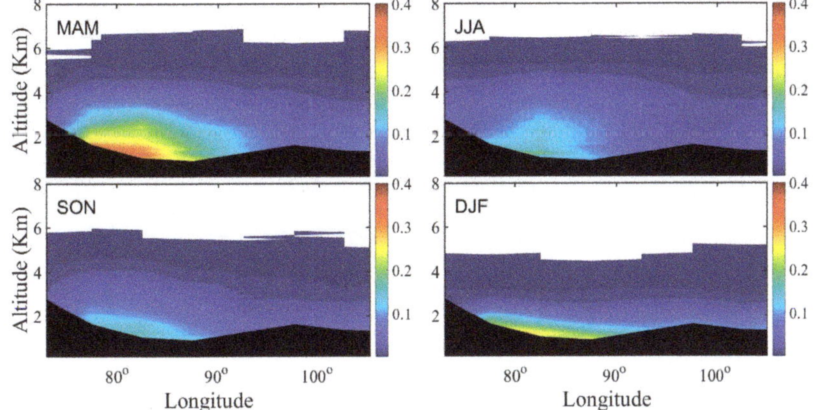

Figure 7. Seasonal cross sections of total dust extinction coefficients (km^{-1}) along a latitude of 40°N and longitudinal range of 73°E to 105° E based on CALIPSO data.

3.5. Horizontal Dust Transport Flux

Figure 8 shows the horizontal dust transport flux for the five homogenous regions of East Asia. Unsurprisingly, horizontal dust transport flux was the highest over the dust source regions: the TD followed by the GD. Seasonally, the horizontal total dust transport flux over all the five regions was, in descending order: DJF, MAM, SON and JJA. The two dust source regions emitted 95.2 Tg, 35.8 Tg, 46.9 Tg and 82 Tg of dust aerosols into the atmosphere during MAM, JJA, SON and DJF, respectively. The NC and SC sources contributed 35.4%, 15.6%, 31.7% and 54.8% to the total in MAM, JJA, SON and DJF, respectively, and the corresponding flux fractions for the KJ amounted to 9.9%, 6.7%, 6.1% and 12.4%. Notably, since a large percent of the dust over the KJ region originated from the TD, the seasonal variation of the horizontal dust transport flux over the KJ resembled that observed over the TD, with higher levels observed in MAM and DJF and lower levels recorded in JJA and SON.

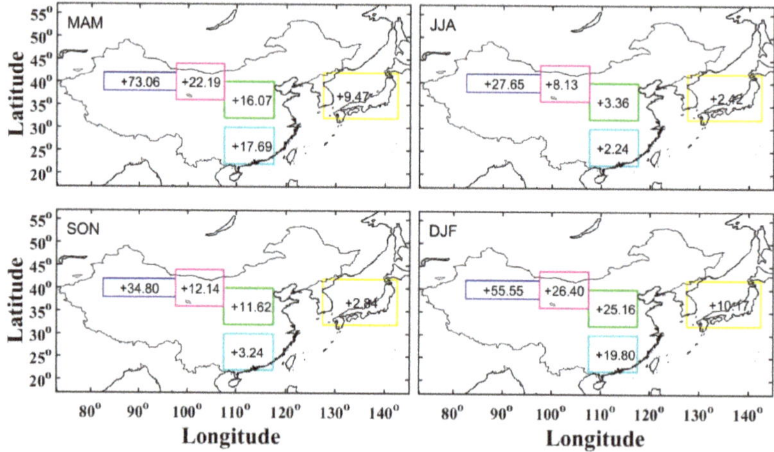

Figure 8. Seasonal distribution of horizontal dust transport flux (Tg) over East Asia (dark blue box: Taklimakan Desert, magenta box: Gobi Desert, green box: Northern China, light blue box: Southern China and yellow box: Korea-Japan).

3.6. Comparison with MERRA-2

Supplementary Figure S1 presents the MERRA-2 AOD and the percent contribution of pure dust aerosol to AOD for our study region. To allow easy comparison, in Figure S2 we present the same quantities from the CALIPSO retrieval. In Figure 5, the percentages of D_AOD include both pure dust and polluted dust. In Figure S2, the percentage values are for pure mineral dust originated from the ground, and so a direct comparison can be made of the MERRA-2 D_AOD percentage values. MERRA-2 can be used as independent evaluation of the CALISPO products because MERRA-2 aerosol variables have been extensively validated against Aerosol Robotic Network (AERONET) measurements [44] and because MERRA-2 does not use CALIPSO to do bias correction. A previous study has shown that the vertical profile of the CALIOP attenuated backscatter coefficient agrees well with that derived by MERRA-2 for Continental USA, South America, Northern and Southern Africa [29]. Here we find that the CALIPSO captured the broad spatial patterns of both AOD and D_AOD for East Asia as calculated by MERRA-2. In the case of AOD, both show two pollution centers, with a heavily polluted region in eastern China and another less pollution region in western China. In the case of D_AOD of pure dust, the highest values were located in the dust source regions (Taklimakan Desert and Gobi Desert) according to both data products.

4. Conclusions

The present study analyzed the seasonal variation of dust aerosols and transport over East Asia using CALIPSO retrievals. To develop deeper insight into the spatiotemporal distribution of dust aerosols, the study domain was divided into five homogenous regions including the Taklimakan Desert (TD), the Gobi Desert (GD), northern China (NC), southern China (SC) and Korea-Japan (KJ) with distinct emissions sources. The frequency of dust occurrence, dust Top Height and extinction coefficients were assessed for each study region.

Our results confirmed that large amounts of dust particles were emitted from the dust regions (TD and GD). The emitted dusts formed a dust belt within a latitudinal range of 25°N to 45°N and extended to eastern China, Japan, Korea and the Pacific Ocean. The dust belt was strongest in the spring outside the rainy season. High frequencies of dust occurrence were found over the dust source regions sequenced in descending order as follows: spring, summer, autumn and winter. FDOs showed to isolated centers over the Taklimakan Desert and Gobi Desert, implying limitations of the diffusion and long-range transport of the dust aerosols over the Taklimakan Desert, meaning that most of uplifted dust aerosols were blocked by the mountains around the Tarim basin and eventually settled back to the ground.

The dust top height over East Asia presents significant seasonal variations with values of roughly 3.5~5.1 km (a.s.e.) over the dust source regions and recorded in descending order as follows: summer, spring, autumn and winter. Over the dust belt, the dust Top Height increased from dust source regions to downwind regions; furthermore, frequent dust storms occurring over East Asia during the spring showed why the dust Top Height was the highest in the spring across the downwind regions.

The AOD was generally characterized by two high AOD centers and two low AOD centers over East Asia. The dust belt was the shortest and had the lowest AOD in the summer with higher dust to AOD percentages (>85%) observed only over the Taklimakan and Gobi Deserts, showing that dust aerosols can hardly travel long distances in the rainy season due to wet deposition.

Pure dust was predominant across all four seasons. Over northern China and in areas direct downwind of the dust source regions, polluted dust generated from anthropogenic activities also played an important role. The dust extinction coefficients over the Taklimakan Desert were sequenced in descending order as follows: spring, winter, summer and autumn. The west-to-east transect of the extinction coefficient further confirmed that only some of the emitted dust particles successfully reached the downwind regions due to topographical obstacles and this blocking effect was especially strong in the autumn and winter.

The main advantages of the CALIPSO system lie in its ability to obtain vertical profiles of aerosols, which are crucial for understanding the distribution and transport of dust. However, the satellite completes a cycle over 16 days, producing data gaps that may miss detection of some dust events. A preliminary comparison with MERRA-2 data product, which is continuous in time, showed that when averaged over multiple years, the horizontal distributions of AOD and D_AOD from the CALIPSO captured the broad patterns calculated by MERRA-2. Future work should use MODIS, MERRA-2 and CALIPSO data to build a 3D structure of the dust trajectory over East Asia to gain STRONGER insight into long-range transport of East Asian dust to other regions of the globe.

Supplementary Materials: The following are available online at http://www.mdpi.com/2072-4292/11/6/701/s1, Figure S1: Seasonal distribution of Aerosol Optical Depth (left panels) and percentage contributions of pure dust to the total AOD (right panels) over East Asia based on MERRA-2 data, Figure S2: Seasonal distribution of Aerosol Optical Depth (left panels) and percentage contributions of pure dust to the total AOD (right panels) over East Asia based on CALIPSO data.

Author Contributions: D.L., T.Z., and S.C. designed the study. D.L. carried out THE research and wrote the manuscript. T.Z., R.B., and S.C. contributed to the preparation of the manuscript through review, editing, and comments. Z.L. wrote the results section. Y.W. and Y.Z. helped polish the manuscript. And all authors were involved in modifying the paper, and the literature review.

Funding: This research was supported by the Foundation for National Natural Science Foundation of China (No. 41830965; No. 91837103).

Remote Sens. **2019**, *11*, 701

Acknowledgments: We thank TC Chakraborty, Yale University, for providing the MERRA-2 data graphs.

Conflicts of Interest: The authors declare no conflict of interest.

References

1. Twomey, S.; Piepgrass, M.; Wolfe, T. An assessment of the impact of pollution on global cloud albedo. *Tellus B* **1984**, *36*, 356–366. [CrossRef]
2. Huang, J.; Fu, Q.; Su, J.; Tang, Q.; Minnis, P.; Hu, Y.; Yi, Y.; Zhao, Q. Taklimakan dust aerosol radiative heating derived from CALIPSO observations using the Fu-Liou radiation model with CERES constraints. *Atmos. Chem. Phys.* **2009**, *9*, 4011–4021. [CrossRef]
3. Chen, S.; Huang, J.; Li, J.; Jia, R.; Jiang, N.; Kang, L.; Ma, X.; Xie, T. Comparison of dust emission, transport, and deposition between the Taklimakan Desert and Gobi Desert from 2007 to 2011. *Sci. China Earth Sci.* **2017**, *60*, 1–18. [CrossRef]
4. Zhang, X.; Arimoto, R.; An, Z. Dust emission from Chinese desert sources linked to variations in atmospheric circulation. *J. Geophys. Res. Atmos.* **1997**, *102*, 28041–28047. [CrossRef]
5. Huang, J.; Patrick, M.; Chen, B.; Huang, Z.; Liu, Z.; Zhao, Q.; Yi, Y.; Ayers, J. Long-range transport and vertical structure of Asian dust from CALIPSO and surface measurements during PACDEX. *J. Geophys. Res. Atmos.* **2008**, *113*. [CrossRef]
6. Xie, X.; Liu, X.; Che, H.; Xie, X.; Wang, H.; Li, J.; Shi, Z.; Liu, Y. Modeling East Asian Dust and Its Radiative Feedbacks in CAM4-BAM. *J. Geophys. Res. Atmos.* **2018**, *123*, 1079–1096. [CrossRef]
7. Duce, R.; Unni, C.; Ray, B.; Prospero, J.; Merrill, J. Long rang atmospheric transport of soil dust from Asia to the Tropical North Pacific: Temporal variability. *Science* **1980**, *209*, 1522–1524. [CrossRef] [PubMed]
8. Zhang, X.; Gong, S.; Zhao, T.; Arimoto, R.; Wang, Y.; Zhou, Z. Sources of Asian dust and role of climate change versus desertification in Asian dust emission. *Geophys. Res. Lett.* **2003**, *30*, 5–8. [CrossRef]
9. Claquin, T.; Schulz, M.; Balkanski, Y.; Boucher, O. Uncertainties in assessing radiative forcing by mineral dust. *Tellus B* **1998**, *50*, 491–505. [CrossRef]
10. Zhang, G.; Zhang, Z.; Liu, J. Spatial distribution of wind erosion and its driving factors in China. *J. Geogr. Sci.* **2001**, *11*, 127–139.
11. Bourgeois, Q.; Ekman, A.; Krejci, R. Aerosol transport over the Andes from the Amazon Basin to the remote Pacific Ocean: A multiyear CALIOP assessment. *J. Geophys. Res. Atmos.* **2015**, *120*, 8411–8425. [CrossRef]
12. Chen, S.; Huang, J.; Qian, Y.; Zhao, C.; Kang, L.; Yang, B.; Wang, Y.; Liu, Y.; Yuan, T.; Wang, T.; et al. An Overview of Mineral Dust Modeling over East Asia. *J. Meteorol. Res.* **2017**, *31*, 633–653. [CrossRef]
13. Xu, X.; Zhou, X.; Weng, Y.; Tian, G.; Liu, Y.; Yan, P.; Ding, G.; Zhang, Y.; Mao, J.; Qiu, H. Study on variational aerosol fields over Beijing and its adjoining areas derived from Terra-MODIS and ground sunphotometer observation. *Chin. Sci. Bull.* **2003**, *48*, 2010–2017. [CrossRef]
14. Qiu, H.; Zhong, J.; Dong, X. Land-use and land-cover changes and dust storms in Tarim Basin, northwest China. *Proc. SPIE Int. Soc. Opt. Eng.* **2003**, *4890*, 652–656.
15. Huang, L.; Jiang, J.; Tackett, J.; Su, H.; Fu, R. Seasonal and diurnal variations of aerosol extinction profile and type distribution from CALIPSO 5-year observations. *J. Geophys. Res. Atmos.* **2013**, *118*, 4572–4596. [CrossRef]
16. Guo, J.; Liu, H.; Wang, F.; Huang, J.; Xia, F.; Lou, M.; Wu, Y.; Jiang, J.; Xie, T.; Zhaxi, Y.; et al. Three-dimensional structure of aerosol in china: A perspective from multi-satellite observations. *Atmos. Res.* **2016**, *178*, 580–589. [CrossRef]
17. Proestakis, E.; Amiridis, V.; Marinou, E.; Georgoulias, A.; Solomos, S.; Kazadzis, S.; Chimot, J.; Che, H.; Alexandri, G.; Binietoglou, I.; et al. 9-year spatial and temporal evolution of desert dust aerosols over South-East Asia as revealed by CALIOP. *Atmos. Chem. Phys.* **2017**, *18*, 1337–1362. [CrossRef]
18. Nan, Y.; Wang, Y. De-coupling interannual variations of vertical dust extinction over the Taklimakan Desert during 2007–2016 using CALIOP. *Sci. Total. Environ.* **2018**, *633*, 608–617. [CrossRef] [PubMed]
19. Yu, X.; Kumar, K.; Lü, R.; Ma, J. Changes in column aerosol optical properties during extreme haze-fog episodes in January 2013 over urban Beijing. *Environ. Pollut.* **2016**, *210*, 217–226. [CrossRef] [PubMed]
20. Winker, D.; Vaughan, M.; Omar, A.; Hu, Y.; Powell, K. Overview of the calipso mission and caliop data processing algorithms. *J. Atmos. Ocean. Technol.* **2009**, *26*, 2310–2323. [CrossRef]

21. Winker, D.; Hunt, W.; Mcgill, M. Initial performance assessment of CALIOP. *Geophys. Res. Lett.* **2007**, *34*, L19803. [CrossRef]

22. Liu, Z.; Vaughan, M.; Winker, D.; Kittaka, C.; Getzewich, B.; Kuehn, R.; Omar, A.; Powell, K.; Trepte, C.; Hostetler, C. The calipso lidar cloud and aerosol discrimination: Version 2 algorithm and initial assessment of performance. *J. Atmos. Ocean. Technol.* **2009**, *26*, 1198–1213. [CrossRef]

23. Liu, Z.; Omar, A.; Vaughan, M.; Hair, J.; Kittaka, C.; Hu, Y.; Powell, K.; Trepte, C.; Winker, D.; Hostetler, C.; et al. Calipso lidar observations of the optical properties of Saharan dust: A case study of long-range transport. *J. Geophys. Res.* **2008**, *113*, D07207.

24. Lolli, S.; Madonna, F.; Rosoldi, M.; Campbell, J.; Welton, E.; Lewis, J.; Gu, Y.; Pappalardo, G. Impact of varying lidar measurement and data processing techniques in evaluating cirrus cloud and aerosol direct radiative effects. *Atmos. Meas. Tech.* **2018**, *11*, 1639–1651. [CrossRef]

25. CALIPSO. CALIPSO Quality Statements Lidar Level 3 Aerosol Profile Monthly Products Version Release 1.00. 2011. Available online: http://eosweb.larc.nasa.gov/PRODOCS/calipso/Quality_Summaries/CALIOP_L3AProProducts_1-00.html (accessed on 15 July 2012).

26. Kaufman, Y.; Koren, I.; Remer, L.; TanrÉ, D.; Ginoux, P.; Fan, S. Dust transport and deposition observed from the terra-moderate resolution imaging spectroradiometer (MODIS) spacecraft over the Atlantic ocean. *J. Geophys. Res. Atmos.* **2005**, *110*, 1–12. [CrossRef]

27. Maring, H. Mineral dust aerosol size distribution change during atmospheric transport. *J. Geophys. Res.* **2003**, *108*, 8592. [CrossRef]

28. Huang, J.; Wang, T.; Wang, W.; Li, Z.; Yan, H. Climate effects of dust aerosols over East Asian arid and semiarid regions. *J. Geophys. Res. Atmos.* **2014**, *119*, 11398–11416. [CrossRef]

29. Buchard, V.; Randles, C.; Silva, A.; Darmenov, A.; Colarco, P.; Govindaraju, R.; Ferrare, R.; Hair, J.; Beyersdorf, A.; Ziemba, L.; et al. The MERRA-2 Aerosol Reanalysis, 1980 Onward. Part II: Evaluation and Case Studies. *J. Clim.* **2017**, *30*, 6851–6872. [CrossRef]

30. Chen, S.; Huang, L.; Kang, H.; Wang, X.; Ma, Y.; He, T.; Yuan, B.; Yang, Z.; Huang, Z.; Zhang, G. Emission, transport and radiative effects of mineral dust from Taklimakan and Gobi Deserts: Comparison of measurements and model results. *Atmos. Chem. Phys.* **2017**, *17*, 2401–2421. [CrossRef]

31. Xu, C.; Ge, J.; Huang, J.; Fu, Q.; Liu, H.; Chen, B. Observations of Dust Aerosol over China Based on CALIPSO Spaceborne Lidar. *J. Desert Res.* **2014**, *34*, 1353–1362.

32. Xu, C.; Ma, Y.; Yang, K.; You, C. Tibetan Plateau Impacts on Global Dust Transport in the Upper Troposphere. *J. Clim.* **2018**, *31*, 4745–4756. [CrossRef]

33. Kang, L.; Huang, J.; Chen, S.; Wang, X. Long-term trends of dust events over Tibetan Plateau during 1961–2010. *Atmos. Environ.* **2016**, *125*, 188–198. [CrossRef]

34. Chen, S.; Huang, J.; Zhao, C.; Qian, Y.; Leung, L.; Yang, B. Modeling the transport and radiative forcing of Taklimakan dust over the Tibetan Plateau: A case study in the summer of 2006. *J. Geophys. Res. Atmos.* **2013**, *118*, 797–812. [CrossRef]

35. Hu, K.; Kumar, K.; Kang, N.; Boiyo, R.; Wu, J. Spatial-temporal characteristics of aerosols and changes in trends over China with recent MODIS Collection 6 satellite data. *Environ. Sci. Pollut. Res.* **2017**, *25*, 6909–6927. [CrossRef] [PubMed]

36. Luo, H.; Han, Y.; Li, Y. Temporal evolution of the boundary layer height and contribution of dust devils to dust aerosols. *China Environ. Sci.* **2017**, *37*, 2438–2442.

37. Han, Y.; Fang, X.; Zhao, T.; Kang, S. Long range trans-pacific transport and deposition of asian dust aerosols. *J. Environ. Sci.* **2008**, *20*, 424–428. [CrossRef]

38. Che, H.; Zhao, H.; Wu, Y.; Xia, X.; Zhu, J.; Wang, H.; Wang, Y.; Sun, J.; Yun, J.; Zhang, X.; et al. Analyses of aerosol optical properties and direct radiative forcing over urban and industrial regions in Northeast China. *Meteorol. Atmos. Phys.* **2015**, *127*, 345–354. [CrossRef]

39. Huang, J.; Liu, J.; Chen, B.; Nasiri, S. Detection of anthropogenic dust using CALIPSO lidar measurements. *Atmos. Chem. Phys.* **2015**, *15*, 11653–11665. [CrossRef]

40. Tian, P.; Zhang, L.; Ma, J.; Tang, K.; Xu, L.; Wang, Y.; Cao, X.; Liang, J.; Ji, Y.; Jiang, J.; et al. Radiative absorption enhancement of dust mixed with anthropogenic pollution over East Asia. *Atmos. Chem. Phys.* **2018**, *18*, 7815–7825. [CrossRef]

41. Chen, S.; Huang, J.; Qian, Y.; Ge, J.; Su, J. Effects of aerosols on autumn precipitation over Mid-eastern China. *J. Trop. Meteorol.* **2014**, *20*, 242–250.

42. Chen, S.; Jiang, N.; Huang, J.; Xu, X.; Zhang, H.; Zang, Z.; Huang, K.; Xu, X.; Wei, Y.; Guan, X.; et al. Quantifying contributions of natural and anthropogenic dust emission from different climatic regions. *Atmos. Environ.* **2018**, *191*, 94–104. [CrossRef]

43. Han, Y.; Wang, K.; Liu, F.; Zhao, T.; Yin, Y.; Duan, J.; Luan, Z. The contribution of dust devils and dusty plumes to the aerosol budget in western China. *Atmos. Environ.* **2016**, *126*, 21–27. [CrossRef]

44. Randles, C.; Sliva, A.; Buchard, V.; Colarco, P.; Darmenov, A.; Govindaraju, R.; Smirnov, A.; Holben, B.; Ferrare, R.; Hair, J.; et al. The MERRA-2 Aerosol Reanalysis, 1980—Onward, Part I: System Description and Data Assimilation Evaluation. *J. Clim.* **2017**, *30*, 6823–6850. [CrossRef] [PubMed]

 *remote sensing*

Letter

Features of the Cloud Base Height and Determining the Threshold of Relative Humidity over Southeast China

Yuzhi Liu *, Yuhan Tang, Shan Hua, Run Luo and Qingzhe Zhu

Key Laboratory for Semi-Arid Climate Change of the Ministry of Education, College of Atmospheric Sciences, Lanzhou University, Lanzhou 730000, China; tangyh14@lzu.edu.cn (Y.T.); huash15@lzu.edu.cn (S.H.); luor14@lzu.edu.cn (R.L.); zhuqzh16@lzu.edu.cn (Q.Z.)
* Correspondence: liuyzh@lzu.edu.cn; Tel.: +86-139-1991-5375

Received: 13 October 2019; Accepted: 3 December 2019; Published: 5 December 2019

Abstract: Clouds play a critical role in adjusting the global radiation budget and hydrological cycle; however, obtaining accurate information on the cloud base height (CBH) is still challenging. In this study, based on Lidar and aircraft soundings, we investigated the features of the CBH and determined the thresholds of the environmental relative humidity (RH) corresponding to the observed CBHs over Southeast China from October 2017 to September 2018. During the observational period, the CBHs detected by Lidar/aircraft were commonly higher in cold months and lower in warm months; in the latter, 75.91% of the CBHs were below 2000 m. Overall, the RHs at the cloud base were mainly distributed between 70 and 90% for the clouds lower than 1000 m, in which the most concentrated RH was approximately 80%. In addition, for the clouds with a cloud base higher than 1000 m, the RH thresholds decreased dramatically with increasing CBH, where the RH thresholds at cloud bases higher than 2000 m could be lower than 60%. On average, the RH thresholds for determining the CBHs were the highest (72.39%) and lowest (63.56%) in the summer and winter, respectively, over Southeast China. Therefore, to determine the CBH, a specific threshold of RH is needed. Although the time period covered by the collected CBH data from Lidar/aircraft is short, the above analyses can provide some verification and evidence for using the RH threshold to determine the CBH.

Keywords: cloud base height; ground-based observations; relative humidity profile; threshold

---

1. Introduction

Clouds can adjust the Earth's energy budget and hydrological cycle through dynamic and thermal processes [1–3] and further drive the climate to change globally [4]. However, considerable uncertainties in cloud properties have been found [5], further contributing to errors in weather forecasting and climate prediction [6]. The immense uncertainties regarding clouds include optical [7–9], microphysical [10,11], and geometrical [12,13] features, the effects on the radiation budget [14], interactions with aerosols [15–21], and impacts on precipitation [22,23]. In particular, the cloud profile is poorly understood at present and remains a primary source of uncertainty in global weather and climate research [24].

The cloud base height (CBH), which is an important parameter of the cloud vertical profile, largely determines the energy exchanges between the clouds and surface. Accordingly, determining the CBHs is extremely critical for weather forecasting and ensuring flight safety [25,26]. Currently, retrieving the CBH generally relies primarily on satellite and ground-based observations. Space-borne active satellite remote sensing (e.g., the cloud profile radar (CPR) mounted on CloudSat and the Cloud-Aerosol Lidar with Orthogonal Polarization (CALIOP) aboard Cloud-Aerosol Lidar and Infrared Satellite Observation (CALIPSO)) has allowed cloud profile information to be obtained globally [27–29]. Some studies have estimated the CBH by applying both satellite-derived cloud optical depth, cloud water path, and some additional parameterizations that connect cloud optical depth with cloud geometrical thickness [30,31]. Other methods have also been used to estimate the CBH [32–34]. For example, Liang et al. [35] estimated the CBH by combining measurements from CloudSat/CALIPSO and Moderate-resolution Imaging Spectroradiometer (MODIS) based on the International Satellite Cloud Climatology Project (ISCCP) cloud-type classification and a weighted algorithm; unfortunately, the calculation of the CBH is dependent on an assumption of the cloud water content [36]. However, a comparison with the ground-based active remote sensing of clouds revealed large uncertainties in the CBH from satellite observations [37]. Therefore, obtaining information on the CBH with high accuracy is urgent.

Compared with satellite observations, ground-based cloud observations can provide CBH measurements with higher accuracy [34]. Some retrievals of the CBH are based on Lidar instruments [33], ceilometers [38], radiosondes [39–42], and total-sky-imager (TSI) [43]. Long-term research on the CBH measurements by radiosondes and ceilometer has been ongoing such as an analysis of 25-year CBHs measured by ceilometer at the Arctic site [44]. Some inter-comparisons among ground-based instruments have been performed [45–49]. In addition, some methods for calculating the CBHs have been reported by some earlier studies. Chernykh and Eskridge [50] proposed a method for judging the position of clouds using the second derivative of temperature and relative humidity (RH) versus the height in the ground-based soundings. For a long time, good agreement between the CBH and the lifting condensation level (LCL) estimated from the surface layer air have been confirmed and applied [51–53]. On this basis, Romps [54] gave the dependence of LCL on the temperature and RH and the conditions for cloud formation at heights such as the lifting deposition level (LDL) and the lifting freezing level (LFL). Wang and Rossow [2] postulated that an RH threshold value of 84% could be used to determine the cloud base location based on rawinsonde data. Additionally, Zhang et al. [42] performed an uncertainty analysis on the sensitivity of the CBH to different RH thresholds (83%, 84%, and 85%). To some degree, RH information at the cloud layer is significant in determining the CBHs from ground-based observations. Thus, an assessment of the RH threshold that can best determine the CBH is needed.

In this study, based on Lidar, pilot balloon, and aircraft soundings, the features of the CBH and the threshold of the environmental RH to determine the cloud base over Southeast China were investigated in detail. First, a comparison among the CBHs derived from three kinds of ground-based observations was performed. The observations with high accuracy were then selected as the reference. Next, combining the RH profiles from ERA-Interim data, the RH thresholds were calculated corresponding to the observed CBHs. Finally, the RH thresholds in different seasons for different CBHs were analyzed.

2. Datasets and Methods

2.1. Ground-Based Observations

At numerous sites throughout Southeast China (blue triangles in Figure 1), we performed observations with cloud Lidar (laser ceilometer), pilot balloon, and aircraft soundings from October 2017 to September 2018. The information on the sites is listed in Table 1.

We derived the cloud height using a cloud Lidar, whose emission source was a InGaAs 905 nm wavelength and 1.76 µJ pulse energy with a pulse repetition frequency of 1000 Hz. The pulse duration was 45 ns, and the beam divergence was less than 3 mrad. The detection range of cloud Lidar spans from 20 m to 7600 m with a vertical resolution of 3.8 m and a temporal resolution of 30 seconds. Lidar can scan the atmosphere with an elevation angle ranging from −30° to 30° and an azimuth in the range of 0–240°. The details about the technical specifications of the cloud Lidar used in this study are shown in Table 2. The retrieved CBH data [55,56] at seven sites from October 2017 to September 2018 were used.

Meanwhile, the CBH observations sounded by aircraft were also used in this study. The CBH detected by the aircraft soundings is obtained when the aircraft is flying upward through the cloud at the observational sites. When the aircraft enters the cloud, the pilot gives an altitude report for that moment, which is considered the height of the cloud base.

Additionally, a pilot balloon together with a GYR1 electronic optical wind theodolite was used to detect the CBH. After releasing a pilot balloon with a fixed rise velocity (ω) from the surface, a theodolite telescope is used to track the balloon. When the balloon starts to enter the clouds, the angular coordinate (elevation angle and azimuth angle) can be recorded. Then, the duration from the time of releasing balloon to the time of entering the cloud (t) can be calculated. Finally, the CBH can be estimated as the product of ω and t, that is, $CBH = \omega \times t$.

Table 1. Information on the observational sites for the cloud base height (CBH).

Site	Location	Elevation (m; Above Sea Level)	Number of Samples		
			Aircraft	Lidar	Pilot Balloon
A	(117° E, 25° N)	397	12	3268	230
B	(116° E, 23° N)	13.8	58	2854	161
C	(115° E, 28° N)	16	60	300	168
D	(120° E, 26° N)	366.6	45	1748	182
E	(118° E, 34° N)	15.7	75	76	19
F	(120° E, 30° N)	4.1	28	177	23
G	(121° E, 31° N)	4.4	50	47	24

Table 2. Technical specifications of the cloud Lidar.

Parameter Name	Parameter Value
Laser	InGaAs (a semiconductor laser)
Wavelength	905 ± 10 nm
Single laser pulse energy	≤ 20 μJ
Pulse width	45 ns ± 10 ns
Scattering angle of laser beam	≤ 3 mrad
Pulse repetition frequency	1 kHz $\pm 15\%$
Effective aperture of the optical system	102 mm
Interferometric filter	910 ± 15 nm

2.2. Clouds and the Earth's Radiant Energy System (CERES)

Cloud fraction and cloud base pressure data (SYN dataset) from October 2017 to September 2018 derived from Clouds and the Earth's Radiant Energy System (CERES) mounted on the Aqua satellite were used in this paper. The temporal and spatial resolutions of these CERES data are hourly and $1° \times 1°$, respectively [57]. Here, the cloud base pressure (CBP) data provided by CERES were used for an intercomparison with the above-mentioned ground-based observations. According to the locations of the ground-based sites and observational time, the CBPs from the CERES data corresponding to the site location were extracted for a comparison with the calculated CBPs from CBHs.

2.3. ERA-Interim Reanalysis Data

ERA-Interim data released by the European Centre for Medium-Range Weather Forecasts (ECMWF) can provide assimilated reanalysis data four times a day (the temporal resolution is 6 h). In this study, RH profiles with a horizontal spatial resolution of $0.25° \times 0.25°$ were used. In the vertical direction, the RH profiles had 37 layers from 1000 hPa to 1 hPa. The RHs employed in this study were extracted from the ERA-Interim data at 20 levels (from 1000 to 300 hPa) with a temporal resolution of 6 hours (0:00, 6:00, 12:00, and 18:00 UTC) [58].

2.4. Method of Conversion from Cloud Base Height (CBH) to Cloud Base Pressure (CBP)

In the cloud base datasets, the cloud base information from the CERES data is scaled by pressure (unit: hPa), while the CBHs detected by cloud Lidar and aircraft are measured as the geometric height (unit: m). To compare the ground-based measurement, a conversion from the CBH to the CBP is needed. Here, according to the pressure-height formula of polytropic atmosphere [59], CBP can be calculated from CBH.

$$CBP = P_{SF} \times \left[1 - \frac{\Gamma \times CBH}{T_{v\_CB}} \right]^{\frac{g}{R\Gamma}} \tag{1}$$

$$T_{v\_SF} = (1 + 0.608 q_{SF}) T_{SF} \tag{2}$$

$$T_{v\_CB} = T_{v\_SF} - \Gamma \times CBH \tag{3}$$

where g is the acceleration of gravity, here, it takes as 9.8 m/s; R is the gas constant, $R = 287.05\ J/kg/K$; and Γ is the temperature lapse rate, taken as 6.5 K/km here. P_{SF} is the atmospheric pressure at the surface. Based on Equation (2), the virtual air temperature at 2 m (T_{v_SF}) can be calculated from specific humidity (q_{SF}) based on ERA-Interim data and observed temperature at the ground-based sites. The virtual air temperature at the cloud base (T_{v_CB}) can be further calculated by Equation (3). Finally, based on the above formulas, the CBP can be calculated.

Based on the RH profiles from the ERA reanalysis data, the RH values at the cloud base can be extracted according to the CBPs calculated according to the above method, and the extracted RH value is regarded as the RH threshold at the cloud base.

3. Results

Figure 1 shows the distribution of observational sites (blue triangles) and the annual mean cloud fraction derived from CERES product in China from October 2017 to September 2018. As shown in Figure 1, during the above period, a high cloud fraction was distributed across South China with a value of approximately 80%. Low-value areas were located in North China, especially Inner Mongolia, with a minimum cloud fraction of 37.35%. Among the seven sites (blue triangles in Figure 1) involved in this study, the annual mean cloud fractions over sites C and D were approximately 70%, while those over sites A, B, and F were approximately 65%; however, the lowest cloud fraction (below 60%) was found over site E.

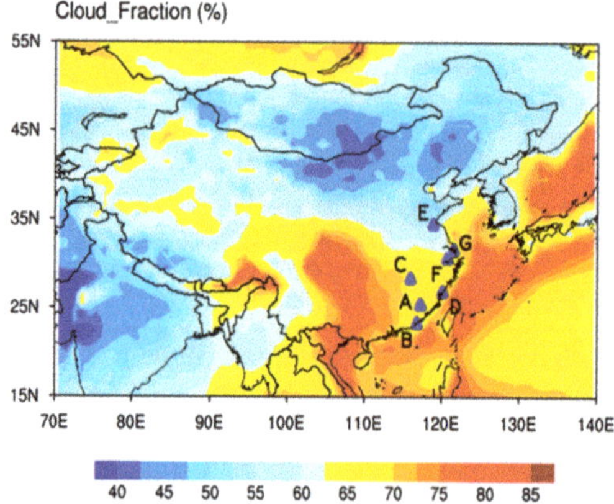

Figure 1. Distribution of the annual mean cloud fraction (unit: %) derived from Clouds and the Earth's Radiant Energy System (CERES) product in the period from October 2017 to September 2018 and the distribution of observational sites for the cloud base height (CBH) in Southeast China. The blue triangles denote the ground-based observational sites for the CBH.

3.1. Intercomparison among the CBHs from Multi-Sourced Data

Compared with satellite measurements, ground-based cloud observations can provide CBH measurements with higher accuracy and a continuous temporal coverage. Moreover, aircraft soundings can provide more accurate CBH information than ground-based observations. Therefore, aircraft-sounded CBHs were considered to be an accurate reference in this study. In the subsequent analyses, the CBHs derived from two kinds of ground-based observations, pilot balloon and cloud Lidar, and from aircraft soundings during the period from October 2017 to September 2018 were analyzed at each of the seven sites (details as shown in Figure 1 and Table 1) to give the features of the CBH over those areas. Figure 2 shows a comparison of the CBHs sounded by aircraft with those detected by cloud Lidar and pilot balloon during the observational period. The results show that the correlation coefficient between the aircraft-sounded and Lidar-observed CBH was 0.86, which indicates good consistency between the CBHs detected by the cloud Lidar and those sounded by the aircraft. In addition, most of the CBHs detected by the Lidar were somewhat higher than those sounded by aircraft when the cloud bases were lower than 2000 m. However, when the cloud bases were higher than 2000 m, most of the Lidar-observed CBHs were slightly lower than the aircraft-soundings. Additionally, comparing the results detected by the pilot balloon with the CBHs sounded by aircraft, most of the CBHs sounded by the former were significantly higher than those sounded by the latter; the correlation coefficient was 0.65. Thus, the CBH data detected by both Lidar and aircraft were regarded as accurate values. In the following analyses, these two sets of data were complementary and used in conjunction to analyze the features of the CBH over Southeast China.

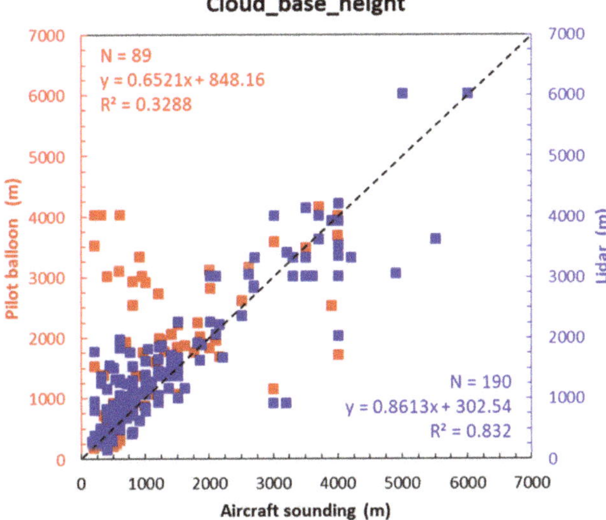

Figure 2. Scatter-grams of the Lidar/pilot-balloon observed and aircraft sounded CBHs at sites in Southeast China. The red and blue squares denote the CBHs from the pilot balloon and cloud Lidar, respectively.

3.2. Features of the CBH over Southeast China

Based on the CBH samples detected by the cloud Lidar and aircraft, the seasonal variations of the CBH during the period from October 2017 to September 2018 at seven sites were analyzed, as shown in Figure 3, 75.91% of the CBH values at the seven sites were primarily below 2000 m. Overall, the CBHs in the summer were consistently lower than those in the other seasons at most of the sites except for site A. In addition, the summer CBHs were below 1000 m at sites D, E, and G, which are located along the oceanic coast. In this sense, at site C, which is farther from the ocean than the other sites, the seasonal variation of the CBH is relatively small throughout the whole year. Therefore, at sites near the ocean, monsoon systems could affect the CBH, which may be the reason for the seasonal CBH discrepancies among the different sites. Here, the error bars reflect the standard errors based on the means of the CBH samples at the seven sites. The standard errors at the sites are small, except for sites E and F, which indicates the reliability of the statistical results with large sample sizes.

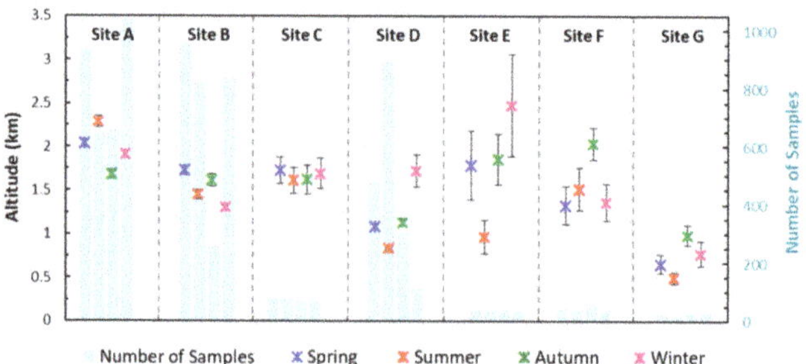

Figure 3. Seasonal variation of the mean CBH detected by cloud Lidar and aircraft during the period from October 2017 to September 2018 at seven ground-based sites. The light cyan bar represents the number of samples at each site in four seasons. Error bars represent the confidence levels of the mean values, assuming independent data. Errors are calculated as $s/(n-2)^{\frac{1}{2}}$, where n is the sample number of CBH measurements within the season and s is the standard deviation.

Furthermore, the monthly and diurnal variations of the CBH were similarly analyzed, as illustrated in Figure 4. Due to the absence of records at some sites where observations are performed only at certain times of day, fewer sites provided diurnal variation information (Figure 4b) than sites that provide monthly variation information (Figure 4a). Relative to the average CBHs at all sites (black line in Figure 4a), the lower CBHs were found in June–August, while the higher CBHs occur in February and April. Overall, most of the CBHs (75.91%) observed at seven sites were below 2000 m; in addition, the higher CBHs appeared in cold months, while the lower ones occurred in warmer months. Specifically, as shown in Figure 4a, the average CBHs over site A in most months (January, March, and May–August) were higher values relatively, which was also reflected by the diurnal variation of the CBH (Figure 4b). Meanwhile, for the whole year, the monthly mean CBHs over site G presented the lowest values (below 1000 m) among the seven sites, which is in agreement with the seasonal mean results in Figure 3. As shown in Figure 4b, overall, the mean values of the CBHs at sites A and B (green line in Figure 4b) were lower in the daytime than the nocturnal ones. Moreover, it was found that the CBHs over site B were much lower than those at site A, especially in the daytime (from 06:00 to 18:00). This phenomenon may be related to the fact that site B is located closer to the ocean than site A (as shown in Figure 1), as the abundance of water vapor from the ocean is beneficial to the formation of water clouds over site B.

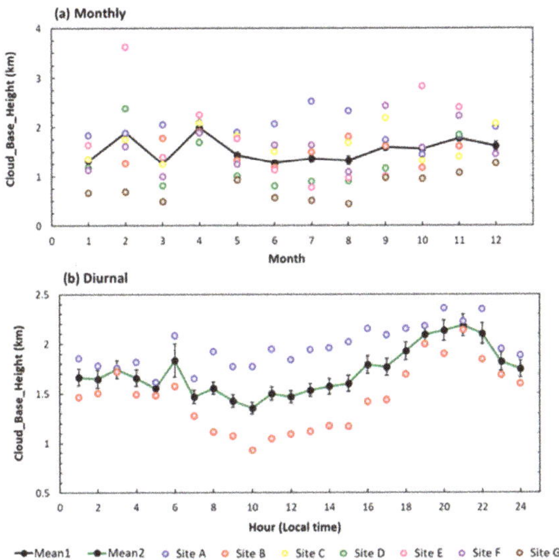

Figure 4. (**a**) Monthly and (**b**) diurnal variations of the CBH during the period from October 2017 to September 2018. Black line in (**a**) represents the average CBH at all sites. Green line in (**b**) denotes the average CBH at sites A and B. Error bars represent the confidence levels of the mean values, assuming independent data. Errors were calculated as $s/(n-2)^{\frac{1}{2}}$, where n is the sample number of CBH measurements within the season and s is the standard deviation.

3.3. Features of the Relative Humidity (RH) Threshold for Determining the CBH over Southeast China

According to Wang et al. [2], the CBH can be determined by the RH, and an RH of 84% was used as a threshold to determine the cloud base location. However, the scarcity of the sounding data limited an in-depth verification of determining the CBH by an RH threshold. Based on the above CBH data detected by cloud Lidar and aircraft, features of the RH thresholds used to determine the CBHs over these observational sites were investigated in the following analysis. The RH profiles derived from ERA reanalysis data together with observed surface air temperature and pressure at the ground-based sites were used to convert CBH to CBP.

Based on the method described in Section 2.4, a conversion from the CBH in meters to the CBP in hPa was performed. The calculated pressures at the cloud bases were compared with those retrieved from CERES observations, as shown in Figure 5. The time series and a comparison of the CBPs derived from the CERES dataset with the CBPs calculated from ground-based observations are given in Figure 5a. The red and blue lines represent the CBPs obtained from CERES observations and those converted from the Lidar/aircraft measurements, respectively. As shown in Figure 5a, the CBPs from the CERES product were consistently smaller (corresponding higher cloud base) than those calculated from the Lidar/aircraft measurements during a large time period; the averaged CBPs from the CERES product and Lidar/aircraft measurements during the period from October 2017 to September 2018 were 806.81 hPa and 860.67 hPa, respectively. Here, the CBH measurements obtained by Lidar/aircraft were relatively accurate. In this sense, the pressures at the cloud bases observed by CERES could overestimate the CBHs over these sites, which relates to the limited detection ability of passive satellite remote sensing for the cloud base location. Furthermore, according to the geometric heights of the cloud bases measured by Lidar/aircraft, the CBPs obtained from the CERES product were compared with those calculated from the Lidar/aircraft measurements, as shown in Figure 5b. For clouds higher than 1100 m, the CBPs observed by CERES and measured by Lidar/aircraft showed great agreement. However, as shown in Figure 5b, the CERES observations slightly overestimated the cloud base locations in reference to the detection results of Lidar/aircraft, especially for clouds lower than 1100 m. Furthermore, the correlation of CBPs between CERES observations with those calculated by Lidar/aircraft measurements was performed (Figure 5c,d). For the samples of all CBHs (represented by green in Figure 5c), CBPs observed by CERES were significantly smaller than those from Lidar/aircraft measurements (especially for the clouds lower than 1100 m), with a correlation coefficient of 0.790 (significant above the 99% confidence level). However, it was found that the correlation coefficient of the CBPs between CERES observations and the ones calculated from Lidar/aircraft measurements for the clouds higher than 1100 m (represented by yellow in Figure 5d) was 0.832 (significant above the 99% confidence level), which indicates the great agreement of the CERES observations with calculations from Lidar/aircraft measurements.

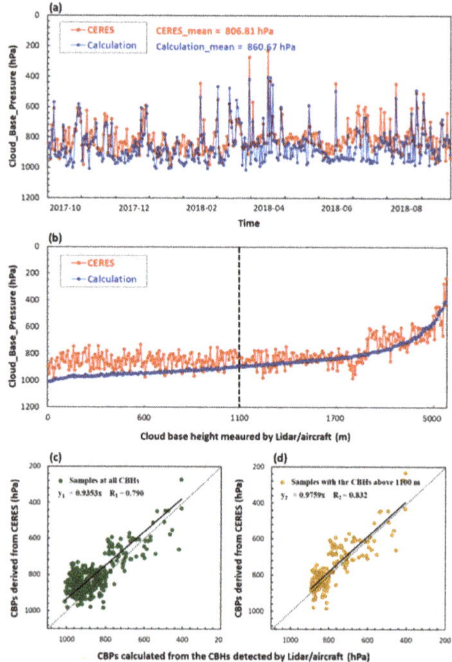

Figure 5. (**a**) Time series and (**b**) comparison of the CBPs derived from CERES observations with those calculated from the CBHs detected by Lidar/aircraft at site B during the period from October 2017 to September 2018. Correlation between the CBPs derived from CERES observations with those calculated from Lidar/aircraft measurements for samples of (**c**) all CBHs and (**d**) CBHs above 1100 m at site B during the period from October 2017 to September 2018.

Furthermore, based on the CBPs calculated from the CBHs detected by Lidar/aircraft, the RH values at the cloud base were further extracted from the ERA data. In a sense, the extracted RH values at the cloud base can be regarded as the thresholds for determining the CBHs. Figure 6 shows the correlation between the RHs at the cloud base and the CBHs detected by Lidar/aircraft at the seven sites during the period from October 2017 to September 2018. Overall, most of the RH values at the cloud base ranged from approximately 70 to 90%, where the CBHs were below 2000 m. As the CBH increased from 2000 m, the RH threshold began to decrease to smaller than 60%. As shown in Figure 6, when the CBH was lower than 1000 m, it corresponded to a stable RH threshold of approximately 80%. When the CBH ranged from 1000 to 2000 m, the RH threshold was below 80% and decreased with increasing CBH. However, when the CBH was higher than 2000 m, the RH threshold decreased dramatically with increasing CBH. Additionally, it was found that the samples showed a large scatter and were sparsely distributed in the region with a RH threshold below 60%. Here, the raw values of the RH calculated from the detected CBHs and the RH profiles from the ERA data are illustrated in Figure 6, which presents an uncertainty induced by the RH profiles provided by the ERA data, especially for the clouds at middle and high levels. However, the phenomenon of a relatively low RH threshold for determining the CBHs for relatively high clouds was revealed.

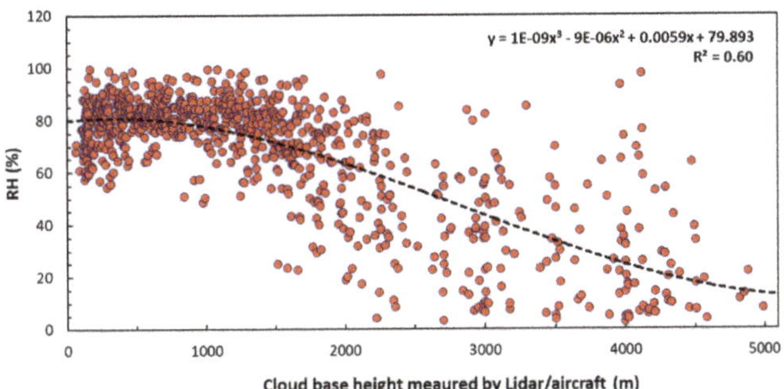

Figure 6. Correlation between the RHs at the cloud base and the CBHs detected by Lidar/aircraft at all sites during the period from October 2017 to September 2018. The black dashed line denotes the fitting result with a cubic polynomial.

The sample percentages for each RH threshold bin were statistically calculated, as shown in Figure 7. For the clouds with base heights ranging from 0 to 1000 m, the samples were dominantly distributed from 50 to 100%, where the maximum percentage of samples (47.43%) was in the bin ranging from 80 to 90%. For the CBHs ranging from 1000 to 2000 m, the samples were mainly in the RH bins from 20 to 100%, where the maximum percentage of samples (31.01%) was distributed in the bin ranging from 80 to 90%. However, for the clouds with base heights exceeding 3000 m, the samples were mainly distributed in the bins with RHs lower than 50%, where the highest sample percentage (49.12%) was distributed in the RH bin below 20% for CBHs larger than 4000 m.

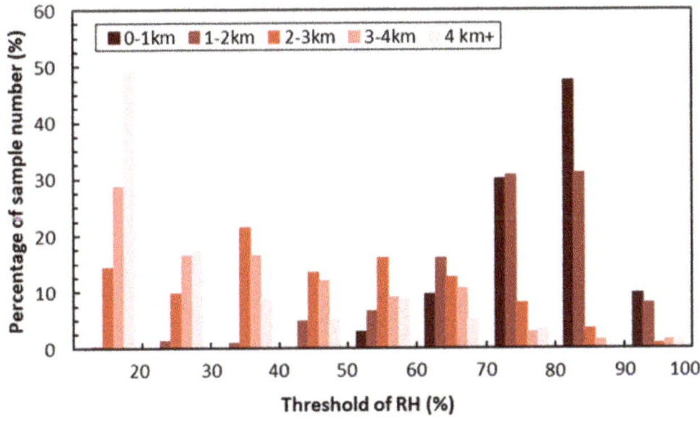

Figure 7. Distribution of the sample number percentage (unit: %) in each RH threshold bin for various CBHs measured by Lidar/aircraft at sites B, C, and F during the period from October 2017 to September 2018.

Among the seven sites, the CBH data detected by Lidar/aircraft at sites B, C, and F had better temporal continuity than the CBH data at the other sites during the period from October 2017 to September 2018. Here, an analysis on the RH thresholds for determining the CBHs was performed based on the above three sites (B, C, and F), as shown in Figure 8 (details are in Table 3). As revealed above, the RH thresholds at the three sites ranged approximately from 70 to 90%, where the CBHs were below 2000 m (as shown in the box in Figure 8); the means of the RH thresholds were approximately 80% except for site F (as shown in black squares in Figure 8). As the CBH increased from 2000 m, the RH threshold began to decrease, especially at site C. Overall, the average RH threshold decreased with the increase in CBH. The maximum RH threshold (with a mean of 79.88% for these three sites) was found for the clouds with base heights ranging from 0 to 1000 m. Then, with an increase in the CBH above 1000 m, the average RH threshold decreased and reached the minimum (28.92%) for the CBHs larger than 4000 m. As shown in Figure 8, a significant difference among the RH thresholds for determining the CBHs among the above three sites was mainly observed for the clouds with base heights exceeding 4000 m.

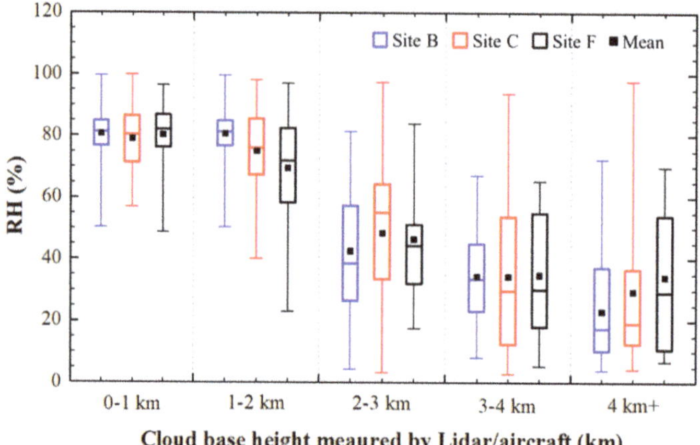

Figure 8. Statistics on the RH threshold for the clouds with different height at sites B, C, and F during the period from October 2017 to September 2018. Whiskers cover the range of RH thresholds. The upper, middle, and lower lines of the box correspond to the first, second, and third quartiles (the 75th, 50th, and 25th percentiles). Black squares denote the means of RH threshold.

Table 3. Characteristics of the relative humidity thresholds of the CBH (unit: %).

Altitude of the Cloud Base	Site B	Site C	Site F	Number of Samples	Mean Threshold
≤1 km	80.54	78.90	80.21	415	79.88
1–2 km	76.04	74.91	72.44	337	74.46
2–3 km	42.56	48.50	46.51	110	45.86
3–4 km	34.36	34.32	34.84	64	34.51
>4 km	23.11	29.53	34.12	56	28.92

As illustrated above, overall, the RH threshold decreased with an increase in the CBH, especially when the CBH was higher than 2000 m. Furthermore, the seasonal variation of the RH thresholds for determining the CBHs during the period from October 2017 to September 2018 was analyzed. The sample percentages for each RH threshold bin in different seasons are shown in Figure 9. For all seasons, the highest samples were dominantly distributed in the bin from 80 to 90%, where the maximum percentage was 40.55% in the summer. This finding is in agreement with the results in Figure 4. In another sense, for the RH threshold bins larger than 70%, most of the samples were obtained in spring and summer, while the fewest samples were obtained in autumn. However, for the RH threshold bins ranging from 30 to 70%, the samples were mainly collected in the autumn. For the RH threshold bins below 30%, most of the samples were detected in the spring (7.81%) and winter (11.93%). Furthermore, the RH thresholds for determining the CBHs at sites B, C, and F in different seasons were statistically calculated, as shown in Figure 10 (details are in Table 4). This figure shows that the maximum RH threshold (with a mean of 72.39% for sites B, C, and F) was found in the summer, which is consistent with the results shown in Figure 9. Moreover, the average RH thresholds in the spring, autumn, and winter were 65.25%, 66.91%, and 63.56%, respectively. Overall, the RHs at site B indicated slightly higher thresholds than the RHs at sites C and F for all seasons. Furthermore, the RH threshold at site C showed obvious variation in the autumn and winter seasons (as shown in the box in Figure 10), which may be related to the water vapor condition over there.

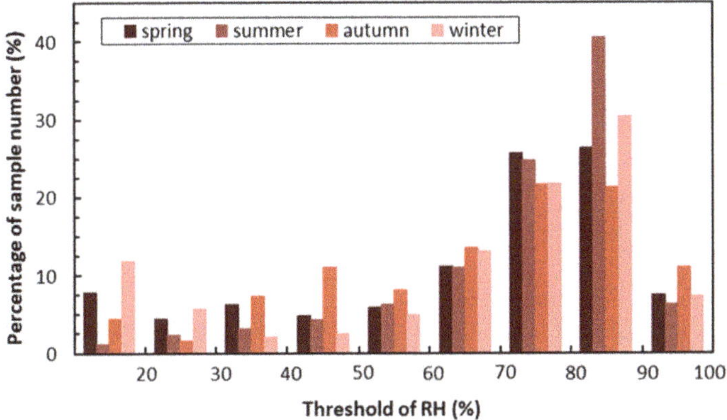

Figure 9. Seasonal distribution of the sample number percentage (unit: %) in each RH threshold bin for various CBHs measured by Lidar/aircraft at sites B, C, and F during the period from October 2017 to September 2018.

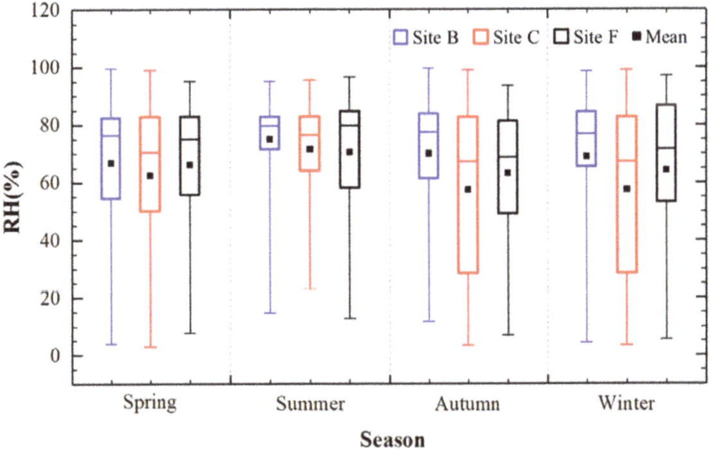

Figure 10. Seasonal statistics on the RH threshold for determining the CBHs at sites B, C, and F during the period from October 2017 to September 2018. Whiskers cover the range of RH thresholds. The upper, middle, and lower lines of the box correspond to the first, second, and third quartiles (the 75th, 50th, and 25th percentiles). Black squares denote the means of RH threshold.

Table 4. Seasonal average of the relative humidity thresholds of the CBH (unit: %).

Seasons	Site B	Site C	Site F	Number of Samples	Mean Threshold
Spring	66.89	62.56	66.31	269	65.25
Summer	75.09	71.58	70.51	254	72.39
Autumn	70.52	64.88	65.32	262	66.91
Winter	68.93	57.46	64.28	243	63.56

4. Conclusions

CBH data detected by Lidar, pilot balloon, and aircraft over Southeast China during the period from October 2017 to September 2018 were analyzed in this study. A comparison among the CBHs detected by Lidar, pilot balloon, and aircraft at seven ground-based sites showed that the CBHs observed by Lidar and aircraft were more consistent with a correlation coefficient of 0.86, and thus, the data from Lidar and aircraft were regarded as an accurate reference. During the observational period, the CBHs were higher in the cold months and lower in the warm months; in the latter, most of the CBHs were primarily below 2000 m.

Combined with the RH profiles provided by ERA-Interim data, the RH thresholds were calculated corresponding to the observed CBHs. Overall, the RH threshold was stable at approximately 80% when the CBH was lower than 1000 m; however, with the increase in CBH, the RH threshold began to decrease dramatically, even below 60%, as the CBH was larger than 2000 m. Seasonally, the maximum (72.39%) and minimum (63.56%) RH thresholds were found in the summer and winter, respectively. In addition, the average RH thresholds in the spring and autumn were 65.25% and 66.91%, respectively.

Although some interesting results were found in this study, some uncertainties in the RH threshold calculation based on the profiles from the ERA reanalysis data may be present in the analyses. A huge uncertainty may be induced by the establishment of humidity profiles from ERA reanalysis data. As pointed out by Chernykh and Aldukhov [60], the profile resolution of the reanalysis data could produce some errors in the gradient calculation that forms part of the cloud base determination. Second, the calculation of the CBH, according to the pressure-height formula of polytropic atmosphere, could introduce inevitable errors. Additionally, the RH thresholds for determining the CBHs varied dramatically with the time and CBH, and thus, using an average RH threshold to determine the CBH may conceal some accurate cloud height information. In the future, by combining ground-based and satellite-based observations of the CBH, an artificial neural network method can be used to obtain more accurate CBHs, which will be significantly beneficial to weather forecasting.

Author Contributions: Y.L. designed the paper; Y.T. wrote the original draft; Y.L. and Y.T. wrote, reviewed, and edited; S.H. and Q.Z. helped in data processing; Y.L., Y.T., and R.L. reviewed and revised the paper.

Funding: This research was funded by the Strategic Priority Research Program of the Chinese Academy of Sciences (Grant No. XDA2006010301) and was jointly supported by the National Natural Science Foundation of China (91744311 and 91737101).

Acknowledgments: We acknowledge the CERES (https://ceres.larc.nasa.gov/) and ECMWF (https://www.ecmwf.int/en/forecasts/datasets/reanalysis-datasets/era-interim) science teams for providing excellent and accessible data products that made this study possible. We are also grateful to the ground-based observations from the numerous sites.

Conflicts of Interest: The authors declare no conflicts of interest.

References

1. Ramanathan, V.; Cess, R.D.; Harrison, E.F.; Minnis, P.; Barkstrom, B.R.; Ahmad, E.; Hartmann, D. Cloud-radiative forcing and Climate: Results from the earth radiation budget experiment. *Science* **1989**, *243*, 57–63. [CrossRef]

2. Wang, J.H.; Rossow, W.B. Determination of cloud vertical structure from upper-air observations. *J. Appl. Meteor.* **1995**, *34*, 2243–2258. [CrossRef]

3. Sun, B.; Groisman, P.Y. Cloudiness variations over the former Soviet Union. *Int. J. Climatol.* **2000**, *20*, 1097–1111. [CrossRef]

4. Naud, C.M.; Muller, J.P.; Clothiaux, E.E. Comparison between active sensor and radiosonde cloud boundaries over the ARM Southern Great Plains Site. *J. Geophys. Res.* **2003**, *108*, 1–12. [CrossRef]

5. Houghton, J.T.; Ding, Y.; Griggs, D.J.; Noguer, M.; van der Linden, P.J.; Dai, X.; Maskell, K.; Johnson, C.A. *Climate Change 2001: The Scientific Basis*; Cambridge University Press: New York, NY, USA, 2001; pp. 1–421.

6. Li, Z.Q.; William, W.K.; Ramanathan, V.; Wu, G.X.; Ding, Y.H.; Madakshira, G.M.; Liu, J.; Qian, Y.F.; Li, J.P.; Zhou, T.J.; et al. Aerosol and Monsoon Climate Interactions over Asia. *Rev. Geophys.* **2016**, *54*, 866–929. [CrossRef]

7. Shang, H.; Letu, H.; Nakajima, T.Y.; Wang, Z.; Ma, R.; Wang, T.; Lei, Y.; Ji, D.; Li, J. Diurnal cycle and seasonal variation of cloud cover over the Tibetan Plateau as determined from Himawari-8 new-generation geostationary satellite data. *Sci. Rep.* **2018**, *8*, 1105. [CrossRef] [PubMed]

8. Letu, H.; Nagao, T.M.; Nakajima, T.Y.; Riedi, J.; Ishimoto, H.; Baran, A.J.; Shang, H.; Sekiguchi, M.; Kikuchi, M. Ice cloud properties from Himawari-8/AHI next-generation geostationary satellite: Capability of the AHI to monitor the DC cloud generation process. *IEEE Trans. Geosci. Remote. Sens.* **2019**, *57*, 3229–3239. [CrossRef]

9. Liu, Y.; Hua, S.; Jia, R.; Huang, J. Effect of aerosols on the ice cloud properties over the Tibetan Plateau. *J. Geophys. Res. Atmos.* **2019**, *124*, 9594–9608. [CrossRef]

10. Huang, J.; Minnis, P.; Lin, B.; Yi, Y.H.; Fan, T.F.; Sun, S.M.; Ayers, J.K. Determination of ice water path in ice-over-water cloud systems using combined MODIS and AMSR-E measurements. *Geophys. Res. Lett.* **2006**, *33*, L21801. [CrossRef]

11. Li, J.; Jian, B.; Huang, J.P.; Hu, Y.; Zhao, C.; Kawamoto, K.; Liao, S.; Wu, M. Long-term variation of cloud droplet number concentrations from space-based Lidar. *Remote. Sens. Environ.* **2018**, *213*, 144–161. [CrossRef]

12. Letu, H.; Nagao, T.M.; Nakajima, T.Y.; Matsumae, Y. Method for validating cloud mask obtained from satellite measurements using ground-based sky camera. *Appl. Opt.* **2014**, *53*, 7523–7533. [CrossRef] [PubMed]

13. Li, J.; Lv, Q.; Zhang, M.; Wang, T.; Kawamoto, K.; Chen, S.; Zhang, B. Effects of atmospheric dynamics and aerosols on the fraction of supercooled water clouds. *Atmos. Chem. Phys.* **2017**, *17*, 1847–1863. [CrossRef]

14. Hua, S.; Liu, Y.; Jia, R.; Chang, S.; Wu, C.; Zhu, Q.; Shao, T.; Wang, B. Role of Clouds in Accelerating Cold-Season Warming During 2000-2015 over the Tibetan Plateau. *Int. J. Climatol.* **2018**, *38*, 4950–4966. [CrossRef]

15. Huang, J.; Minnis, P.; Lin, B.; Wang, T.; Yi, Y.; Hu, Y.; Sun-Mack, S.; Ayers, K. Possible influences of Asian dust aerosols on cloud properties and radiative forcing observed from MODIS and CERES. *Geophys. Res. Lett.* **2006**, *33*, L06824. [CrossRef]

16. Liu, Y.; Huang, J.; Shi, G.; Takamura, T.; Khatri, P.; Bi, J.; Shi, J.; Wang, T.; Wang, X.; Zhang, B. Aerosol optical properties and radiative effect determined from sky-radiometer over Loess Plateau of Northwest China. *Atmos. Chem. Phys.* **2011**, *11*, 11455–11463. [CrossRef]

17. Li, Z.; Guo, J.; Ding, A.; Liao, H.; Liu, J.; Sun, Y.; Wang, T.; Xue, H.; Zhang, H.; Zhu, B. Aerosol and boundary-layer interactions and impact on air quality. *Natl. Sci. Rev.* **2017**, *4*, 810–833. [CrossRef]

18. Li, J.; Huang, J.; Stamnes, K.; Wang, T.; Lv, Q.; Jin, H. A global survey of cloud overlap based on CALIPSO and CloudSat measurements. *Atmos. Chem. Phys.* **2015**, *15*, 519–536. [CrossRef]

19. Chen, S.; Jiang, N.; Huang, J.; Zang, Z.; Guan, X.; Ma, X.; Luo, Y.; Li, J.; Zhang, X.; Zhang, Y. Estimations of indirect and direct anthropogenic dust emission at the global scale. *Atmos. Environ.* **2018**, *200*, 50–60. [CrossRef]

20. Guo, J.; Liu, H.; Li, Z.; Rosenfeld, D.; Jiang, M.; Xu, W.; Jiang, J.; He, J.; Chen, D.; Min, M.; et al. Aerosol-induced changes in the vertical structure of precipitation: A perspective of TRMM precipitation radar. *Atmos. Chem. Phys.* **2018**, *18*, 13329–13343. [CrossRef]

21. Zhu, Q.; Liu, Y.; Jia, R.; Hua, S.; Shao, T.; Wang, B. A numerical simulation study on the impact of smoke aerosols from Russian forest fires on the air pollution over Asia. *Atmos. Environ.* **2018**, *182*, 263–274. [CrossRef]

22. Guo, J.; Deng, M.; Lee, S.S.; Wang, F.; Li, Z.; Zhai, P.; Liu, H.; Lv, W.; Yao, W.; Li, X. Delaying precipitation and lightning by air pollution over the Pearl River Delta. Part I: Observational analyses. *J. Geophys. Res. Atmos.* **2016**, *121*, 6472–6488. [CrossRef]

23. Liu, Y.; Zhu, Q.; Huang, J.; Hua, S.; Jia, R. Impact of dust-polluted convective clouds over the Tibetan Plateau on downstream precipitation. *Atmos. Environ.* **2019**, *209*, 67–77. [CrossRef]

24. Stephens, G. Cloud feedbacks in the climate system: A critical review. *J. Clim.* **2005**, *18*, 237–273. [CrossRef]
25. Leyton, S.M.; Fritsch, J.M. The impact of high-frequency surface weather observations on short-term probabilistic forecasts of ceiling and visibility. *J. Appl. Meteorol.* **2004**, *43*, 145–156. [CrossRef]
26. Inoue, M.; Fraser, A.D.; Phillips, H.E. An assessment of numerical weather prediction–derived low-cloud-base height forecasts. *Wea. Forecast.* **2015**, *30*, 486–497. [CrossRef]
27. Costa-Surós, M.; Calbó, J.; González, J.A.; Martin-Vide, J. Behavior of cloud base height from ceilometer measurements. *Atmos. Res.* **2013**, *127*, 64–76. [CrossRef]
28. L'Ecuyer, T.S.; Jiang, J. Touring the atmosphere aboard the A-Train. *Phys. Today* **2010**, *63*, 36–41. [CrossRef]
29. Leeuw, G.; Kokhanovsky, A.; Cermak, J. Remote sensing of aerosols and clouds: Techniques and applications (editorial to special issue in Atmospheric Research). *Atmos. Res.* **2012**, *113*, 40–42. [CrossRef]
30. Hutchison, K.D. The retrieval of cloud base heights from MODIS and three-dimensional cloud fields from NASA's EOS Aqua mission. *Int. J. Remote. Sens.* **2002**, *23*, 5249–5265. [CrossRef]
31. Hutchison, K.D.; Wong, E.; Ou, S.C. Cloud base heights retrieved during night-time conditions with MODIS data. *Int. J. Remote. Sens.* **2006**, *27*, 2847–2862. [CrossRef]
32. Kuji, M.; Nakajima, T.Y.; Mukai, S. Retrieval of cloud geometrical properties using optical remote sensing data. *Proc. SPIE* **2000**. [CrossRef]
33. Borg, L.A.; Holz, R.E.; Turner, D.D. Investigating cloud radar sensitivity to optically thin cirrus using collocated Raman lidar observations. *Geophys. Res. Lett.* **2011**, *38*, L05807. [CrossRef]
34. Sharma, S.; Vaishnav, R.; Shukla, M.V.; Kumar, P.; Thapliyal, P.K.; Lal, S.; Acharya, Y.B. Evaluation of cloud base height measurements from Ceilometer CL31 and MODIS satellite over Ahmedabad, India. *Atmos. Meas. Technol.* **2015**, *8*, 11729–11752. [CrossRef]
35. Liang, Y.; Sun, X.; Miller, S.D.; Li, H.; Zhou, Y.; Zhang, R.; Li, S. Cloud Base Height Estimation from ISCCP Cloud-Type Classification Applied to A-Train Data. *Adv. Meteorol.* **2017**. [CrossRef]
36. Oh, S.B.; Kim, Y.H.; Cho, C.H.; Lim, E. Verification and correction of cloud base and top height retrievals from Ka-band cloud radar in Boseong, Korea. *Adv. Atmos. Sci.* **2016**, *33*, 73–84. [CrossRef]
37. Zhang, J.Q.; Xia, X.A.; Chen, H.B. A comparison of cloud layers from ground and satellite active remote sensing at the Southern Great Plains ARM site. *Adv. Atmos. Sci.* **2017**, *34*, 347–359. [CrossRef]
38. Martucci, G.; Milroy, C.; O'Dowd, C.D. Detection of cloud-base height using Jenoptik CHM15K and Vaisala CL31 ceilometers. *J. Atmos. Ocean. Technol.* **2010**, *27*, 305–318. [CrossRef]
39. Poore, K.D.; Wang, J.; Rossow, W.B. Cloud layer thicknesses from a combination of surface and upper-air observations. *J. Clim.* **1995**, *8*, 550–568. [CrossRef]
40. Yan, W.; Han, D.; Lu, W.; Lei, X. Research of cloud base height retrieval based on COSMIC occultation sounding data. *Chin. J. Geophys.* **2012**, *55*, 1–15. [CrossRef]
41. Zhang, J.; Chen, H.; Li, Z.; Fan, X.; Peng, L.; Yu, Y.; Cribb, M. Analysis of cloud layer structure in Shouxian, China using RS92 radiosonde aided by 95 GHz cloud radar. *J. Geophys. Res.* **2010**, *115*, D00K30. [CrossRef]
42. Zhang, Y.; Zhang, L.; Guo, J.; Feng, J.; Cao, L.; Wang, Y.; Zhou, Q.; Li, L.; Li, B.; Xu, H.; et al. Climatology of cloud-base height from long-term radiosonde measurements in China. *Adv. Atmos. Sci.* **2018**, *35*, 158–168. [CrossRef]
43. Kassianov, E.I.; Long, C.N.; Christy, J. Cloud-Base-Height Estimation from Paired Ground-Based Hemispherical Observations. *J. Appl. Meteorol.* **2005**, *44*, 1221–1233. [CrossRef]
44. Maturilli, M.; Ebell, K. Twenty-five years of cloud base height measurements by ceilometer in Ny-Ålesund, Svalbard. *Earth Syst. Sci. Data* **2018**, *10*, 1451–1456. [CrossRef]
45. Wang, Z.; Wang, Z.H.; Cao, X. Consistency analysis for cloud vertical structure derived from millimeter cloud radar and radiosonde profiles. *Acta. Meteorol. Sin.* **2016**, *74*, 815–826.
46. Forsythe, J.; Haar, T.V.; Reinke, D. Cloud-base height estimates using a combination of meteorological satellite imagery and surface reports. *J. Appl. Meteorol.* **2000**, *39*, 2336–2347. [CrossRef]
47. Barker, H.W. Estimating cloud field albedo using one-dimensional series of optical depth. *J. Atmos. Sci.* **1996**, *53*, 2826–2837. [CrossRef]
48. Berg, L.; Stull, R. Accuracy of point and line measures of boundary layer cloud amount. *J. Appl. Meteor.* **2002**, *41*, 640–650. [CrossRef]
49. Kassianov, E.I.; Long, C.; Ovtchinnikov, M. Cloud sky cover versus cloud fraction: Whole-sky simulations and observations. *J. Appl. Meteor.* **2005**, *44*, 86–98. [CrossRef]

50. Chernykh, I.V.; Eskridge, R.E. Determination of cloud amount and level from radiosonde soundings. *J. Appl. Meteorol.* **1996**, *35*, 1362–1369. [CrossRef]
51. Craven, J.P.; Jewell, R.E.; Brooks, H.E. Comparison between observed convective cloud-base heights and lifting condensation level for two different lifted parcels. *Wea. Forecast.* **2002**, *17*, 885–890. [CrossRef]
52. Stull, R.B.; Eloranta, E. A case study of the accuracy of routine, fair-weather cloud-base reports. *Natl. Wea. Dig.* **1985**, *10*, 19–24.
53. Zhang, Y.; Klein, S.A. Factors controlling the vertical extent of fair-weather shallow cumulus clouds over land: Investigation of diurnal-cycle observations collected at the ARM Southern Great Plains site. *J. Atmos. Sci.* **2013**, *70*, 1297–1315. [CrossRef]
54. Romps, D.M. Exact expression for the lifting condensation level. *J. Atmos. Sci.* **2017**, *74*, 3891–3900. [CrossRef]
55. Kleet, J.D. Stable analytical inversion solution for processing lidar returns. *Appl. Opt.* **1981**, *20*, 211–220. [CrossRef] [PubMed]
56. Collis, R.T.H.; Russell, P.B. Lidar measurement of particles and gases by elastic backscateringand differential absorption. In *Laser Monitoring of the Atmosphere*; Springer: Berlin/Heidelberg, Germany, 1976; pp. 71–151.
57. Chambers, L.H.; Lin, B.; Young, D.F. Examination of new CERES data for evidence of tropical iris feedback. *J. Clim.* **2002**, *15*, 3719–3726. [CrossRef]
58. Dee, D.P.; Uppala, S.M.; Simmons, A.J.; Berrisford, P.; Poli, P.; Kobayashi, S.; Andrae, U.; Balmaseda, M.A.; Balsamo, G.; Bauer, P.; et al. The ERA-Interim reanalysis: Configuration and performance of the data assimilation system. *Quart. J. R. Meteor. Soc.* **2011**, *137*, 553–597. [CrossRef]
59. Iribarne, J.V.; Cho, H.-R. *Atmospheric Physics*; Reidel: Dordrecht, The Netherlands, 1980; ISBN 90-277-1033-3.
60. Chernykh, I.; Aldukhov, O. Vertical distribution of cloud layers from atmospheric radiosounding data. *Izv. Atmos. Ocean. Phys.* **2004**, *40*, 41–53.

Technical Note

Evaluation of Terra-MODIS C6 and C6.1 Aerosol Products against Beijing, XiangHe, and Xinglong AERONET Sites in China during 2004–2014

Muhammad Bilal [1], Majid Nazeer [2,3], Janet Nichol [4], Zhongfeng Qiu [1,*], Lunche Wang [5], Max P. Bleiweiss [6], Xiaojing Shen [7], James R. Campbell [8] and Simone Lolli [9,10]

[1] School of Marine Sciences, Nanjing University of Information Science and Technology, Nanjing 210044, China; muhammad.bilal@connect.polyu.hk

[2] Key Laboratory of Digital Land and Resources, East China University of Technology, Nanchang 330013, China; majid.nazeer@comsats.edu.pk

[3] Earth and Atmospheric Remote Sensing Lab (EARL), Department of Meteorology, COMSATS University Islamabad, Islamabad 45550, Pakistan

[4] Department of Geography, School of Global Studies, University of Sussex, Brighton BN19RH, UK; janet.nichol@connect.polyu.hk

[5] Department of Geography, School of Earth Sciences, China University of Geosciences, Wuhan 430074, China; wang@cug.edu.cn

[6] Department of Entomology, Plant Pathology and Weed Science, New Mexico State University (NMSU), Las Cruces, NM 88003, USA; maxb@nmsu.edu

[7] School of Atmospheric Science at Nanjing University of Information Science and Technology, Nanjing 210044, China; shenxj@nuist.edu.cn

[8] Naval Research Laboratory, Monterey, CA 93943, USA; james.campbell@nrlmry.navy.mil

[9] Institute of Methodologies for Environmental Analysis, CNR, 85050 Tito Scalo (PZ), Italy; simone.lolli@imaa.cnr.it

[10] Joint Center for Earth Systems Technology, University of Maryland Baltimore County, Baltimore, MD 21221, USA

[*] Correspondence: zhongfeng.qiu@nuist.edu.cn; Tel.: +86-025-5869-5696

Received: 24 December 2018; Accepted: 21 February 2019; Published: 27 February 2019

Abstract: In this study, Terra-MODIS (Moderate Resolution Imaging Spectroradiometer) Collections 6 and 6.1 (C6 & C6.1) aerosol optical depth (AOD) retrievals with the recommended high-quality flag (QF = 3) were retrieved from Dark-Target (DT), Deep-Blue (DB) and merged DT and DB (DTB) level–2 AOD products for verification against Aerosol Robotic Network (AERONET) Version 3 Level 2.0 AOD data obtained from 2004–2014 for three sites located in the Beijing-Tianjin-Hebei (BTH) region. These are: Beijing, located over mixed bright urban surfaces, XiangHe located over suburban surfaces, and Xinglong located over hilly and vegetated surfaces. The AOD retrievals were also validated over different land-cover types defined by static monthly NDVI (Normalized Difference Vegetation Index) values obtained from the Terra-MODIS level-3 product (MOD13A3). These include non-vegetated surfaces (NVS, NDVI < 0.2), partially vegetated surfaces (PVS, $0.2 \leq$ NDVI ≤ 0.3), moderately vegetated surfaces (MVS, $0.3 <$ NDVI < 0.5) and densely vegetated surfaces (DVS, NDVI ≥ 0.5). Results show that the DT, DB, and DTB-collocated retrievals achieve a high correlation coefficient of ~ 0.90–0.97, 0.89–0.95, and 0.86–0.95, respectively, with AERONET AOD. The DT C6 and C6.1 collocated retrievals were comparable at XiangHe and Xinglong, whereas at Beijing, the percentage of collocated retrievals within the expected error ($\leftrightarrow$EE) increased from 21.4% to 35.5%, the root mean square error (RMSE) decreased from 0.37 to 0.24, and the relative percent mean error (RPME) decreased from 49% to 27%. These results suggest significant relative improvement in the DT C6.1 product. The percentage of DB-collocated AOD retrievals $\leftrightarrow$EE was greater than 70% at Beijing and Xinglong, whereas less than 66% was observed at XiangHe. Similar to DT AOD, DTB AOD retrievals performed well at XiangHe and Xinglong compared with Beijing. Regionally, DB C6 and C6.1-collocated retrievals performed better than DT and DTB in terms of good quality retrievals and relatively small errors. For diverse vegetated surfaces, DT-collocated retrievals reported small errors and good quality retrievals only for NVS and DVS, whereas larger errors were reported for PVS. MVS. DB contains good quality AOD retrievals over PVS, MVS, and DVS compared with NVS. DTB C6.1 collocated retrievals were better than C6 over NVS, PVS, and DVS. C6.1 is substantially improved overall, compared with C6 at local and regional scales, and over diverse vegetated surfaces.

Keywords: MOD04; Dark-Target; Deep-Blue; AERONET; LiDAR; AOD; Beijing; China

---

1. Introduction

Aerosol optical depth (AOD) is used for understanding the impact of aerosol on the Earth's climate system [1], human health [2–4], atmospheric visibility [5], and air quality [6–10]. In order to perform continuous in-situ measurements of AOD, a large number of sun photometers have been deployed worldwide under the Aerosol Robotic Network (AERONET) [11,12], which provides AOD at relatively high spectral and temporal resolutions, though at specific point-based locations. Therefore, to expand upon this framework, global AOD observations are required for better understanding of aerosol distributions and their impacts on regional and larger scales.

The spatial distribution of AOD can be examined from passive radiometric satellite sensors, but the accuracy of AOD retrievals depends on instrument calibration, cloud screening fidelities, estimates of background surface reflectance, and available spectral aerosol models to support requisite radiance inversions [13]. Specifically, AOD over land can be obtained from space-borne sensors such as the AVHRR (Advanced Very High Resolution Radiometer) [14,15], SeaWiFS (Sea-viewing Wide Field of view Sensor) [16], MISR (Multiangle Imaging Spectroradiometer) [17,18], TOMS (Total Ozone Mapping Spectroradiometer) [19], OMI (Ozone Monitoring Instrument) [20], the MERIS (Medium Resolution Imaging Spectroradiometer) [21], the VIIRS (Visible Infrared Imaging Radiometer Suite) [22,23], and the MODIS (MODerate resolution Imaging Spectroradiometer) [24,25]. The accuracy of available land surface reflectance, however, mostly limits the application of over-land AOD retrievals compared with over water [25,26]. Improvements to over-land retrieval algorithms as a whole, therefore, are important to increase data availability globally.

MODIS, aboard the NASA Terra and Aqua satellites respectively, features over-land AOD retrievals globally based on the Dark-Target (DT) [25] and the Deep-Blue (DB) algorithms [24]. For DT, pixels for dense vegetated surfaces are selected for a top-of-atmosphere (TOA) reflectance between 0.01 and 0.25 and corrected for gaseous absorption at 500 m spatial resolution. The selected pixels are arranged in a retrieval window of 20×20 pixels (400 pixels) and screened for clouds, snow/ice, and other bright surfaces. The remaining pixels are separated from land and water surfaces, and the 50% brightest pixels and the 20% darkest pixels are discarded to perform aerosol retrievals. The newly-released Collection 6.1 (C6.1) DT AOD product features the following improvements/changes: (a) the quality of AOD retrievals is degraded to zero for greater than 50% (20%) of coastal pixels (water pixels) within the 20×20 pixel window of, and (b) surface reflectance ratios for urban areas are updated using the MOD09 surface reflectance product [26]. Improvements/changes for the DT C6.1 algorithm can be found at [27]. The expected error (EE) of the DT algorithm is $\pm (0.05 + 0.15 \times \mathrm{AOD})$ [25], which represents about 66% of retrievals within EE on a global scale [28].

For DB, first the pixels are screened for clouds and snow/ice surfaces and the surface reflectance is estimated for the remaining pixels. Unlike DT, the DB algorithm retrieves AOD over dark as well as bright surfaces at a spatial resolution of 1 km and aggregates the retrievals at a spatial resolution of 10 km. The newly-released DB C6.1 AOD product features the following improvements/changes compared to C6: (a) artifact correction for heterogeneous terrain, (b) seasonal and regional aerosol models have been updated, (c) improvements have been made for the estimation of surface reflectance over elevated terrain, (d) metadata for Ångström Exponent have been updated, (e) updated EE, and (f) updated internal smoke detection masks. Improvements/changes for the DB C6.1 algorithm can be found at [27]. DB EE depends on geometry. That is, DB has different error characteristics, and the EE of DT is used in this study [24,29].

The MODIS C6 and C6.1 aerosol products include a merged DT and DB (DTB) AOD product for better spatial coverage of AOD over land (i.e., to include pixels from DT (DB) for areas where DB (DT) fails to retrieve) [25,30]. The DTB AOD product was developed using NDVI (Normalized Difference Vegetation Index) thresholds (i.e., the DT [DB] AOD is selected for surfaces with NDVI > 0.3 [NDVI < 0.2], and for surfaces with NDVI between 0.2 and 0.3, an average of DB and DT AOD or AOD available either from DB or DT with recommended high-quality flag is selected) [25].

Beijing-Tianjin-Hebei (BTH) is the national capital and largest urbanized metropolis region in northern China. It undergoes severe air pollution events in the form of haze composed from fine particles and dust storms consisting of larger coarse particles. To control and monitor air pollution over this region, accurate spatio-temporal satellite observations are required. Therefore, the main objectives of the present study are to compare and validate the C6 and C6.1 DT, DB and DTB AOD products from 2004–2014 (i) over the three Aerosol Robotic Network (AERONET) sites located in Beijing-Tianjin-Hebei (BTH) region (Beijing, XiangHe, and Xinglong), to highlight the performance difference in C6.1 compared with C6, over (ii) diverse vegetated surfaces to understand the generation of DTB AOD product by the relative contributions of the DB and DT retrievals.

2. Data Sets

Terra-MODIS DT, DB, and DTB C6 and C6.1 aerosol products at 10 km spatial resolution for the BTH region from 2004 to 2014 were downloaded from "the Level-1 and Atmosphere Archive & Distribution System (LAADS) Distributed Active Archive Center (DAAC)". The Terra-MODIS C6 level-three monthly NDVI product (MOD13A3) at 1 km resolution is used to define different land surface types. For the validation of the MOD04 AOD product, AERONET [11,12] cloud–screened and quality–assured (Version 3, Level 2.0) [31] data were downloaded from the AERONET website for three sites. Specifically, the Beijing site is located over mixed bright urban surfaces, the XiangHe site is located over suburban surfaces, and Xinglong is located over hilly and vegetated surfaces (Figure 1). AERONET provides measurements three-to-five times more accurate than satellite data [32] in seven channels (0.340–1.020μm) every 15 min with an uncertainty of ~0.01–0.02 [11] in the absence of thin cirrus cloud contamination [33]. AERONET data are available from Beijing, XiangHe, and Xinglong sites from March 2001 to June 2018, September 2004 to May 2017, and February 2006 to October 2014, respectively. A summary of the dataset is provided in Table 1.

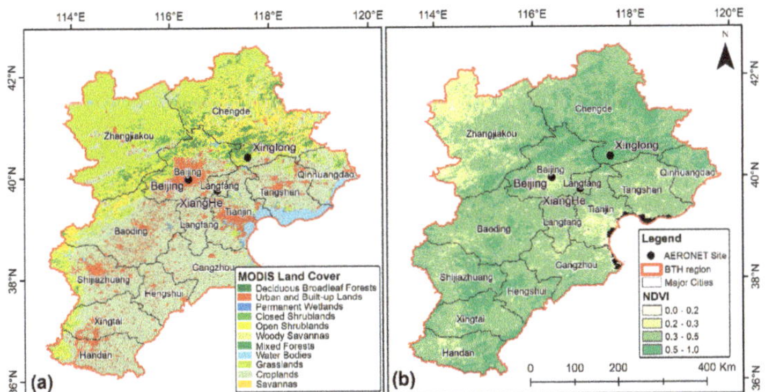

Figure 1. Land cover and yearly-averaged Normalized Difference Vegetation Index (NDVI) maps of the study area with overlaid Aerosol Robotic Network (AERONET) sites. (a) MODIS land cover data, (b) AERONET sites and NDVI values

Table 1. Summary of the data sets used.

Data	Scientific Data Set (SDS)	AOD
AERONET	Version 3 Level 2.0	AOD
MOD04 C6 and C6.1	Optical_Depth_Land_And_Ocean	DT AOD over land and ocean
	Deep_Blue_Aerosol_Optical_Depth_550_Land	DB AOD over land
	Deep_Blue_Aerosol_Optical_Depth_550_Land_QA_flag	Indicate quality of pixel
	AOD_550_Dark_Target_Deep_Blue_Combined	DT, DB or their average AOD
	AOD_550_Dark_Target_Deep_Blue_Combined_QA_Flag	Indicate quality of pixel
MOD13A3 C6	1 km NDVI	Monthly NDVI

3. Methodology

The performance of the Terra–MODIS DT, DB and DTB C6 and C6.1 AOD retrievals at 10 km spatial resolution is evaluated on a local scale against three AERONET stations located at urban (Beijing), suburban (XiangHe), and hilly and vegetated surfaces (Xinglong) for a period of 11 years (2004–2014). The step-by-step methodology is as follows: (i) DT AOD retrievals were obtained from the scientific data set (SDS) "Optical_Depth_Land_And_Ocean" containing the recommended high-quality flag (QF = 3) AOD retrievals over land, and the DB AOD retrievals were obtained from the SDS "Deep_Blue_Aerosol_Optical_Depth_550_Land_Best_Estimate" containing the recommended high-quality flag (QF $\geq$ 2). DTB AOD retrievals were obtained from the SDS "AOD_550_Dark_Target_Deep_Blue_Combined" and the recommended high-quality flag (QF = 3) AOD retrievals were filtered using the SDS "AOD_550_Dark_Target_Deep_Blue_Combined_QA_Flag". (ii) DT, DB and DTB AOD retrievals were validated and compared by combining all available collocations for each AERONET site. (iii) The DT, DB, and DTB AOD retrievals were filtered according to the categorized four land surface types: non-vegetated surfaces (NVS, NDVI < 0.2), partially vegetated surfaces (PVS, $0.2 \leq$ NDVI ≤ 0.3), moderately vegetated surfaces (MVS, $0.3 <$ NDVI < 0.5) and densely vegetated surfaces (DVS, NDVI ≥ 0.5) [27], as defined by MOD13A3 NDVI static values. The accuracy of the DTB_{C6} AOD retrievals can be improved by using dynamic values of NDVI Bilal and Nichol [34], but static NDVI values were used as the DTB product was developed using these values [25,30]. (iv) As AOD data at 550 nm were not available from AERONET measurements, an Ångström exponent 440–675 nm ($\alpha_{440-675}$) was used to interpolate AOD at 550 nm. (v) To increase the number of collocations and the temporal coverage of AOD retrievals, at least 2 out of 9 pixels were considered within a spatial window of 3 $\times$ 3 pixels centered on the AERONET site and at least two values of AERONET AOD were considered within $\pm$ 1hr of the Terra overpass. (vi) Errors in collocated AOD retrievals were reported using the relative percent mean error (RPME, Equation (1)), the root mean square error (RMSE, Equation (2)), and the expected error (EE, Equation (3)) of the DT algorithm over land, defined as

$$\text{RPME} = \left(\frac{\overline{AOD}_{(MODIS)} - \overline{AOD}_{(AERONET)}}{\overline{AOD}_{(AERONET)}} \right) \times 100 \tag{1}$$

$$\text{RMSE} = \sqrt{\frac{1}{n} \sum_{i=1}^{n} \left(AOD_{(MODIS)i} - AOD_{(AERONET)i} \right)^2} \tag{2}$$

$$\text{EE} = \pm \left(0.05 + 0.15 \times AOD_{(AERONET)} \right) \tag{3}$$

4. Results and Discussion

4.1. Validation of MODIS AOD Retrievals at Local Scale

In general, variance in the collocated AOD retrievals is mostly influenced by either the surface reflectance or the aerosol model used in the inversion process [9,35–37]. Variance during high aerosol loading events is usually due to an error in the aerosol scheme. In contrast, variance in AOD retrievals during low aerosol loading events are usually due to the error in the estimated surface reflectance [6–10,27,35,37–46]. High (low) agreement between satellite retrievals and AERONET AOD indicate that satellite retrievals follow (or do not follow) the aerosol variation as measured by the sun photometers [27].

For retrieval verification with AERONET, total numbers of DT C6 collocations were 721, 1060, and 60 for Beijing, XiangHe and Xinglong, respectively. Small numbers of collocations for the Xinglong site compared to the other two sites may be due to the limitation of the algorithm to retrieve AOD over the high elevated site (899 m above sea level), as a total of 909 AOD measurements were available within ± 1 h of the Terra overpass. In Figure 2, red lines represent the EE envelope and the black line represents the 1:1 line. Verification shows that the DT C6 AOD retrievals (Figure 2a–c) were correlated well with AOD measurements for each site, as the range of correlation coefficient (R) was between 0.90-0.96. This indicates that the DT algorithm has the ability to represent the aerosol variation measured by AERONET [27]. DT C6 AOD retrievals performed better over Xinglong and XiangHe, as 78.3% and 67.7% of the retrievals were within (↔) the EE compared with Beijing (↔EE = 21.4%), respectively.

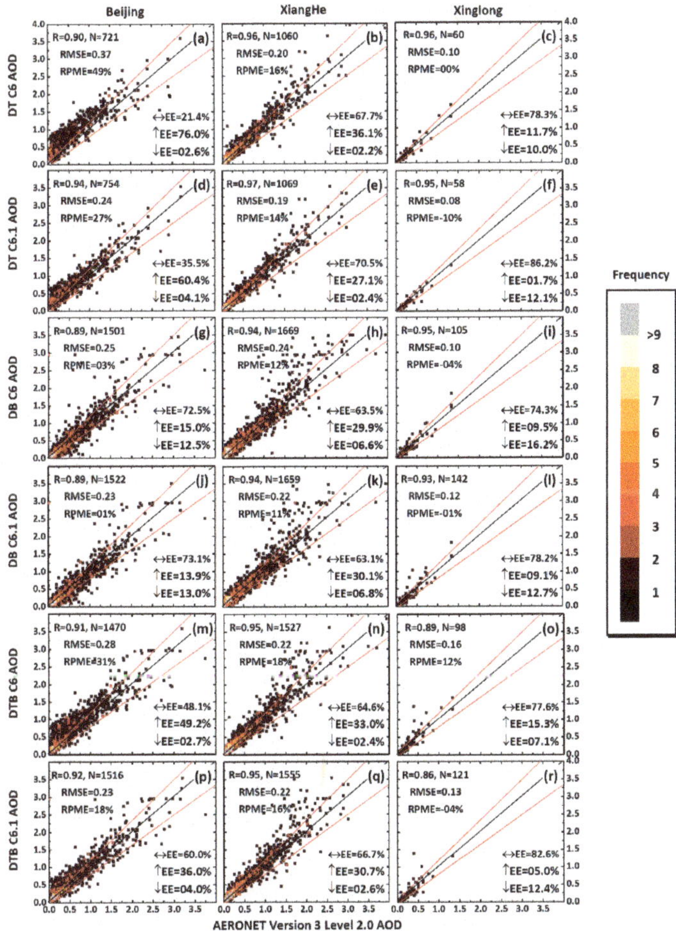

Figure 2. Validation of MODerate resolution Imaging Spectroradiometer (MODIS) aerosol optical depth (AOD) retrievals from 2004–2014 at the Beijing, XiangHe and Xinglong AERONET sites, including DT C6 (**a–c**), DT C6.1 (**d–f**), DB C6 (**g–i**), DB C6.1 (**j–l**), DTB C6 (**m–o**), and DTB C6.1 (**p–r**). The black line is the 1:1 line, and red lines are the expected error envelope.

Large errors were observed at Beijing, which might be due to the errors in the estimated surface reflectance and aerosol scheme during low and high aerosol loadings, respectively that led to 76% of the retrievals above (↑) the EE, RMSE of 0.37 and RPME of 49%. Similar results were reported by previous studies over the region [8,40,41]. Similar to C6, DT C6.1 performed better over Xinglong and XiangHe compared to Beijing. Modifications in DT C6.1 increased the percentage of retrievals ↔EE and reduced the RMSE and RPME. Specifically at Beijing, where the percentage of retrievals ↔EE increased from 21.4% to 35.5%, the percentage of retrievals ↑EE decreased from 76% to 60.4% and the RMSE and RPME decreased from 0.37 to 0.24 and from 49% to 27%, respectively. These relatively significant improvements at Beijing may be due to the modified surface reflectance ratios for urban areas based on the MODIS operational surface reflectance (MOD09) [26].

DB C6 and C6.1 AOD collocations were 75–108% and 101–144% greater than the DT C6 and C6.1, respectively. DB AOD retrievals performed well over Beijing and Xinglong sites, compared to XiangHe, in terms of larger percentage of retrievals ↔EE, a smaller percentage of retrievals ↑EE, and smaller RPME. In contrast, a large number of retrievals were below (↓) the EE at the Beijing and Xinglong sites compared to XiangHe. DB includes more collocations than DT over Xinglong, although DT is designed to retrieve AOD over such regional surface. Also, the number of collocations of DB C6.1 was greater than DB C6 over these hillier areas, which might be due to improved surface reflectance modeling for elevated terrain in C6.1. Significant improvements in DB C6.1 compared to C6 were not observed. Overall, the performance of C6 and C6.1 were comparable.

The numbers of collocations for DTB AOD retrievals (Figure 2m–r) were greater than DT (Figure 2a–f) but less than DB AOD. Similar statistics were reported by previous studies [27,39,42]. Similar to DT AOD, the performance of DTB retrievals was better over the Xinglong and XiangHe sites compared with Beijing, which was due to more contributions of DT retrievals in the DTB AOD dataset. At Beijing, a larger percentage of retrievals ↔EE, a smaller percentage of retrievals ↑EE, and small RMSE and RPME were found compared to DT AOD. This plausibly occurred due to relative contributions of the DB AOD retrievals (Figure 2g–l), as they performed better than DT AOD over this site. Modifications and improvements in DT and DB C6.1 increased the percentage of DTB AOD retrievals ↔EE from 48.1% to 60%, decreased the percentage of retrievals ↑EE from 49.2% to 36%, and reduced the RMSE and RPME from 0.28 to 0.23, and 31% to 18%, respectively at Beijing, though performance at the other sites was comparable with C6.

4.2. Evaluation of MODIS AOD Retrievals at Regional Scale

The MODIS DT, DB and DTB AODs from each site were combined together for regional verification (Figure 3). This exercise shows the high correlative agreement of 0.93 for the DT (Figure 3a,d) collocated AOD, which indicates that DT reproduces aerosol variation regionally. DT C6 and C6.1 have collocation totals and R, but improvements and modifications in C6.1 significantly improve AOD quality, as the percentage of retrievals (↑) ↔EE (decreased) increased from (47.5%) 49.9% to (39.7%) 59.9%, RMSE and RPME decreased from 0.28 to 0.21 and 29% to 19%, respectively, compared with C6. Overall, significant improvements and modifications are still required for DT to improve over BTH.

The number of DB C6.1 (C6) collocations was 77% (80%) greater than DT C6.1 (C6) at the regional scale. Similar to DT, the DB algorithm has the ability to follow the actual variation in aerosol concentrations as measured by AERONET, as R between the DB AOD retrievals and AERONET AOD measurements was > 0.90. The DB AOD has a large percentage of retrievals ↔EE and small RPME compared with DT AOD over BTH. Overall, the performance of DB AOD was reasonable, as 68% of the retrievals fell ↔EE. However, no significant improvements were observed in C6.1 compared with C6.

Similar to the local scale, DTB AOD retrievals (Figure 3c,f) were greater in numbers than DT AOD (Figure 3a,c), but less than DB AOD (Figure 3b,d). DTB AOD were more influenced by the DT AOD than DB AOD, as the correlation coefficient, the percentage of retrievals ↓, ↑, and ↔EE, RMSE, and RPME were comparable with DT AOD at the regional scale. Overall, the performance of DTB

AOD was poor compared with DB AOD, and this can be improved by considering more DB AOD retrievals in DTB as suggested by our previous studies [27,39,42].

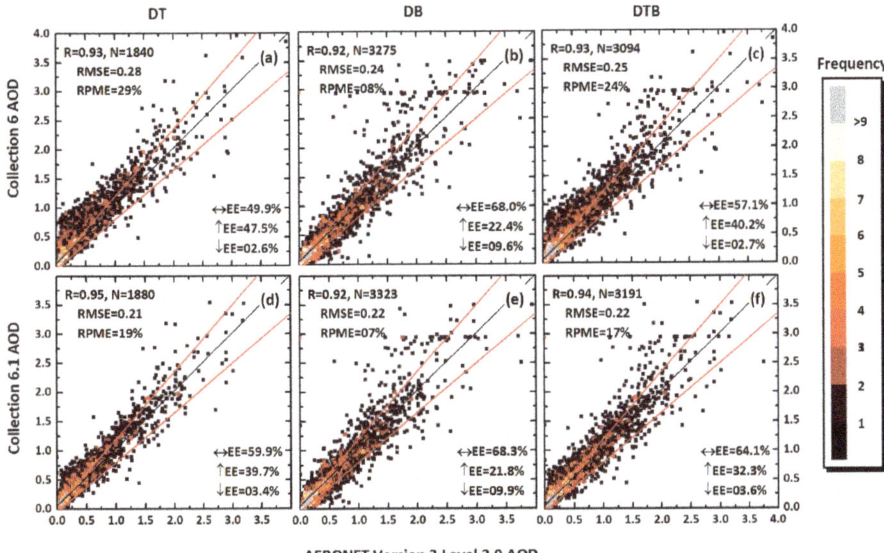

Figure 3. Validation of MODIS AOD retrievals from 2004–2014 at the regional scale: DT C6 (**a**), DB C6 (**b**), DTB C6 (**c**), DT C6.1 (**d**), DB C6.1 (**e**), and DTB C6.1 (**f**). The black line is the 1:1 line and red lines are the expected error envelope.

Figure 4 shows differencing scatter plots between C6 – AEORNET and C6.1 – AERONET for DT- (Figure 4a), DB- (Figure 4b) and DTB- collocated (Figure 4c) AOD observations to determine the exact temporal differences between C6 and C6.1 AOD. For this purpose, an equal number of C6 and C6.1-collocated AOD observations for the same time and site were obtained. Results show that differences between C6 – AEROENT and C6.1 – AERONET for DB were well correlated with each other, as the correlation (R ~ 0.99) was higher than DT (R ~ 0.87) and DTB (R ~ 0.90) AOD differences. Negative values of RPME suggest that DT, DB, and DTB C6.1 AOD have 2.7%, 7.3%, and 27.6%, respectively, less error compared with C6 AOD. Overall, these results suggest that the DB C6 and C6.1 AOD were comparable over the region due to high values of R, a slope close to 1 (0.997) and small RPME, compared to DT and DTB AOD. Further, significant differences were observed between DTB C6 and C6.1 AOD in terms of RPME.

In order to evaluate the temporal performance of the AOD products against AERONET over BTH, collocated AOD observations from AERONET and MODIS were averaged for each month from 2004—2014 (Figure 5). Results show different temporal trends of DT (Figure 5a) and DB (Figure 5b) AOD over the region. High DT and DB AOD were observed in July and June, respectively, which might be due to different numbers of collocated observations of DT and DB for each month. No significant improvements in DT and DB C6.1 were observed on a monthly basis compared with C6, as monthly averaged observations were the same for both datasets. In the case for DT, AOD was overestimated for C6 and C6.1 from March to October compared with AERONET. Whereas, for DB, AOD from both collections was overestimated from October to March, compared to AERONET. Figure 5c shows that the DTB AOD have the same temporal trend as DT from April to October and the same trend as DB from October to March due to more contributions from the respective DT and DB retrievals. Similar results were reported in a previous study [34]. Overall, the DB AOD retrievals performed better and compared more favorably with AERONET measurements in terms of small RPME (C6 = 5.3, C6.1 = 4.8), compared to DT (C6 = 17.5, C6.1 = 17.0) and DTB (C6 = 18.3, C6.1 = 17.5) AOD.

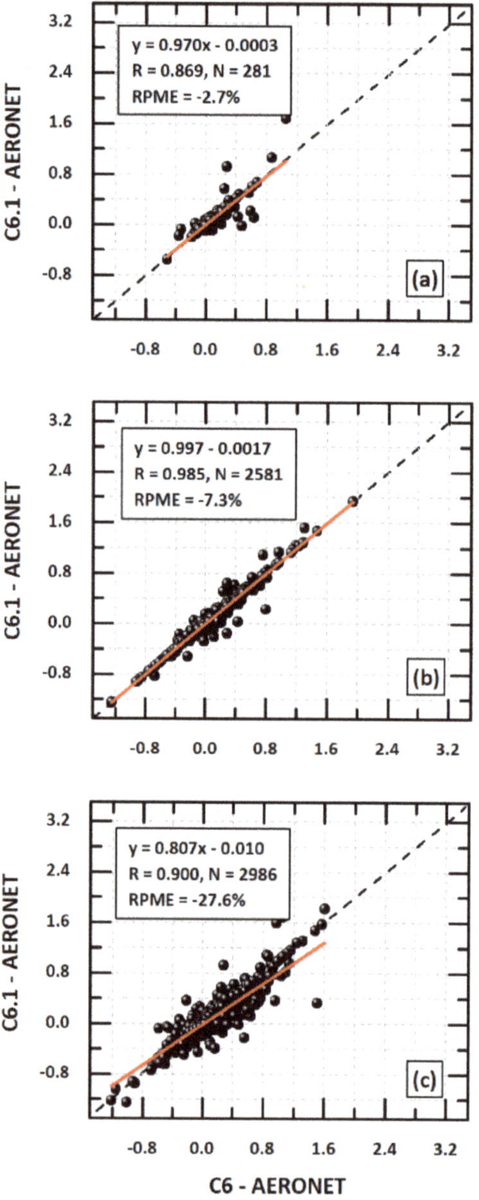

Figure 4. Differencing scatter plots between (C6 − AERONET)- and (C6.1 − AERONET)-collocated AOD retrievals, for, (**a**) DT AOD, (**b**) DB AOD, and (**c**) DTB AOD. The red and dotted black lines represent the regression and 1:1 lines, respectively.

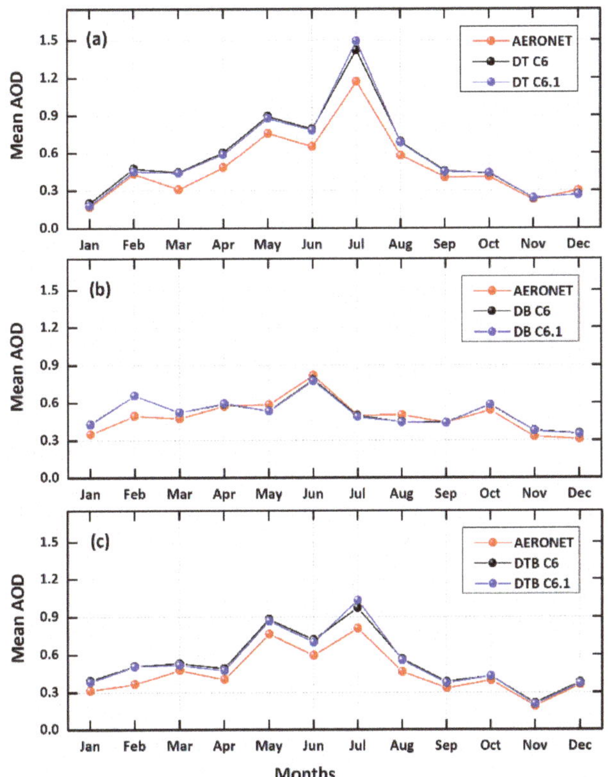

Figure 5. Temporal trends of monthly averaged collocated observations of DT AOD (**a**), DB AOD (**b**), and DTB AOD (**c**) retrievals from 2004–2014 over BTH.

C6 and C6.1 AOD were also compared spatially over BTH as seen in Figure 6. For this purpose, daily AOD retrievals from 2004 to 2014 were averaged for each collection and product. Distinct layers of AOD were observed over the region (Figure 6) and AOD varied from north to south (hilly to urban areas). The DT algorithm retrieved high AOD compared to DB, which contributed to the DTB AOD retrievals over this region. Similar findings were observed in the previous section and other studies [27,39]. Overall, the range of differences between C6 and C6.1 was found from −0.89 to 0.30, and large differences were observed between DT and DTB C6 and C6.1 AOD over certain areas compared to DB AOD. These results suggest that DB C6 and C6.1 AOD retrievals exhibit the same spatial pattern and values compared to DT and DTB AOD.

In order to evaluate DT and DB AOD over different land surfaces and understand their relative contributions to DTB AOD, each as validated over diverse vegetated surfaces (Figure 7) defined by NDVI: non-vegetated surfaces (NVS, NDVI < 0.2), partially-vegetated surfaces (PVS, $0.2 \leq$ NDVI ≤ 0.3), moderately-vegetated surfaces (MVS, 0.3 < NDVI < 0.5) and densely-vegetated surfaces (DVS, NDVI ≥ 0.5) [27]. For DT C6.1 (C6), the 52 (65), 535 (503), 981 (961), and 312 (311) collocations were available for NVS, PVS, MVS, and DVS, respectively. The RMSE (RPME) of DT C6.1 AOD was reduced from 0.12 (22%) to 0.09 (19%) for NVS, 0.28 (50%) to 0.17 (25%) for PVS, and 0.30 (28%) to 0.23 (19%) for MVS, and no significant improvements were observed for DVS. DT C6 and C6.1 collocated retrievals did not fulfill the requirements of EE for PVS and MVS, as the percentage of retrievals ↔EE was less than 66% [28]. Relatively good performance from the DT algorithm was in fact expected, as it was designed and developed to retrieve accurate AOD for vegetated surfaces.

Only DT AOD was considered in DTB for MVS, which introduced the same errors as observed in DT as shown in Figure 7s,w. Overall, the improvements observed in C6.1 compared to C6 were mainly due to the modified estimated surface reflectance.

The numbers of DB collocations were greater in numbers than DT for all types of surfaces except DVS, where they were comparable. In general, it is expected that the DB algorithm performs better for NVS than PVS, MVS, and DVS, and the DT algorithm exhibits better performance over vegetated surfaces than DB. Based on this concept, DB AOD retrievals were ignored in DTB over surfaces where NDVI > 0.3, and only considered over the surfaces where 0.3 < NDVI [25,27,30,34,39,42]. However, this study found that the performance and accuracy of DB- collocated retrievals were good over PVS, MVS and DVS compared to NVS (Figure 7i–p) in terms of (small) large percentages of retrievals (↑) ↔EE, and small RMSE and RPME. The DB AOD retrievals were also better than DT-collocated retrievals for NDVI > 0.2 surfaces. In contrast, the DT AOD performance was better than DB AOD for NDVI < 0.2 surfaces in terms of large (small) percentage of retrievals ↔ (↑)EE, and small RMSE, though they are in small numbers. Similar to our previous studies [27,39,42], this study also suggests that to improve the quality for DTB, DT AOD can be included in the DTB product for NDVI < 0.2 and DB AOD can be included for NDVI > 0.3 surfaces, as these retrievals were ignored in DTB [25]. Improvements/changes in the DB C6.1 algorithm did not show significant improvements, as C6- (Figure 7i–l) and C6.1- (Figure 7m–p) collocated retrievals were comparable for all surfaces.

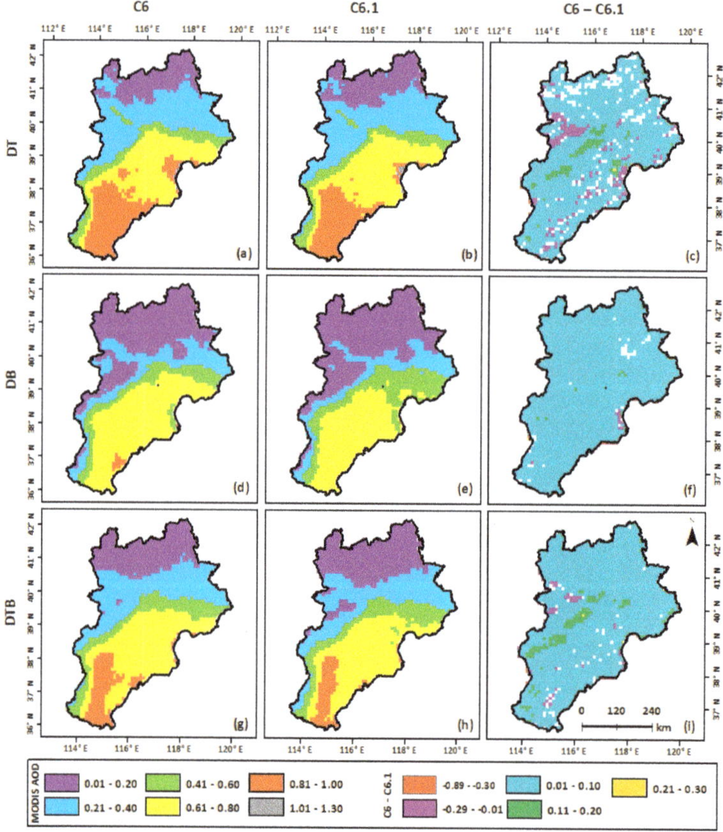

Figure 6. Spatial comparison between C6 and C6.1 AOD retrievals from 2004 to 2014 over the BTH region, where (**a**), (**b**) and (**c**) represent DT C6, C6.1 and C6 – 6.1, (**d**), (**e**) and (**f**) represent DB C6, C6.1 and C6 – 6.1, and (**g**), (**h**) and (**i**) represent DTB C6, C6.1 and C6 – 6.1, respectively.

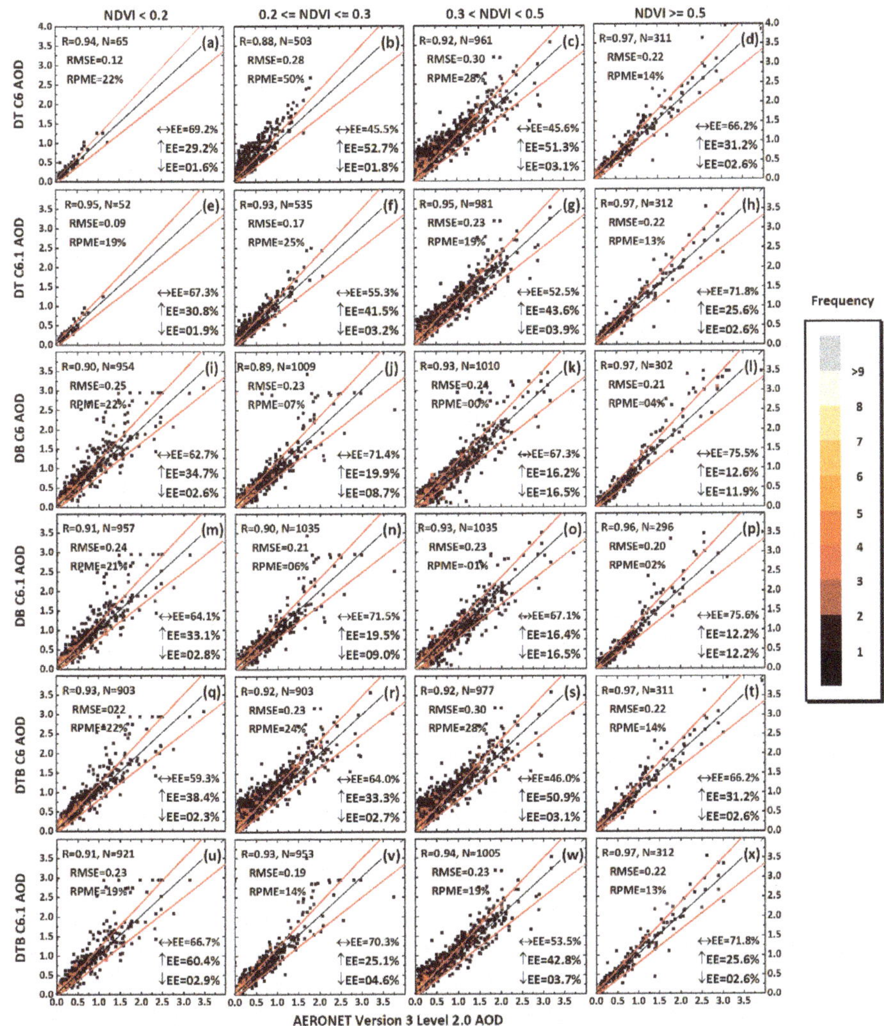

Figure 7. Validation of MODIS AOD retrievals from 2004–2014 over NVS, PVS, MVS, and DVS, where DT C6 (**a–d**), DB C6 (**e–h**), DTB C6 (**i–l**), DT C6.1 (**m–p**), DB C6.1 (**q–t**), and DTB C6.1 (**u–x**). The black line is the 1:1 line, and red lines are the expected error envelope.

DTB AOD retrievals for NVS and MVS should be the same as DB and DT AOD retrievals, respectively, but a mismatch was found. This might be due to using different NDVI data. This study used the monthly NDVI L3 operational product, whereas NDVI climatological data were used for the generation of DTB AOD retrievals [25,30]. The performance of the DTB AOD over NVS, and MVS and DVS was almost the same as DB and DT AOD, respectively, which was expected. For these surfaces, only DB and DT AOD were considered, respectively. However, DTB AOD performance over PVS was much better than DT AOD due to more contributions of DB. DTB AOD performance over NVS, and MVS and DVS can thus be improved by considering DT and DB AOD, respectively [27,39]. Overall, the DTB C6.1 AOD retrievals were better than C6 due to modifications and improvements in the respective retrievals.

Figure 8 shows a trend in the percentage of AOD retrievals ↔EE, RMSE, and RPME over diverse vegetated surfaces. Over NVS (NDVI < 0.2), the statistical performance of DT C6 and C6.1 was better than DB and DTB in terms of a larger percentage of retrievals ↔EE, and a very small RMSE, though RPME is comparable for all retrievals. The reasonable performance of DT suggests that the surface reflectance scheme used in the inversion works well over NVS. Over PVS ($0.2 \leq$ NDVI ≤ 0.3), DT C6 retrievals performed poorly compared with other retrievals for all of the statistical parameters, and this might be due to the underestimation in the estimated surface reflectance and error in the aerosol scheme as can be seen in Figure 7. DT performance was improved in C6.1 compared to C6, likely due to the modified surface reflectance scheme. However, it was still worst compared with DB.

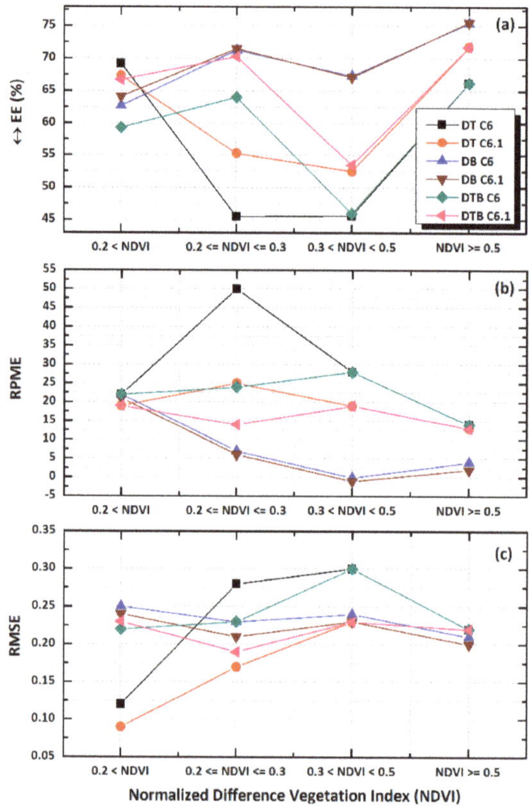

Figure 8. A trend of the percentage of retrievals ↔EE (**a**), RMSE (**b**), and RPME (**c**) in AOD retrievals over diverse vegetated surfaces of the BTH region.

Over MVS (0.3 < NDVI < 0.5) and DVS (DVS, NDVI $\geq$ 0.5), statistical performance of DB C6 and C6.1 was better than the DTB retrievals in terms of the larger percentage of retrievals ↔EE and a very small RPME, though RMSE was comparable with DT. Overall, the statistical performance of DT C6 showed its worst results over PVS and MVS, which might be due to errors in the estimated surfaces reflectance. The new modified surface reflectance scheme introduced in the DT C6.1 has likely improved such performance. These results suggest that DT-collocated AOD were good for surfaces with NDVI < 0.2 and DB-collocated AOD were good for surfaces with NDVI > 0.3, and that these retrievals can be considered in the DTB combined AOD product to improve overall performance.

5. Conclusions

The primary objective of this study is to verify and compare Terra-MODIS Collection (C6) and C6.1 Dark-Target (DT), Deep-Blue (DB) and merged DTB AOD retrievals versus AERONET sun photometer AOD derived over mixed urban surfaces (Beijing), suburban surfaces (XiangHe) and hilly and vegetated surfaces (Xinglong). For this, high-quality assurance flag MODIS AOD retrievals were obtained from 2004–2014. These retrievals were also validated and compared over diverse land surface types, including non-vegetated surfaces (NVS, NDVI < 0.2), partially vegetated surfaces (PVS, $0.2 \leq$ NDVI ≤ 0.3), moderately vegetated surfaces (MVS, $0.3 <$ NDVI < 0.5) and densely vegetated surfaces (DVS, NDVI ≥ 0.5), as categorized by static values of monthly NDVI obtained from the Terra-MODIS L3 monthly NDVI product (MOD13A3 C6). The main outcomes of this research are:

(1) DT, DB, and DTB C6 and C6.1 collocated AOD retrievals correlated well with AERONET AOD measurements at the three designated sites.

(2) DT C6.1 collocated AOD retrievals were better than C6 due to use of modified surface reflectance ratios, and, overall, both had a large error at Beijing.

(3) DB has more collocations than DT over Xinglong, although DT is designed to retrieve AOD over these surfaces.

(4) DB C6 and C6.1 AOD retrievals performed equally, as no significant changes in DB C6.1 compared to C6 were observed.

(5) The percentage of DTB-collocated AOD retrievals $\leftrightarrow$EE increased, and the RMSE and RPME decreased due to improvements/changes in the DT C6.1 at Beijing.

(6) At the regional scale, DB C6 and C6.1 AOD retrievals performed better than DT and DTB C6 and C6.1 AOD.

(7) For diverse vegetated surfaces, the percentage of DT C6 and C6.1-collocated AOD was less than 68% for MVS surfaces, where the reasonable performance of DT was expected since it was designed and developed to retrieve accurate AOD for vegetated surfaces.

(8) DT and DB-collocated AOD retrievals performed better than each other over surfaces with NDVI < 0.2 and NDVI > 0.2, respectively, in terms of small RMSE and a large percentage of collocated retrievals $\leftrightarrow$EE.

This study also concludes that DB and DT AOD retrievals should be considered for surfaces with NDVI > 0.3 and NDVI < 0.2, respectively, in the new integrated product, to improve the quality of DTB AOD overall.

Author Contributions: M.B. and J.N. designed the paper; J.N., M.B., Z.Q., J.C., and S.L. reviewed and revised the paper; M.N., L.W., and S.X. helped in data processing.

Funding: This work was jointly supported by the National Key Research and Development Program of China (No.2016YFC1400904), Jiangsu Chair Professor, and the Startup Foundation for Introduction Talent of NUIST (2017r107).

Acknowledgments: The authors would like to acknowledge NASA Goddard Space Flight Center for MODIS data, and Principal Investigators of AERONET sites. We are thankful to Devin White (Oak Ridge National Laboratory) for MODIS Conversion Tool Kit (MCTK).

Conflicts of Interest: The authors declare no conflict of interest

References

1. Kaufman, Y.J.; Tanré, D.; Boucher, O. A satellite view of aerosols in the climate system. *Nature* **2002**, *419*, 215–223. [CrossRef] [PubMed]
2. Pope, C.A.; Ezzati, M.; Dockery, D.W. Fine-particulate air pollution and life expectancy in the United States. *N. Engl. J. Med.* **2009**, *360*, 376–386. [CrossRef] [PubMed]
3. Pope, C.A.; Dockery, D.W. Health effects of fine particulate air pollution: Lines that connect. *J. Air Waste Manag. Assoc. (1995)* **2006**, *56*, 709–742. [CrossRef]

4. Pope, C.A.; Burnett, R.T.; Thun, M.J.; Calle, E.E.; Krewski, D.; Ito, K.; Thurston, G.D. Lung cancer, cardiopulmonary mortality, and long-term exposure to fine particulate air pollution. *Jama J. Am. Med. Assoc.* **2002**, *287*, 1132–1141. [CrossRef]

5. Cheung, H.-C.; Wang, T.; Baumann, K.; Guo, H. Influence of regional pollution outflow on the concentrations of fine particulate matter and visibility in the coastal area of southern China. *Atmos. Environ.* **2005**, *39*, 6463–6474. [CrossRef]

6. Bilal, M.; Nichol, J.; Spak, S. A New Approach for Estimation of Fine Particulate Concentrations Using Satellite Aerosol Optical Depth and Binning of Meteorological Variables. *Aerosol Air Qual. Res.* **2017**, *11*, 356–367. [CrossRef]

7. Zou, B.; Pu, Q.; Bilal, M.; Weng, Q.; Zhai, L.; Nichol, J.E. High-Resolution Satellite Mapping of Fine Particulates Based on Geographically Weighted Regression. *IEEE Geosci. Remote Sens. Lett.* **2016**, *13*, 495–499. [CrossRef]

8. Bilal, M.; Nichol, J.E. Evaluation of MODIS aerosol retrieval algorithms over the Beijing-Tianjin-Hebei region during low to very high pollution events. *J. Geophys. Res. Atmos.* **2015**, *120*, 7941–7957. [CrossRef]

9. Bilal, M.; Nichol, J.E.; Chan, P.W. Validation and accuracy assessment of a Simplified Aerosol Retrieval Algorithm (SARA) over Beijing under low and high aerosol loadings and dust storms. *Remote Sens. Environ.* **2014**, *153*, 50–60. [CrossRef]

10. Bilal, M.; Nichol, J.E.; Bleiweiss, M.P.; Dubois, D. A Simplified high resolution MODIS Aerosol Retrieval Algorithm (SARA) for use over mixed surfaces. *Remote Sens. Environ.* **2013**, *136*, 135–145. [CrossRef]

11. Holben, N.; Tanr, D.; Smirnov, A.; Eck, T.F.; Slutsker, I.; Newcomb, W.W.; Schafer, J.S.; Chatenet, B.; Lavenu, F.; Kaufman, J.; et al. An emerging ground-based aerosol climatology: Aerosol optical depth from AERONET. *J. Geophys. Res. Atmos.* **2001**, *106*, 12067–12097. [CrossRef]

12. Holben, B.N.; Eck, T.F.; Slutsker, I.; Tanré, D.; Buis, J.P.; Setzer, A.; Vermote, E.; Reagan, J.A.; Kaufman, Y.J.; Nakajima, T.; et al. AERONET—A Federated Instrument Network and Data Archive for Aerosol Characterization. *Remote Sens. Environ.* **1998**, *66*, 1–16. [CrossRef]

13. Li, Z.; Zhao, X.; Kahn, R.; Mishchenko, M.; Remer, L.; Lee, K.-H.; Wang, M.; Laszlo, I.; Nakajima, T.; Maring, H. Uncertainties in satellite remote sensing of aerosols and impact on monitoring its long-term trend: A review and perspective. *Ann. Geophys.* **2009**, *27*, 2755–2770. [CrossRef]

14. Riffler, M.; Popp, C.; Hauser, A.; Fontana, F.; Wunderle, S. Validation of a modified AVHRR aerosol optical depth retrieval algorithm over Central Europe. *Atmos. Meas. Tech.* **2010**, *3*, 1255–1270. [CrossRef]

15. Hauser, A.; Oesch, D.; Foppa, N.; Wunderle, S. NOAA AVHRR derived aerosol optical depth over land. *J. Geophys. Res.* **2005**, *110*, D08204. [CrossRef]

16. Sayer, A.M.; Hsu, N.C.; Bettenhausen, C.; Jeong, M.-J.; Holben, B.N.; Zhang, J. Global and regional evaluation of over-land spectral aerosol optical depth retrievals from SeaWiFS. *Atmos. Meas. Tech.* **2012**, *5*, 1761–1778. [CrossRef]

17. Kahn, R.A.; Gaitley, B.J.; Garay, M.J.; Diner, D.J.; Eck, T.F.; Smirnov, A.; Holben, B.N. Multiangle Imaging SpectroRadiometer global aerosol product assessment by comparison with the Aerosol Robotic Network. *J. Geophys. Res.* **2010**, *115*, D23209. [CrossRef]

18. Kahn, R.A.; Gaitley, B.J.; Martonchik, J.V.; Diner, D.J.; Crean, K.A.; Holben, B. Multiangle Imaging Spectroradiometer (MISR) global aerosol optical depth validation based on 2 years of coincident Aerosol Robotic Network (AERONET) observations. *J. Geophys. Res.* **2005**, *110*, D10S04. [CrossRef]

19. Torres, O.; Bhartia, P.K.; Herman, J.R.; Sinyuk, A.; Ginoux, P.; Holben, B. A Long-Term Record of Aerosol Optical Depth from TOMS Observations and Comparison to AERONET Measurements. *J. Atmos. Sci.* **2002**, *59*, 398–413. [CrossRef]

20. Torres, O.; Tanskanen, A.; Veihelmann, B.; Ahn, C.; Braak, R.; Bhartia, P.K.; Veefkind, P.; Levelt, P. Aerosols and surface UV products from Ozone Monitoring Instrument observations: An overview. *J. Geophys. Res.* **2007**, *112*, D24S47. [CrossRef]

21. Vidot, J.; Santer, R.; Aznay, O. Evaluation of the MERIS aerosol product over land with AERONET. *Atmos. Chem. Phys.* **2008**, *8*, 7603–7617. [CrossRef]

22. Liu, H.; Remer, L.A.; Huang, J.; Huang, H.-C.; Kondragunta, S.; Laszlo, I.; Oo, M.; Jackson, J.M. Preliminary evaluation of S-NPP VIIRS aerosol optical thickness. *J. Geophys. Res. Atmos.* **2014**, *119*, 3942–3962. [CrossRef]

23. Jackson, J.M.; Liu, H.; Laszlo, I.; Kondragunta, S.; Remer, L.A.; Huang, J.; Huang, H.-C. Suomi-NPP VIIRS aerosol algorithms and data products. *J. Geophys. Res. Atmos.* **2013**, *118*, 12673–612689. [CrossRef]

24. Hsu, N.C.; Jeong, M.-J.; Bettenhausen, C.; Sayer, A.M.; Hansell, R.; Seftor, C.S.; Huang, J.; Tsay, S.-C. Enhanced Deep Blue aerosol retrieval algorithm: The second generation. *J. Geophys. Res. Atmos.* **2013**, *118*, 9296–9315. [CrossRef]

25. Levy, R.C.; Mattoo, S.; Munchak, L.A.; Remer, L.A.; Sayer, A.M.; Patadia, F.; Hsu, N.C. The Collection 6 MODIS aerosol products over land and ocean. *Atmos. Meas. Tech.* **2013**, *6*, 2989–3034. [CrossRef]

26. Gupta, P.; Levy, R.C.; Mattoo, S.; Remer, L.A.; Munchak, L.A. A surface reflectance scheme for retrieving aerosol optical depth over urbansurfaces in MODIS Dark Target retrieval algorithm. *Atmos. Meas. Tech.* **2016**, *9*, 3293–3308. [CrossRef]

27. Bilal, M.; Nazeer, M.; Qiu, Z.; Ding, X.; Wei, J. Global Validation of MODIS C6 and C6.1 Merged Aerosol Products over Diverse Vegetated Surfaces. *Remote Sens.* **2018**, *10*, 475. [CrossRef]

28. Levy, R.C.; Remer, L.A.; Kleidman, R.G.; Mattoo, S.; Ichoku, C.; Kahn, R.; Eck, T.F. Global evaluation of the Collection 5 MODIS dark-target aerosol products over land. *Atmos. Chem. Phys.* **2010**, *10*, 10399–10420. [CrossRef]

29. Sayer, A.M.; Hsu, N.C.; Bettenhausen, C.; Jeong, M.-J. Validation and uncertainty estimates for MODIS Collection 6 "Deep Blue" aerosol data. *J. Geophys. Res. Atmos.* **2013**, *118*, 7864–7872. [CrossRef]

30. Sayer, A.M.; Munchak, L.A.; Hsu, N.C.; Levy, R.C.; Bettenhausen, C.; Jeong, M.J. MODIS Collection 6 aerosol products: Comparison between Aqua's e-Deep Blue, Dark Target, and "merged" data sets, and usage recommendations. *J. Geophys. Res. Atmos.* **2014**, *119*, 13965–13989. [CrossRef]

31. Giles, D.M.; Sinyuk, M.S.; Sorokin, J.S.; Schafer, A.; Smirnov, I.; Slutsker, T.F.; Eck, B.N.; Holben, J.R.; Lewis, J.R.; Campbell, E.J.; et al. Advancements in the Aerosol Robotic Network (AERONET) Version 3 database–automated near real-time quality control algorithm with improved cloud screening for Sun photometer aerosol optical depth measurements. *Atmos. Meas. Tech.* **2019**, *12*, 169–209. [CrossRef]

32. Remer, L.A.; Chin, M.; DeCola, P.; Feingold, G.; Halthore, R.; Kahn, R.A.; Quinn, P.K.; Rind, D.; Schwarts, S.E.; Streets, D.; et al. Executive Summary, in Atmospheric Aerosol Properties and Climate Impacts. In *A Report by the U.S. Climate Change Science Program and the Subcommittee on Global Change Research*; Chin, M., Kahn, R.A., Schwartz, S.E., Eds.; National Aeronautics and Space Administration: Washington, DC, USA, 2009.

33. Chew, B.N.; Campbell, J.R.; Reid, J.S.; Giles, D.M.; Welton, E.J.; Salinas, S.V.; Liew, S.C. Tropical cirrus cloud contamination in sun photometer data. *Atmos. Environ.* **2011**, *45*, 6724–6731. [CrossRef]

34. Bilal, M.; Nichol, J. Evaluation of the NDVI-Based Pixel Selection Criteria of the MODIS C6 Dark Target and Deep Blue Combined Aerosol Product. *IEEE J. Sel. Top. Appl. Earth Obs. Remote Sens.* **2017**, *10*, 3448–3453. [CrossRef]

35. Li, Z.; Niu, F.; Lee, K.-H.; Xin, J.; Hao, W.M.; Nordgren, B.L.; Wang, Y.; Wang, P. Validation and understanding of Moderate Resolution Imaging Spectroradiometer aerosol products (C5) using ground-based measurements from the handheld Sun photometer network in China. *J. Geophys. Res.* **2007**, *112*. [CrossRef]

36. He, Q.; Li, C.; Tang, X.; Li, H.; Geng, F.; Wu, Y. Validation of MODIS derived aerosol optical depth over the Yangtze River Delta in China. *Remote Sens. Environ.* **2010**, *114*, 1649–1661. [CrossRef]

37. Xie, Y.; Zhang, Y.; Xiong, X.; Qu, J.J.; Che, H. Validation of MODIS aerosol optical depth product over China using CARSNET measurements. *Atmos. Environ.* **2011**, *45*, 5970–5978. [CrossRef]

38. Bilal, M.; Nazeer, M.; Nichol, J.E. Validation of MODIS and VIIRS derived aerosol optical depth over complex coastal waters. *Atmos. Res.* **2017**, *186*, 43–50. [CrossRef]

39. Bilal, M.; Nichol, J.; Wang, L. New customized methods for improvement of the MODIS C6 Dark Target and Deep Blue merged aerosol product. *Remote Sens. Environ.* **2017**, *197*, 115–124. [CrossRef]

40. He, L.; Wang, L.; Lin, A.; Zhang, M.; Bilal, M.; Wei, J. Performance of the NPP-VIIRS and aqua-MODIS Aerosol Optical Depth Products over the Yangtze River Basin. *Remote Sens.* **2018**, *10*, 117. [CrossRef]

41. Wei, J.; Sun, L.; Huang, B.; Bilal, M.; Zhang, Z.; Wang, L. Verification, improvement and application of aerosol optical depths in China Part 1: Inter-comparison of NPP-VIIRS and Aqua-MODIS. *Atmos. Environ.* **2018**, *175*, 221–233. [CrossRef]

42. Bilal, M.; Qiu, Z.; Campbell, J.R.; Spak, S.; Shen, X.; Nazeer, M. A New MODIS C6 Dark Target and Deep Blue Merged Aerosol Product on a 3 km Spatial Grid. *Remote Sens.* **2018**, *10*, 463. [CrossRef]

43. Wei, J.; Huang, B.; Sun, L.; Zhang, Z.; Wang, L.; Bilal, M. A Simple and Universal Aerosol Retrieval Algorithm for Landsat Series Images Over Complex Surfaces. *J. Geophys. Res. Atmos.* **2017**. [CrossRef]

44. Nichol, J.; Bilal, M. Validation of MODIS 3 km Resolution Aerosol Optical Depth Retrievals Over Asia. *Remote Sens.* **2016**, *8*, 328. [CrossRef]

45. Bilal, M.; Nichol, J.E.; Nazeer, M. Validation of Aqua-MODIS C051 and C006 Operational Aerosol Products Using AERONET Measurements Over Pakistan. *IEEE J. Sel. Top. Appl. Earth Obs. Remote Sens.* **2016**, *9*, 2074–2080. [CrossRef]

46. Sun, L.; Wei, J.; Bilal, M.; Tian, X.; Jia, C.; Guo, Y.; Mi, X. Aerosol optical depth retrieval over bright areas using Landsat 8 OLI images. *Remote Sens.* **2016**, *8*, 23. [CrossRef]

 *remote sensing*

Technical Note

Vertically Resolved Precipitation Intensity Retrieved through a Synergy between the Ground-Based NASA MPLNET Lidar Network Measurements, Surface Disdrometer Datasets and an Analytical Model Solution

Simone Lolli [1,2,*], Leo Pio D'Adderio [3,4], James R. Campbell [5], Michaël Sicard [6,7], Ellsworth J. Welton [8], Andrea Binci [9], Alessandro Rea [10], Ali Tokay [2], Adolfo Comerón [6], Ruben Barragan [6,7], Jose Maria Baldasano [11], Sergi Gonzalez [12,13], Joan Bech [12], Nicola Afflitto [1], Jasper R. Lewis [2] and Fabio Madonna [1]

[1] CNR-IMAA, Consiglio Nazionale delle Ricerche, Contrada S. Loja snc, Tito Scalo, 85050 Potenza, Italy; nicola.afflitto@imaa.cnr.it (N.A.); fabio.madonna@imaa.cnr.it (F.M.)

[2] JCET-UMBC, University of Maryland Baltimore County, Baltimore, MD 21228, USA; ali.tokay-1@nasa.gov (A.T.); jasper.r.lewis@nasa.gov (J.R.L.)

[3] CNR-ISAC, Consiglio Nazionale delle Ricerche, Roma, Via del Fosso del Cavaliere 100, 00133 Roma, Italy; leopio.dadderio@artov.isac.cnr.it

[4] Dipartimento di Fisica, Università degli Studi di Ferrara, Via Saragat 1, 44121 Ferrara, Italy

[5] Naval Research Laboratory, Monterey, CA 93940, USA; james.campbell@nrlmry.navy.mil

[6] CommSensLab, Department of Signal Theory and Communications, Universitat Politècnica de Catalunya, 08034 Barcelona, Spain; msicard@tsc.upc.edu (M.S.); comeron@tsc.upc.edu (A.C.); ruben.barragan@tsc.upc.edu (R.B.)

[7] Ciències i Tecnologies de l'Espai—Centre de Recerca de l'Aeronàutica i de l'Espai/Institut d'Estudis Espacials de Catalunya (CTE-CRAE / IEEC), Universitat Politècnica de Catalunya, 08034 Barcelona, Spain

[8] NASA GSFC, Code 612, Greenbelt, MD 20771, USA; ellsworth.j.welton@nasa.gov

[9] Dipartimento Di Matematica, Università degli Studi di Roma II, Via della Ricerca Scientifica 1, 00133 Rome, Italy; abinci@gmail.com

[10] GRASI S.r.l., Via Tumoli, 03100 Frosinone, Italy; alessandrorea@me.com

[11] Environmental Modeling Laboratory, Universitat Politècnica de Catalunya, 08007 Barcelona, Spain; jose.baldasano@upc.edu

[12] Department of Applied Physics—Meteorology, University of Barcelona, 08007 Barcelona, Spain; sergi.gonzalez@meteo.ub.edu (S.G.); joan.bech@ub.edu (J.B.)

[13] Meteorological Surveillance and Forecasting Group, DT Catalonia, Agencia Estatal de Meteorología (AEMET), DT Catalonia. Arquitecte Sert, 1, 08003 Barcelona, Spain

[*] Correspondence: simone.lolli@cnr.it; Tel.: +39-0971-427250

Received: 18 May 2018; Accepted: 9 July 2018; Published: 11 July 2018

Abstract: In this paper, we illustrate a new, simple and complementary ground-based methodology to retrieve the vertically resolved atmospheric precipitation intensity through a synergy between measurements from the National Aeronautics and Space Administration (NASA) Micropulse Lidar network (MPLNET), an analytical model solution and ground-based disdrometer measurements. The presented results are obtained at two mid-latitude MPLNET permanent observational sites, located respectively at NASA Goddard Space Flight Center, USA, and at the Universitat Politècnica de Catalunya, Barcelona, Spain. The methodology is suitable to be applied to existing and/or future lidar/ceilometer networks with the main objective of either providing near real-time (3 h latency) rainfall intensity measurements and/or to validate satellite missions, especially for critical light precipitation (<3 mm h^{-1}).

Remote Sens. **2018**, *10*, 1102

Keywords: rainfall; lidar; disdrometer; evaporation; meteorology; climate change; latent heat; precipitation

---

1. Introduction

Rain and precipitation fundamentally influence life on Earth. With respect to the Earth-Atmosphere system, they play a role in pairing water and energy cycles, serving as a proxy for latent heat in the atmosphere. In fact, precipitation, modulating the latent heat content in the atmosphere [1], also modifies atmospheric column thermodynamics, affecting cloud lifetime [2]. Moreover, the hydrological cycle, which characterizes the continuous exchange of water in all its three phases, below and above the earth surface, is strongly dependent on precipitation. As a result, characterizing rainfall intensity and its variability at a global scale, is crucial not only to improving our knowledge of the hydrological cycle, but also to reducing uncertainties of global climate change model predictions for future environmental scenarios. Understanding rainfall accumulation paths, together with their spatial variability, besides helping in identifying world regions subject to drought and flooding, is of fundamental importance in reducing global climate models uncertainty to forecasting global temperature change [3]. In this context and thanks to the technological progress in satellite remote sensing techniques, the National Aeronautics and Space Administration (NASA) launched jointly with the Japan Aerospace Exploration Agency (JAXA) the Tropical Rainfall Measuring Mission (TRMM) followed by the Global Precipitation Measurement (GPM) [1]. The main objective of TRMM missions was to monitor and study precipitation with satellite measurements in the tropics where two-thirds of global precipitations occurs.

GPM further extended the measurement range towards the polar regions, (i.e., up to 69°N/S). NASA is at the forefront for retrievals for vertically resolved microphysical rain drop properties from ground-based multi-wavelength lidar measurements [4] and their improvement through comparison with an analytical model solution that uses disdrometer and radiosonde data as inputs [5]. Taking advantage of the experience gained in these previous studies, we develop in this paper a new methodology to retrieve range-resolved rainfall intensity through a synergy between elastic lidar measurements, disdrometer data and an analytical model solution. Measurements obtained with this simple method, if implemented globally through existing or future lidar Level 2 algorithms and output datasets, will fill a gap to help validate satellite data for light precipitation (intensity $< 3 \text{ mm h}^{-1}$) for which current global climate model predictions are in disagreement [1]. Results obtained from two mid-latitude NASA Micro Pulse Lidar NETwork (MPLNET [6]) permanent observational sites, one located at Goddard Space Flight Center (GSFC), USA, and the other at the Universitat Politècnica de Catalunya (UPC), Spain, are presented.

2. Materials and Methods

2.1. MPLNET Lidar Data Measurements

The ground-based lidar systems used in this study are the elastic polarization-sensitive micro pulse lidar (P-MPL v. 4B, Sigma Space Corp., now LEICA Geosystems, Lanham, MD, USA), which are deployed at two permanent MPLNET lidar network observational sites. The purpose of the NASA MPLNET network [6], active since 1999, is to retrieve automatically and continuously the geometrical and optical aerosol and cloud properties under most meteorological conditions and to the limit of laser signal attenuation. Measurements and retrievals obtained from worldwide deployed permanent stations are publicly available at MPLNET website [7]. Multi-year network data were previously analyzed to assess cloud [7–10] and aerosol [2,11,12] radiative effects.

The P-MPL samples the atmosphere with a relatively high frequency (2500 Hz) using a low-energy (7 μJ) Nd:YAG (neodymium-doped yttrium aluminum garnet) laser at 532 nm. The P-MPL acquisition settings at the two sites focused upon in this study follow the NASA MPLNET temporal and spatial specifications (60 s integration time and 75 m vertical resolution for GSFC and 60 s and 30 m for UPC). Polarization capabilities rely on the collection of two-channel measurements (i.e., the signal measured in the so-called 'co-polar' and 'cross-polar' channels of the instrument, respectively denoted as P_{co} (z) and P_{cr} (z). For reference, these channels are not to be confused with traditional linear depolarization measurements, where co- and cross-polar channels represent those linear states with respect to the linearly-polarized laser source (e.g., [13]). The MPL uses a nematic liquid crystal switching between two states [14,15]. In one of them, the crystal behaves like an isotropic medium, not having an effect on the wave propagating through it. In the other state, the crystal behaves as a quarter wave plate with principal axes at $45°$ with respect to the polarization direction of the transmitted electric field. The total power, P, is reconstructed as $P = P_{co} + 2P_{cr}$ [16,17]). The signal, P, multiplied by the squared range is the basis for retrieving all of the different Level 2 cloud and aerosol products [18,19]. Since the P-MPL is a single wavelength lidar, however, the retrieval of the vertically-resolved microphysical and optical aerosol properties are subject to stronger assumptions with respect to multi-wavelength lidars [20].

Among the newly available MPLNET Version 3 (V3) release products, we specifically consider here the Level 1 V3 Cloud algorithm (beta product) [21], which automatically retrieves the cloud base height that is used to correctly compute the precipitating drop size distribution from the ground. The MPLNET systems used are those of the Universitat Politècnica de Catalunya (UPC), Barcelona, Spain, (41.38N, 2.11E, 115 m a.s.l.) and of the Goddard Space Flight Center (GSFC), USA, (38.99N, 76.84W, 50 m a.s.l.).

2.2. Disdrometer

The disdrometer is an in situ measurement device designed to measure the drop size distribution (DSD; [22]), represented as the number of drops per unit of volume and per unit of raindrop diameter. Disdrometers can be based on different measurement principles (high-speed cameras, Doppler effect, laser-optical, impact, etc.). Two different versions of the Parsivel laser-optical disdrometer manufactured by OTT [23] are installed at UPC and GSFC, namely the first generation Parsivel (Parsivel^1) and the second generation Parsivel (Parsivel^2), respectively. Parsivel systems were originally developed by PM Tech Inc., now OTT Hydromet, Kempten, Germany. The instrument has a laser diode (emitting wavelength of 780 nm) generating a horizontal flat beam. The measurement area is nominally 48 cm^2 for the first generation Parsivel and 54 cm^2 for Parsivel^2.

When a hydro-meteor passes through the laser beam, it produces attenuation proportional to its size. A relationship between the laser beam occlusion by the falling particle is applied to estimate the particle size. Parsivel instruments can measure particle diameters up to about 25 mm classifying them in 32 size classes of different width. The instrument also estimates the hydro-meter fall velocity by measuring the time necessary for the particle to pass through the laser beam, and thus it stores particles in 32×32 matrices. The disdrometers high temporal resolution (60 s for this work) permits study in great detail of physical precipitation variability.

2.3. The Analytical Model Solution

In its original version of the subject model [24], the analytical model solution, based on molecular diffusivity of water vapor in air, permits calculating the *evaporation power* of a generic atmospheric layer in stationary thermodynamic conditions through the variable D^*, which is the initial diameter of a raindrop that evaporates completely after traveling a certain distance in the incident atmospheric layer. Conversely, for our purposes, instead of D^*, we reconstruct backwards the vertically-resolved profile of the raindrop diameter, starting from D_0, measured at the surface by the disdrometer, up to the cloud base at the radiosonding (or the atmospheric model) vertical resolution. The analytical model solution has been previously validated in [5].

In more detail, if the atmosphere is not saturated with respect to the water vapor, a raindrop evaporates through diffusion, which is assumed to be proportional to the water vapor gradient between raindrop surface and the environment [24]. The raindrop mass changes with time t following Equation (1):

$$\frac{dm}{dt} = 4\pi r D_v f_v \Delta \rho_v,$$ (1)

where m is the mass of the raindrop having radius r; and D_v is the water vapor diffusivity in air, f_v is the vapor diffusion ventilation coefficient and $\Delta \rho_v$ is the gradient of water vapor density between raindrop surface and the atmosphere. f_v can be expressed through v, the air kinematic viscosity, the diffusivity D_v, raindrop terminal fall velocity V and raindrop diameter D [24]:

$$f_v = 0.78 + 0.308 \left(\frac{v}{D_v}\right)^{\frac{1}{3}} \left(\frac{VD}{v}\right)^{\frac{1}{2}}.$$ (2)

In turn, D_v and v are depending on atmosphere thermodynamics as pressure and temperature. The water vapor density difference $\Delta \rho_v$ can be determined from the atmospheric sounding and raindrop temperature, which is determined from heat balance raindrop equation:

$$L\frac{dm}{dt} = 4\pi r K f_h \Delta T,$$ (3)

where L is the water vaporization latent heat, K is the air thermal conductivity, f_h is the heat ventilation coefficient, and ΔT the temperature difference between the environment and the raindrop. As stated in [24], the error assuming equality between the diffusion ventilation coefficient for vapor and heat is small and justified by the other approximations. Equation (1) can be rewritten as:

$$VD\frac{dD}{dh} = \frac{4}{\rho_w} D_v f_v \Delta \rho_v,$$ (4)

where h is the vertical coordinate measured from a certain reference level downward, ρ_w is the water density. Ventilation coefficient and diffusivity show a very low variability with height. It is then possible to represent them by their midlevel values in the considered layer as D_{vm} and f_{vm}. Using those values and representing the terminal fall velocity as $V = V_m \left(\frac{\rho_m}{\rho}\right)^{0.4}$ (m subscript indicates density and velocity midlevel values) [24], Equation (4) becomes:

$$\frac{V_m D dD}{f_{vm}} = \left(\frac{\rho_m}{\rho}\right)^{0.4} \frac{4}{\rho_w} D_{vm} \Delta \rho_v dh.$$ (5)

In Equation (5), the right side is only height h dependent, while the left side is a function of raindrop diameter D only. Likewise, $\Delta \rho_v$ is just depending on h and the atmospheric thermodynamics. Following the approach of Li [24], the left side integral of Equation (5) can be fit using a quadratic formula as:

$$F(D) = \int_0^D \frac{V_m D}{f_{vm}}\, dD \simeq c_1 D + c_2 D^2,$$ (6)

where c_1 and c_2 are the best-fit values. As an example, at 800 hPA (midlevel point) corresponding to a temperature T = 283 K, $c_1 = 2.008$ cm^2 s^{-1} and $c_2 = 30.146$ cm s^{-1}. The two coefficients should be calculated on a case-by-case basis depending on midlevel point and temperature. Our methodology computes the DSD from surface (measured by the disdrometer) up to cloud base at the same spatial resolution of the sounding or a complementary atmospheric model. If a generic raindrop exhibits a diameter D_0 (measured by the disdrometer) at the surface, at the top of the first considered layer (at height h_1), its diameter will be D_1. In general, if a raindrop exhibits a diameter of D_1 at range h_1 (bottom of the layer) through the analytical model solution, it is possible to compute the value of the raindrop diameter D_2 at height h_2 (top of the layer) as:

$$(c_1 D_1 + c_2 D_1^2) - (c_1 D_2 + c_2 D_2^2) = -\int_{h_1}^{h_2} \left(\frac{\rho_m}{\rho}\right)^{0.4} \frac{4}{\rho_w} D_{vm} \Delta \rho_v dh \equiv E(h_1, h_2). \tag{7}$$

Again, the function $E(h_1, h_2)$ defined as the integral of the right side of Equation (7) is fully determined by the vertical distribution of the thermodynamic variables in the considered layer (i.e., temperature, pressure and water vapor). Consequently, starting from raindrop diameter measurements at the surface, it is possible to estimate the raindrop diameter profile up to the cloud base just knowing the atmosphere thermodynamics. The cloud base height is retrieved using the operational MPLNET lidar product [21]. If the sounding data are unavailable or too far from the measurement site, the atmospheric thermodynamics variables can be obtained from NASA Goddard Modeling and Assimilation Office, version 5.9.1 reanalysis; GEOS-5 [21]), available every three hours and collocated at each MPLNET station.

For each range bin, at the radiosonde or GEOS-5 model resolution, the atmosphere is assumed to be steady. The primary limitation of the analytical model solution is that it does not take into account processes that affect raindrop diameter, such as coalescence and collision. For this reason, this method is more suitable for light intensity rainfall, where those processes are not significant. Moreover, since the methodology further depends on lidar/ceilometer measurements, the rain intensity will affect the instrument signal-to-noise Ratio (SNR). Thus, the lidar/ceilometer signal will only be available up to the cloud base in light intensity rainfall given the potential limits of signal attenuation in heavier showers.

3. Results and Discussion

3.1. Seasonal Differences at UPC

The UPC permanent observational site is located on the Remote Sensing Lab (RSlab) building in Barcelona, Spain. The disdrometer is deployed 600 m away from the lidar at the meteorological observatory of the Applied Physics Department of the University of Barcelona. For this kind of application, such a short distance is not relevant in lighter rainfall and both instruments can be assumed as co-located. We analyzed the variability in seasonal rainfall intensity over 2016 where disdrometer and co-located MPLNET observations were simultaneously available. The largest rainfall events were found during the spring (March–April–May; MAM; 2801 min) and fall (September–October–November; SON, 1278 min) seasons. Rainfall intensity was analyzed at three different levels: 300 m, 800 m and 1300 m above ground level (agl).

During spring (Figure 1a), the peak of the distribution is shifted towards higher rainfall intensities (around 1.5 mm h^{-1}), while, in fall (Figure 1b), the bulk of rainfall intensity is around 0.6 mm h^{-1}. This seasonal difference may be explained with different rain processes taking place (i.e., convective vs. stratiform events). The plots at three different quotes show similar trends, but the highest occurrence peaks of the rainfall intensity probability density function are shifted with respect to the altitude: at 300 m the highest peak is centered at 1.58 mm h^{-1}, at 800 m 1.99 mm h^{-1} and 2.51 mm h^{-1} at 1300 m (Figure 1a). Figure 1b shows less pronounced shift during SON: the highest occurrence peaks of rainfall intensity is 0.50 mm h^{-1} at 300 m while there is basically no difference at 800 m and 1300 m with 0.63 mm h^{-1}. Due to the lower sample size measurements, the same analysis has not been performed at GSFC.

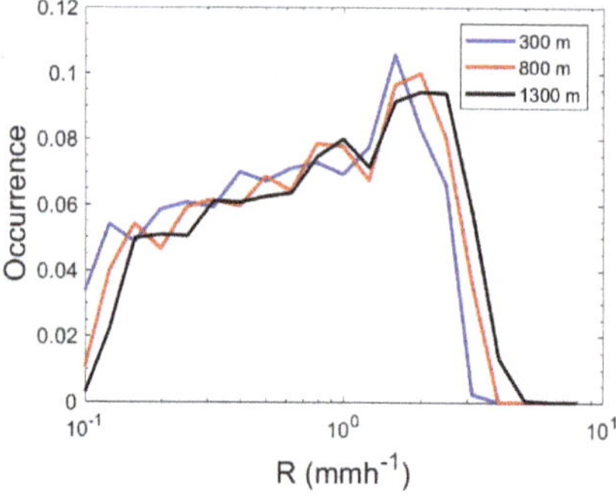

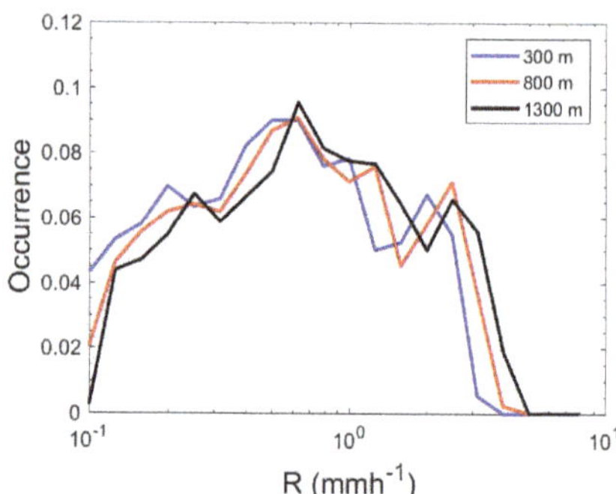

(a) Probability Density Function for rainfall events detected on 2016, Spring (March, April, May; MAM)

(b) Probability Density Function for rainfall events detected on 2016, Fall (September, October, November; SON)

Figure 1. Probability Distribution Function (PDF) for rainfall intensities at three vertical levels (300 m, 800 m, 1300 m) during Spring (**a**) and Fall (**b**) 2016 at the MPLNET Barcelona permanent observation station.

3.2. Case Study Analysis

Two case studies of the analytical model application at UPC and GSFC are presented and discussed in terms of vertically-resolved precipitation temporal evolution.

3.2.1. Retrieval of DSD profiles at UPC

On 4 April 2016, Figure 2a shows the composite plot of the depolarized channel signal, where precipitation contours are visible at around 9:00 a.m. UT and from 4:00 p.m. UT to 7:00 p.m. UT. Figure 2b shows the V3 L1 cloud algorithm cloud base height retrieval used in the inversion. Figure 2c depicts rainfall vertical intensity from 7:40 p.m. UT to 7:50 p.m. UT. Combining local radiosonde data (not showed here) and lidar data, we can state that rain originates from melting ice (cold rain process), with the freezing level detected at 2250 m AGL, just a few tens of meters below the cloud base. This is also confirmed by the GEOS-5 model (Figure 2a), where 0 °C isotherm is in very close agreement with radiosonding. In this rainfall event, the steepest gradient of intensity is 0.03 mm h^{-1} km^{-1}, which is much smaller than the GSFC case study (see Section 3.2.2).

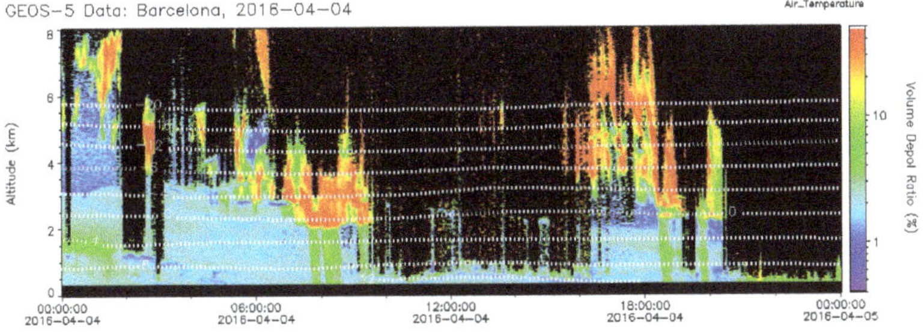

(**a**) Composite MPLNET V3 cross-polar channel with superimposed GEOS-5 isotherms

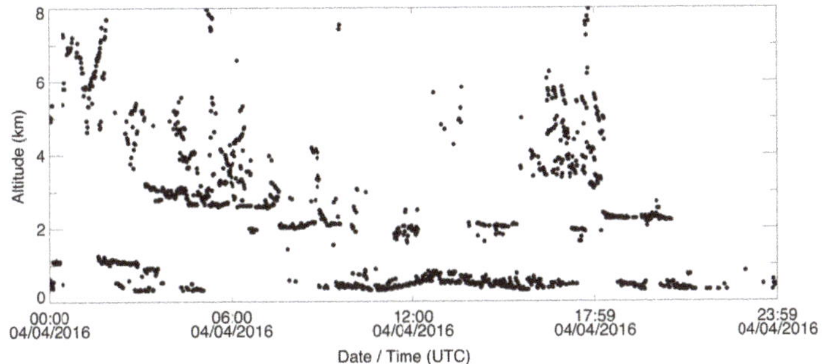

(**b**) MPLNET V3 L1 cloud base height retrieval product

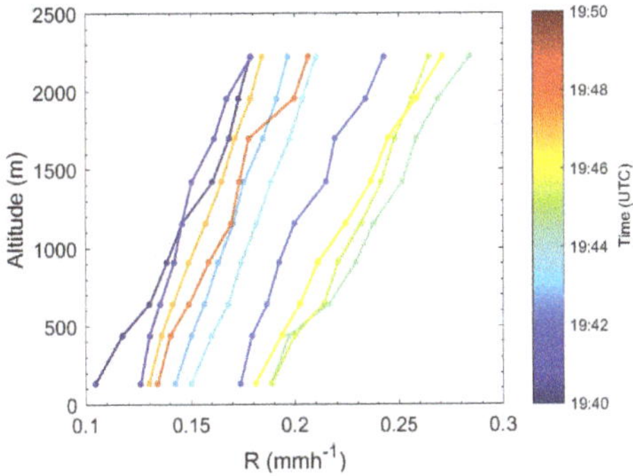

(**c**) Rainfall Intensity on 04 April 2016

Figure 2. Vertically-resolved rainfall intensity computations at different measurement times for UPC MPLNET station on 4 April 2016. (**a**) MPL cross-polar channel signal; (**b**) cloud base height automatically retrieved by V3 L1 Cloud algorithm; (**c**) vertically-resolved rainfall intensities, computed with the analytical model solution using disdrometer data and V3 L1 cloud base height retrieval, from 7:40 p.m. UT to 7:50 p.m. UT.

3.2.2. Retrieval of DSD profiles at GSFC

GSFC disdrometer and co-located lidar measurements were analyzed from November 2015 to April 2016. The vertical profiles of rainfall intensity, after applying the analytical model solution from 5:27 p.m. UT to 5:54 p.m. UT on 22 April 2016, are shown Figure 3. Depicted in Figure 3a is the composite plot of the depolarization channel signal obtained from the lidar on 22 April 2016. The core of the precipitation is clearly visible at around 5:45 p.m. UT. Figure 3b shows the cloud base height retrieval from V3 L1 MPLNET cloud algorithm. In Figure 3c, we can observe that rainfall intensity is weak, but increasing with time. The steepest gradient with respect to altitude is recorded at 5:47 p.m. UT with 0.22 mm h^{-1} km^{-1}.

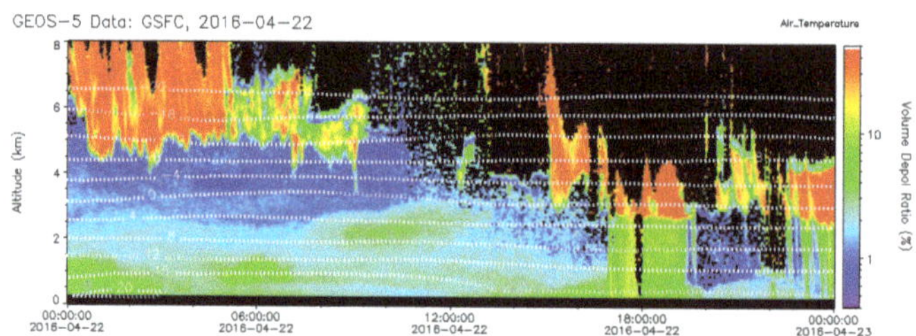

(**a**) Composite MPLNET V3 cross-polar channel with superimposed GEOS-5 isotherms

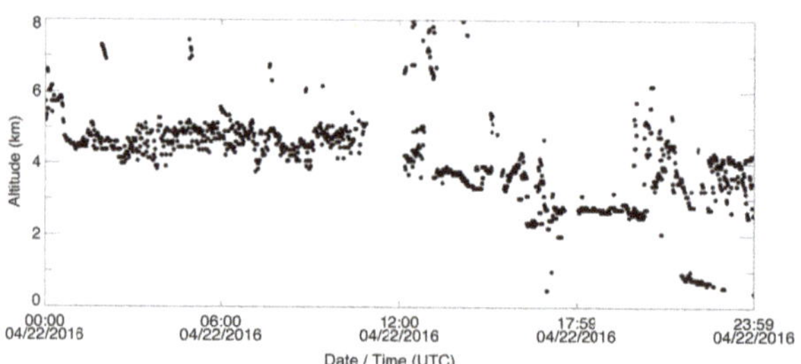

(**b**) MPLNET V3 L1 cloud algorithm cloud base retrieval product

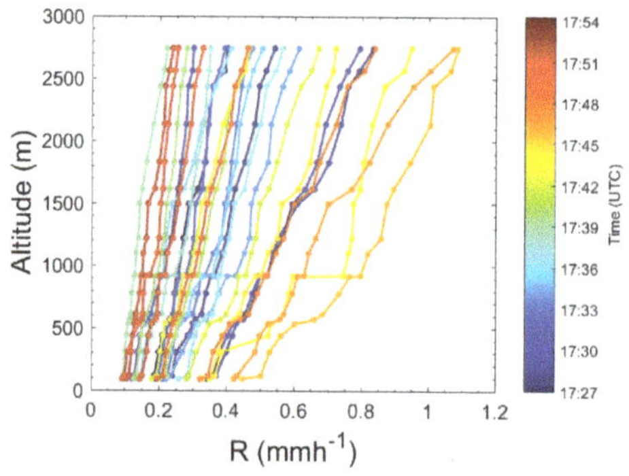

(**c**) Rainfall Intensity on 22 April 2016

Figure 3. Vertically-resolved rainfall intensity computations at different measurement times for the GSFC MPLNET station on 22 April 2016. (**a**) MPL cross-polar channel signal; (**b**) cloud base height automatically retrieved by V3 L1 Cloud algorithm; (**c**) vertically-resolved rainfall intensities, computed with the analytical model solution using disdrometer data and V3 L1 cloud base height retrieval, from 5:27 p.m. UT to 5:54 p.m. UT.

3.3. Evaporation Characteristics at UPC

In order to generalize the rain evaporation properties, UPC data measurements were analyzed as a function of rain parameter differences (i.e., R, the rain rate and Z, an equivalent radar reflectivity) between the cloud base and the ground. This indicates the impact of evaporation on rain integral parameters as R and Z. Figure 4a reports the analysis of R. The evaporation results are more marked (greater ΔR) for higher cloud base heights and increasing R values at the ground. For relatively high cloud bases (higher than 3000 m), the R difference with the ground reaches values as high as 6 mm h^{-1}. For lower R values and low cloud bases, ΔR is roughly constant, never exceeding 1 mm h^{-1}, regardless of the cloud base height. For lower cloud base heights (below 1000 m), ΔR is rain intensity insensitive at the ground and does not exceed 0.6–0.8 mm h^{-1}. For cloud base height over 2000 m, rainfall rate relative difference with respect to the ground exceeds 100%. Figure 4b shows Z properties, calculated as the sixth moment of the DSD. In contrast with R, the plot highlights that ΔZ is dependent only on the cloud base height, with a decrease of a factor two (ΔZ >= 3 dB) between the cloud base and the ground (cloud base higher than 2000 m). This can be explained, from a microphysical point of view, because of the small drop sizes collected in the analyzed data. That is, the lower the rain intensity, the smaller the drop diameters composing the DSD.

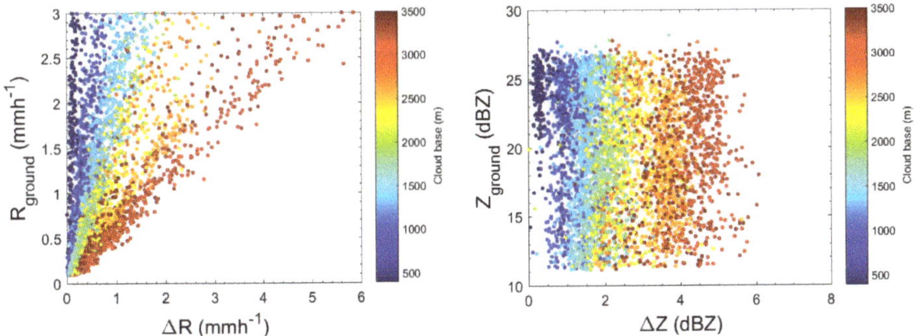

(**a**) Trend of ΔR as function of R at ground and cloud base height

(**b**) Trend of ΔZ as function of R at ground and cloud base height

Figure 4. Trend of the difference between the cloud base and the ground of the rain parameters R and Z as a function of parameter values measured at ground and at cloud base height.

4. Conclusions

We introduce a methodology for computing vertically-resolved rain parameters (i.e., rain intensity) through a synergy between ground-based lidar, in situ disdrometer measurements and an analytical model solution paired with thermodynamic variables measured by atmospheric radiosondes (if unavailable, atmosphere thermodynamic variables can be inferred from a NASA GEOS-5 model). The methodology, applied at two permanent mid-latitude NASA Micro Pulse Lidar Network datasets, the Goddard Space Flight Center (GSFC) and Universitat Politècnica de Catalunya in Barcelona, Spain (UPC), is particularly suited for measurements of low intensity precipitation (rainfall rate, R, <3 mm h^{-1}). If implemented operationally in the network, the methodology can generate near real-time rainfall intensity as a standard Level 2 product. Low-intensity precipitation measurements are crucial for better understanding the hydrological cycle and for validating satellite missions, like the Global Precipitation Mission experiment (GPM) [1].

The analysis of a complete year (2016) of precipitation at UPC (4079 min of data) permitted assessing rainfall intensity seasonal variability for different cloud base altitude ranges. Slightly different rain intensity distributions were observed during spring (MAM) and fall (SON), with a higher occurrence of a relatively high rain rate during spring (1.5 vs. 0.6 mm h^{-1} mean R), and a lower rainfall intensity associated with lower altitudes. This implies rainfall evaporation with consequent atmospheric column cooling. Yearly analysis of UPC MPLNET data shows that the effect of the evaporation on the rainfall rate has different impacts, depending on both rain intensity at ground and cloud base height. On the other hand, the radar reflectivity shows a dependence only on the cloud base height. The comparison between UPC and GSFC indicates that, for approximately the same rain intensity at the ground, the rain intensity gradients observed in GSFC (0.22 mm h^{-1} km^{-1}) are larger than the ones observed at UPC (0.03 mm h^{-1} km^{-1}). This result shows that, for this case study, the GSFC atmosphere is in general drier with respect to UPC.

Both analyzed case studies demonstrate the analytical model capability for reconstructing DSD from ground to cloud base. This also permits computing all of the significant distribution moments (i.e., radar reflectivity, liquid water content, mean mass diameter, etc.) besides rain reflectivity. Future research will focus on assessing light precipitation inter-annual intensity variability from long-term (>15 years) MPLNET stations, especially in polluted regions, to quantifying for the first time the aerosol indirect effects on drizzle reduction/suppression.

Author Contributions: Conceptualization, S.L.; Methodology, S.L., L.P.D., M.S.; Software, S.L., L.P.D., A.R., A.B. and J.R.L.; Validation, S.L., A.R. and A.B.; Formal Analysis, S.L., L.P.D., M.S., J.R.C., F.M.; Investigation, S.L., E.J.W., J.R.C., M.S., L.P.D.; Resources, N.A., A.C., J.B.; Data Curation, S.L., A.T., J.R.L., E.J.W., S.G., J.B.; Writing—Original Draft Preparation, S.L.; Writing—Review & Editing, J.R.C., L.P.D., M.S., J.B., S.G., J.R.L. and N.A.; Visualization, S.L., F.M., N.A.; Supervision, S.L., N.A.; Project Administration, N.A., A.C., J.B., E.J.W., J.R.C.; Funding Acquisition, J.R.C., E.J.W., A.C., J.B. and S.G.

Funding: The MPL measurements and processing in Barcelona are supported by the European Union (H2020, grant 654109, ACTRIS-2), the European Fund for Regional Development, the Spanish Government (grants TEC2015-63832-P, CGL2015-65627-C3-2-R, CGL2016-81828-REDT and CGL2017-90884-REDT) and the Catalan Government (grant 2014 SGR 583). CommSensLab is a Unidad de Excelencia María de Maeztu (grant MDM-2016-0600) funded by the Agencia Estatal de Investigación.

Acknowledgments: The NASA Micro-Pulse Lidar Network (MPLNET) are supported by the NASA Radiation Sciences Program (H. Maring).

Conflicts of Interest: The authors declare no conflict of interest.

References

1. Hou, A.Y.; Kakar, R.K.; Neeck, S.; Azarbarzin, A.A.; Kummerow, C.D.; Kojima, M.; Oki, R.; Nakamura, K.; Iguchi, T. The Global Precipitation Measurement Mission. *Bull. Am. Meteorol. Soc.* **2014**, *95*, 701–722. [CrossRef]

2. Lolli, S.; Campbell, J.R.; Lewis, J.R.; Welton, E.J.; Di Girolamo, P.; Fatkhuroyan, F.; Gu, Y.; Marquis, J.W. Assessment of cirrus cloud and aerosol radiative effect in South-East Asia by ground-based NASA MPLNET lidar network data and CALIPSO satellite measurements. In *Remote Sensing of Clouds and the Atmosphere XXII*; International Society for Optics and Photonics: Bellingham, WA, USA, 2017; Volume 10424, p. 1042405.

3. Bosilovich, M.G.; Schubert, S.D.; Walker, G.K. Global Changes of the Water Cycle Intensity. *J. Clim.* **2005**, *18*, 1591–1608. [CrossRef]

4. Lolli, S.; Welton, E.J.; Campbell, J.R. Evaluating light rain drop size estimates from multiwavelength micropulse lidar network profiling. *J. Atmos. Ocean. Technol.* **2013**, *30*, 2798–2807. [CrossRef]

5. Lolli, S.; Di Girolamo, P.; Demoz, B.; Li, X.; Welton, E. Rain Evaporation Rate Estimates from Dual-Wavelength Lidar Measurements and Intercomparison against a Model Analytical Solution. *J. Atmos. Ocean. Technol.* **2017**, *34*, 829–839. [CrossRef]

6. Welton, E.J.; Campbell, J.R.; Spinhirne, J.D.; Scott, V.S. *Global Monitoring of Clouds and Aerosols Using a Network of Micropulse Lidar Systems*; SPIE: Bellingham, WA, USA, 2001; Volume 4153, pp. 151–159. [CrossRef]

7. Campbell, J.R.; Lolli, S.; Lewis, J.R.; Gu, Y.; Welton, E.J. Daytime cirrus cloud top-of-the-atmosphere radiative forcing properties at a midlatitude site and their global consequences. *J. Appl. Meteorol. Climatol.* **2016**, *55*, 1667–1679. [CrossRef]

8.   Lolli, S.; Campbell, J.R.; Lewis, J.R.; Gu, Y.; Marquis, J.W.; Chew, B.N.; Liew, S.C.; Salinas, S.V.; Welton, E.J. Daytime Top-of-the-Atmosphere Cirrus Cloud Radiative Forcing Properties at Singapore. *J. Appl. Meteorol. Climatol.* **2017**, *56*, 1249–1257. [CrossRef]

9.   Campbell, J.R.; Peterson, D.A.; Marquis, J.W.; Fochesatto, G.J.; Vaughan, M.A.; Stewart, S.A.; Tackett, J.L.; Lolli, S.; Lewis, J.R.; Oyola, M.I.; et al. Unusually deep wintertime cirrus clouds observed over the alaskan sub-arctic. *Bull. Am. Meteorol. Soc.* **2017**, *99*, 27–32. [CrossRef]

10.   Lolli, S.; Campbell, J.R.; Lewis, J.R.; Gu, Y.; Welton, E.J. Fu–Liou–Gu and Corti–Peter model performance evaluation for radiative retrievals from cirrus clouds. *Atmos. Chem. Phys.* **2017**, *17*, 7025–7034. [CrossRef]

11.   Pani, S.K.; Wang, S.H.; Lin, N.H.; Tsay, S.C.; Lolli, S.; Chuang, M.T.; Lee, C.T.; Chantara, S.; Yu, J.Y. Assessment of aerosol optical property and radiative effect for the layer decoupling cases over the northern South China Sea during the 7-SEAS/Dongsha Experiment. *J. Geophys. Res. Atmos.* **2016**, *121*, 4894–4906. [CrossRef]

12.   Tosca, M.G.; Campbell, J.; Garay, M.; Lolli, S.; Seidel, F.C.; Marquis, J.; Kalashnikova, O. Attributing accelerated summertime warming in the southeast united states to recent reductions in aerosol burden: Indications from vertically-resolved observations. *Remote Sens.* **2017**, *9*, 674. [CrossRef]

13.   Campbell, J.R.; Hlavka, D.L.; Welton, E.J.; Flynn, C.J.; Turner, D.D.; Spinhirne, J.D.; Scott, V.S.S., III; Hwang, I.H. Full-Time, Eye-Safe Cloud and Aerosol Lidar Observation at Atmospheric Radiation Measurement Program Sites: Instruments and Data Processing. *J. Atmos. Ocean. Technol.* **2002**, *19*, 431–442. [CrossRef]

14.   Welton, E.J.; Campbell, J.R. Micropulse Lidar Signals: Uncertainty Analysis. *J. Atmos. Ocean. Technol.* **2002**, *19*, 2089–2094. [CrossRef]

15.   Ciofini, M.; Lapucci, A.; Lolli, S. Diffractive optical components for high power laser beam sampling. *J. Opt. A Pure Appl. Opt.* **2003**, *5*, 186. [CrossRef]

16.   Welton, E.J.; Voss, K.J.; Gordon, H.R.; Maring, H.; Smirnov, A.; Holben, B.; Schmid, B.; Livingston, J.M.; Russell, P.B.; Durkee, P.A.; et al. Ground-based lidar measurements of aerosols during ACE-2: Instrument description, results, and comparisons with other ground-based and airborne measurements. *Tellus B* **2000**, *52*, 636–651. [CrossRef]

17.   Welton, E.J.; Voss, K.J.; Quinn, P.K.; Flatau, P.J.; Markowicz, K.; Campbell, J.R.; Spinhirne, J.D.; Gordon, H.R.; Johnson, J.E. Measurements of aerosol vertical profiles and optical properties during INDOEX 1999 using micropulse lidars. *J. Geophys. Res. Atmos.* **2002**, *107*, INX2 18-1–INX2 18-20. [CrossRef]

18.   Lolli, S.; Madonna, F.; Rosoldi, M.; Campbell, J.R.; Welton, E.J.; Lewis, J.R.; Gu, Y.; Pappalardo, G. Impact of varying lidar measurement and data processing techniques in evaluating cirrus cloud and aerosol direct radiative effects. *Atmos. Meas. Tech.* **2018**, *11*, 1639–1651. [CrossRef]

19.   Sassen, K.; Knight, N.C.; Takano, Y.; Heymsfield, A.J. Effects of ice-crystal structure on halo formation: cirrus cloud experimental and ray-tracing modeling studies. *Appl. Opt.* **1994**, *33*, 4590–4601. [CrossRef] [PubMed]

20.   Flynn, C.J.; Mendoza, A.; Zheng, Y.; Mathur, S. Novel polarization-sensitive micropulse lidar measurement technique. *Opt. Express* **2007**, *15*, 2785–2790. [CrossRef] [PubMed]

21.   Lewis, J.R.; Campbell, J.R.; Welton, E.J.; Stewart, S.A.; Haftings, P.C. Overview of MPLNET Version 3 Cloud Detection. *J. Atmos. Ocean. Technol.* **2016**, *33*, 2113–2134. [CrossRef]

22.   D'Adderio, L.; Porcù, F.; Tokay, A. Evolution of drop size distribution in natural rain. *Atmos. Res.* **2018**, *200*, 70–76. [CrossRef]

23.   Löffler-Mang, M.; Joss, J. An Optical Disdrometer for Measuring Size and Velocity of Hydrometeors. *J. Atmos. Ocean. Technol.* **2000**, *17*, 130–139. [CrossRef]

24.   Li, X.; Srivastava, R.C. An Analytical Solution for Raindrop Evaporation and Its Application to Radar Rainfall Measurements. *J. Appl. Meteorol.* **2001**, *40*, 1607–1616 . [CrossRef]

MDPI

St. Alban-Anlage 66

4052 Basel

Switzerland

Tel. +41 61 683 77 34

Fax +41 61 302 89 18

www.mdpi.com

Remote Sensing Editorial Office

E-mail: remotesensing@mdpi.com

www.mdpi.com/journal/remotesensing